QUALITY AND RELIABILITY IN ENGINEERING

Quality and Reliability in Engineering provides an integrated approach to quality specification, quality control and monitoring, and reliability. Examples and exercises stress practical engineering applications. Steps in the development of the theory are implemented in the form of complete, self-contained computer programs. The book serves as a textbook for upper-level undergraduate courses in quality and reliability in mechanical engineering, manufacturing engineering, and industrial engineering programs. It can be used as a supplement to upper-level capstone design courses, short courses for quality training, and as a learning resource for practicing engineers.

Tirupathi R. Chandrupatla has been a professor and Founding Chair of Mechanical Engineering at Rowan University, Glassboro, New Jersey, since 1995. He received his M.S. degree in design and manufacturing from the Indian Institute of Technology (IIT), Bombay, and his Ph.D. from the University of Texas at Austin. He began his career as a design engineer with Hindustan Machine Tools (HMT), Bangalore. His first teaching post was at IIT, Bombay. He has also taught at the University of Kentucky, Lexington, and GMI Engineering and Management Institute (now called Kettering University), Flint, Michigan, before joining Rowan. Chandrupatla is the author of *Introduction to Finite Elements in Engineering, Optimization Concepts and Applications in Engineering,* and *Finite Element Analysis for Engineering and Technology.* The first book has been translated into Spanish, Korean, Greek, and Chinese. Chandrupatla's research interests include design, optimization, manufacturing engineering, finite element analysis, and quality and reliability. He has published widely in these areas and serves as a consultant to industry. Chandrupatla is a registered Professional Engineer and also a Certified Manufacturing Engineer. He is a member of the American Society for Engineering Education (ASEE), the American Society of Mechanical Engineers (ASME), the Society of Automotive Engineers (SAE), and the Society of Manufacturing Engineers (SME).

Quality and Reliability in Engineering

Tirupathi R. Chandrupatla
Rowan University

CAMBRIDGE
UNIVERSITY PRESS

32 Avenue of the Americas, New York NY 10013-2473, USA

Cambridge University Press is part of the University of Cambridge.

It furthers the University's mission by disseminating knowledge in the pursuit of education, learning and research at the highest international levels of excellence.

www.cambridge.org
Information on this title: www.cambridge.org/9780521515221

© Tirupathi R. Chandrupatla 2009

This publication is in copyright. Subject to statutory exception and to the provisions of relevant collective licensing agreements, no reproduction of any part may take place without the written permission of Cambridge University Press.

First published 2009

A catalogue record for this publication is available from the British Library

Library of Congress Cataloguing in Publication data

Chandrupatla, Tirupathi R., 1944–
Quality and reliability in engineering / Tirupathi R. Chandrupatla.
  p. cm.
Includes bibliographical references and index.
ISBN 978-0-521-51522-1 (hardback)
1. Reliability (Engineering)   2. Quality assurance.   I. Title.
TA169.C45   2009
620'.00452 – dc22   2008027686

ISBN 978-0-521-51522-1 Hardback

Additional resources for this publication at www.cambridge.org/9780521515221

Cambridge University Press has no responsibility for the persistence or accuracy of URLs for external or third-party internet websites referred to in this publication, and does not guarantee that any content on such websites is, or will remain, accurate or appropriate.

To Henry M. Rowan

"A detailed attention to quality and reliability is vital to the growth and success of a company."

Henry M. Rowan

Henry M. Rowan is the founder and chairman of Inductotherm Industries, Inc., located in Rancocas, New Jersey. He built his company's first furnace in his backyard; from that humble beginning, Inductotherm has become the world's largest designer and manufacturer for induction melting, heat treating, and welding. Inductotherm is currently a global enterprise of more than eighty companies with facilities in fifteen nations around the world.

A native of Raphine, Virginia, Rowan grew up in Ridgewood, New Jersey. After serving as a pilot in the Army Air Corps during World War II, he earned his B.S. in electrical engineering from Massachusetts Institute of Technology. Rowan took major steps in customer service by creating a highly mobile service team organized around a fleet of company-owned aircraft to ensure maximum uptime of each customer's installation. In 1992, Rowan endowed Glassboro State College with $100 million. It is now known as Rowan University. Rowan's triumphs and tribulations are presented in his autobiography, *The Fire Within* (Penton Publishing, 1995).

Contents

Preface *page* xi
Acknowledgments xv

1 Quality Concepts ... 1

1.1 Introduction — 1
1.2 Quality and Reliability Defined — 1
1.3 Historical Development — 2
1.4 Quality Philosophies — 3
1.5 Conclusion — 6
QUESTIONS FOR DISCUSSION — 7

2 Tolerances and Fits ... 8

2.1 Introduction — 8
2.2 Preferred Numbers — 8
2.3 Tolerances and Fits — 9
2.4 Manufacturing Processes and Tolerances — 15
2.5 Tolerance Selection in Assemblies — 15
2.6 Summary — 21
EXERCISE PROBLEMS — 22

3 Geometric Tolerances ... 24

3.1 Introduction — 24
3.2 Geometric Tolerances – Some Basic Ideas — 24
3.3 Tolerances of Form — 29
3.4 Profile Tolerances — 39
3.5 Orientation Tolerances — 39

vii

3.6	Tolerances of Location	43
3.7	Tolerances of Runout	47
3.8	Summary	49
	EXERCISE PROBLEMS	49

4 Elements of Probability and Statistics 54

4.1	Introduction	54
4.2	Probability	54
4.3	Statistics	58
4.4	Probability Distribution Definitions	64
4.5	Discrete Probability Distributions	68
4.6	Continuous Distributions	73
4.7	Summary	84
	EXERCISE PROBLEMS	84

5 Sampling Concepts 88

5.1	Introduction	88
5.2	The Central Limit Theorem	88
5.3	Confidence Interval Estimation	90
5.4	Confidence Interval for the Mean of a Normal Population	90
5.5	Confidence Interval for the Difference between Two Means	93
5.6	Confidence Interval for a Proportion	94
5.7	Confidence Interval for the Variance	96
5.8	Confidence Interval for the Ratio of Two Variances	97
5.9	Hypothesis Testing	98
5.10	Type I and Type II Errors	102
5.11	Sample Size Determination	106
5.12	Summary	107
	EXERCISE PROBLEMS	107

6 Data Presentation: Graphs and Charts 110

6.1	Introduction	110
6.2	Stem-and-Leaf Plots	110
6.3	Histograms	111
6.4	Cause-and-Effect Diagrams	112
6.5	Pareto Charts	114
6.6	Box Plots	115
6.7	Normal Probability Plot	116
6.8	Run Charts	118
6.9	Summary	119
	EXERCISE PROBLEMS	120

7 Statistical Process Control 122

7.1 Introduction 122
7.2 Order Statistics and Other Preliminaries 122
7.3 Causes of Variation 124
7.4 Statistical Process Control Concepts 125
7.5 Control Charts for Variables 129
7.6 Control Charts for Attributes 146
7.7 Operating Characteristic (OC) Curves 150
7.8 Summary 154
EXERCISE PROBLEMS 154

8 Process Capability Analysis 158

8.1 Introduction 158
8.2 Process Capability 158
8.3 Measurement System Analysis – Gage Repeatability and Reproducibility Study 166
8.4 Propagation of Errors 172
8.5 Prediction and Tolerance Intervals for Normal Distribution 173
8.6 Summary 179
EXERCISE PROBLEMS 179

9 Acceptance Sampling 182

9.1 Introduction 182
9.2 Acceptance Sampling for Attributes 182
9.3 Sampling Plans for Variables 195
9.4 Summary 205
EXERCISE PROBLEMS 205

10 Experimental Design 207

10.1 Introduction 207
10.2 Basic Concepts 207
10.3 Factorial Experiments 209
10.4 2^k Factorial Experiments 223
10.5 Summary 232
EXERCISE PROBLEMS 232

11 Reliability Concepts 236

11.1 Introduction 236
11.2 Reliability Functions 236
11.3 Failure Distributions 238

11.4 System Reliability 243
   11.5 *K*-of-*N* Systems 248
   11.6 Standby Systems 250
   11.7 Summary 251
   EXERCISE PROBLEMS 252

12 **Reliability Testing** **255**
   12.1 Introduction 255
   12.2 Weibull Distribution Parameter Estimation 255
   12.3 Lognormal Distribution Parameter Estimation 260
   12.4 Exponential-Based Life Testing and Confidence Intervals 260
   12.5 Sampling Procedures for Life Testing (Exponential-Based) 263
   12.6 Summary 270
   EXERCISE PROBLEMS 271

APPENDIX 273

Answers to Selected Problems 295

Bibliography 301

Index 305

Preface

The text material evolved out of teaching a course on quality and reliability in an undergraduate program in mechanical engineering. This is a required course in the second semester of our junior year. I have taught the course every year for the past ten years. I received positive feedback from the students who took the course and from the managers in industry who employed them. These positive interactions provided the motivation to develop the course material into a book. This book is a culmination of more than forty years of my experience as a design and manufacturing engineer, teacher, researcher, and consultant.

The underlying philosophy of the book is that a quality product results from the specification of quality at the design stage; measurement, monitoring, and control of quality at the production stage; and quality performance at the final stage. A course on quality and reliability covering all three aspects is needed in every mechanical engineering or manufacturing engineering program. Practicing engineers in design, manufacturing, and quality engineering need to have this material handy in one place. Industrial engineering students also need an exposure to quality specification. There are many excellent books on each of the three areas, but books integrating the three areas are not available.

This book is intended as a textbook for an upper-level course in mechanical engineering, manufacturing engineering, and industrial engineering programs. The book can be used as a reference book for upper-level capstone design courses, and also as a learning resource for practicing engineers. Each chapter introduces the underlying concepts and attempts to explain the origin of some of the data in the tables. As an example, the estimation of the standard deviation in terms of the sample range used in various process control charts is shown to come from order statistics. These and other relationships have been implemented in generating the tables available in the Appendix. The corresponding tables have active formulas and are available for download from **www.cambridge.org/9780521515221**.

Complete computer programs that implement and parallel the theory have been provided. These programs are in Microsoft Excel and are available on the book's web site. Several full-fledged programs have spreadsheet simplicity. Pressing Alt+F11 will show the modules and functions that have been developed. Several programs have interactive features using the spin buttons in Microsoft Excel. All tables give in the Appendix are available on the book's web site in the form of spreadsheets or programs.

The book is organized as follows: Chapter 1 gives definitions of quality and reliability providing a brief historical development. Quality philosophies are presented. Chapter 2 develops the concept of preferred numbers before introducing the international tolerance system. The relationship of manufacturing processes and tolerances is presented. Tolerance selection and tolerance allocation decisions are also discussed. Chapter 3 gives an overview of geometric dimensioning and tolerancing. Tolerances of form, profile, orientation, location, and runout are discussed. Evaluation aspects of form tolerances – straightness, flatness, circularity, and cylindricity – are discussed, and several computer programs are included.

Chapters 4, 5, and 6 provide the key concepts of probability and statistics, sampling concepts, and data presentation tools. Chapter 7 introduces the concepts of order statistics and other preliminaries and goes on to present various control charts for variables and attributes. Operating characteristic curves are given for both variables and attributes. Chapter 8 discusses process capability, measurement system analysis, error propagation, and tolerance intervals. Chapter 9 presents acceptance sampling for attributes and variables. Interactive programs are provided for the design of sampling plans for both attributes and variables.

Chapter 10 gives concepts of experimental design. Completely randomized single-factor experiments, randomized block experiments, two-factor factorial experiments, and 2^k factorial experiments are discussed. Chapter 11 introduces reliability concepts, and various failure distributions are presented. The evaluation of system reliability of series and parallel systems, K-of-N systems, and standby systems are discussed. Chapter 12 discusses parameter estimation aspects for Weibull and lognormal distributions and sampling procedures for reliability life testing.

Programs in mechanical engineering and manufacturing engineering are expected to cover all chapters. Chapter 6 may be covered through some discussion followed by assignments. Some topics in Chapters 7, 8, 9, and 12 may be left as reading material. A course on quality improvement in industrial engineering programs may use Chapter 2 as optional reading material and skip Chapter 3. Needed material from Chapters 4 and 5 may be reviewed. Chapters 6, 7, 8, 9, and 10 should be covered in their entirety. Chapters 11 and 12 may be used as needed. A first course on reliability may cover Chapters 1, 4, 5, 6, 10, 11, and 12, and other chapters may be used as needed.

Junior- or senior-level capstone design and project-based courses in mechanical and manufacturing engineering may use this book as a study reference, with students expected to study Chapters 1, 2, 3, 4, 5, 6, 10, and 11. Some testing aspects of Chapter 12 may also be used.

Training programs in the areas of quality and reliability may use relevant chapters and programs for short courses. Practicing professionals should find the book useful for self-learning.

The use of computer programs must be stressed; the included Excel programs should serve well for this purpose. I would like to get your feedback concerning the included software (you may contact me at Chandrupatla@rowan.edu). Use of software such as MINITAB and other commercial software is encouraged.

Acknowledgments

I would like to thank several reviewers who gave many constructive suggestions. In particular, I would like to thank Dr. Srinivas Chakravarty, Professor of Operations Research and Statistics, Department of Industrial and Manufacturing Engineering, Kettering University, Flint, Michigan, who has been a continuing source of encouragement for many years as a colleague and friend. I cherish all the interesting discussions on statistics, quality, and reliability that we had. I am deeply indebted to Dr. Prabhaker R. Gangasani, Technical Director, Dura-Bar, a division of Wells Manufacturing Company, Woodstock, Illinois, for providing me with valuable insight through his review. I express my sincere thanks to Dr. Ashok D. Belegundu, Professor, Department of Mechanical Engineering, Pennsylvania State University, who, as coauthor of my previous books, always encouraged me to write this book. I thank all the students who took the course and provided many valuable suggestions.

I express my deep gratitude to Mr. Henry M. Rowan of the Inductotherm Group of companies, who has been a constant source of inspiration. I also thank Dr. Dianne Dorland, Dean, College of Engineering at Rowan University, for her encouragement and support in this endeavor. The encouragement received from the entire faculty in the College of Engineering at Rowan is highly appreciated.

I express my sincere thanks to Mr. Peter Gordon, Engineering Editor at Cambridge University Press. He handled the project efficiently and with great speed. I would like to express my thanks to the copyeditor, Ms. Heather Phillips, the project manager, Ms. Peggy M. Rote, and the production team for the fine job that they accomplished.

I would like to thank my wife, Suhasini; my sons, Sreekanth and Hareesh; my daughter-in-law, Vandana; and my grandchildren, Sumanth and Sriya. They all turned this major undertaking into a pleasant chore.

Tirupathi R. Chandrupatla
Rowan University

1
Quality Concepts

1.1 Introduction

Quality is perceived differently by different people. Yet, everyone understands what is meant by "quality." In a manufactured product, the customer as a user recognizes the quality of fit, finish, appearance, function, and performance. The quality of service may be rated based on the degree of satisfaction by the customer receiving the service. The relevant dictionary meaning of quality is "the degree of excellence." However, this definition is relative in nature. The ultimate test in this evaluation process lies with the consumer. The customer's needs must be translated into measurable characteristics in a product or service. Once the specifications are developed, ways to measure and monitor the characteristics need to be found. This provides the basis for continuous improvement in the product or service. The ultimate aim is to ensure that the customer will be satisfied to pay for the product or service. This should result in a reasonable profit for the producer or the service provider. The relationship with a customer is a lasting one. The reliability of a product plays an important role in developing this relationship.

1.2 Quality and Reliability Defined

There are many definitions of quality available in the literature. A definition attributed to quality guru Crosby states the following:

Quality is conformance to requirements.

The preceding definition assumes that the specifications and requirements have already been developed. The next thing to look for is conformance to these requirements. Another frequently used definition comes from Juran:

Quality is fitness for use.

This definition stresses the importance of the customer who will use the product.

W. Edwards Deming defined quality as follows:

Good quality means a predictable degree of uniformity and dependability with a quality standard suited to the customer.

The underlying philosophy of all definitions is the same – consistency of conformance and performance, and keeping the customer in mind.

Another definition that is widely accepted is

Quality is the degree to which performance meets expectations.

This definition provides a means to assess quality using a relative measure.

We provide here the definition adopted by the American Society for Quality (ASQ):

Quality denotes an excellence in goods and services, especially to the degree they conform to requirements and satisfy customers.

This definition assimilates the previous ones and is our definition of choice.

Reliability implies dependability – reliability introduces the concept of failure and time to failure:

Reliability is the probability that a system or component can perform its intended function for a specified interval under stated conditions.

Quality and reliability go hand in hand. The customer expects a product of good quality that performs reliably.

1.3 Historical Development

The history of quality is as old as civilization. The Harappans of the ancient Indus Valley civilization (3000 BC) achieved high precision in the measurement of length, mass, and time. The smallest division, which is marked on an ivory scale from Lothal, was approximately 1.704 millimeters, recorded in the Bronze Age. The dimensions of the pyramids, built around 2500 BC, show a high degree of accuracy. However, the use of tolerancing systems for the specification of quality and statistical principles to monitor quality are of recent origin. The quality movement may be traced back to medieval Europe. Craftsmen began organizing into unions called guilds in the late thirteenth century. Manufacturing in the industrialized world followed the craftsmanship model throughout the eighteenth century. The factory system, with its emphasis on product inspection, started in Great Britain in the mid-1750s and grew into the Industrial Revolution in the early nineteenth century. In 1798 Eli Whitney introduced the concept of producing interchangeable parts to simplify assembly.

Objective methods of measuring and ensuring dimensional consistency evolved in the mid-1800s with the introduction of go gages. A go gage for a hole checks for its lower limit (maximum material condition). No-go gages, which are used to check the upper limit for a hole, were introduced much later. Frederick W. Taylor introduced the principles of scientific management around 1900 and emphasized the division of labor with a focus on productivity. There was a significant rise in productivity but it had a negative effect on quality. Henry Ford's moving automobile assembly line was introduced in 1913. This required that consistently good-quality parts were available so that the production assembly line would not be forced to slow down. In 1924 Walter A. Shewhart introduced the basic ideas of the statistical process control chart, which signaled the beginning of the era of statistical quality control. By the mid-1930s, statistical quality control methods were widely used at Western Electric, a manufacturing arm of the Bell system.

World War II brought increased recognition of quality in manufacturing industries and military applications. The American Society for Quality Control was formed in 1946. (Eventually it shortened its name to ASQ in 1997.) A quality revolution in Japan followed World War II: the Japanese began applying the lessons learned in producing military goods produced for export. Quality stalwarts W. Edwards Deming and Joseph M. Juran lectured extensively in Japan. As a result, the Japanese became leaders in quality by the 1970s. Japanese manufacturers began increasing their share in American markets, resulting in widespread economic effects in the United States. The U.S. response emphasized not only statistics but approaches that embraced the entire organization – a movement that became known as Total Quality Management. Several other quality initiatives followed. The ISO 9000 quality system standards were published in 1987. The Baldrige National Quality Program and the Malcolm Baldrige National Quality Award were established by the U.S. Congress in the same year. The quality philosophies that introduced the modern concepts of quality are presented in the next section.

1.4 Quality Philosophies

Several individuals made significant contributions to quality control and improvement. We take a closer look at the approach and philosophies of W. Edwards Deming, Joseph M. Juran, Philip B. Crosby, and Armand V. Feigenbaum.

W. Edwards Deming

W. Edwards Deming is perhaps the best-known quality expert in the world. He was instrumental in the post-war industrial revival of Japan. Subsequently his ideas were increasingly adopted in industry in the United States and other countries. Deming received his electrical engineering degree from the University of Wyoming and his Ph.D. in mathematical physics. He worked for the Western Electric Company with Walter A. Shewhart, the developer of the control chart. Deming then worked with the U.S. Department of Agriculture and the U.S. Census Bureau. Starting in 1950 he delivered a series of lectures to top management in Japan on statistical process control. Japanese industry adopted his methods which resulted in a significant improvement in quality. Deming firmly believed that quality is the responsibility of the management. The Deming philosophy is summarized in the following fourteen points,[1] which were included in his monumental work *Out of the Crisis*.

The fourteen points apply to both small and large organizations, to the service industry as well as to manufacturing. They also apply to a division within a company. The fourteen points are presented here.

1. *Create constancy of purpose for improvement of product and service.* The point stresses the need for a mission statement which must be understood by all employees, suppliers, and customers. The strategic plan should look for the long-term payback.

2. *Adopt the new philosophy.* Management must learn the responsibilities and take on leadership for change. Poor workmanship, defective products, or bad service are not acceptable.

3. *Cease dependence on mass inspection.* Eliminate the need for inspection on a mass basis by building quality into the product in the first place. Statistical methods of quality control are more efficient.

4. *End the practice of awarding business on the basis of price tag alone.* Instead, minimize the total cost. The aim in the purchase of new tools and other equipment should be to minimize the net cost per hour of operation or per piece produced. Move toward a single supplier for any one item, on a long-term relationship of loyalty and trust.

5. *Improve constantly and forever the system of production and service.* The improvement of product and service is an ongoing process. The Deming cycle involves the four-step process of plan, do, check, act. At the *plan* stage, the opportunities for improvement are identified. The theory is tested on a small scale at the *do* stage, the results of the test are analyzed at the *check* stage, and the results are implemented in the *act* stage.

6. *Institute training.* On-the-job training must be provided for all employees. Employees must be encouraged to implement the knowledge developed through training.

7. *Adopt and institute leadership.* The aim of supervision should be to help people to do a better job using machines. Supervision must create an environment where the workers take leadership roles in accomplishing their work.

---

[1] Deming, W. Edwards, *Out of the Crisis*, pp. 23–24, © 2000, Massachusetts Institute of Technology, by permission of the MIT Press.

1.4 Quality Philosophies

8. *Drive out fear.* Management must create an environment where workers are encouraged to ask questions and make suggestions. A climate of innovation leads to progress.

9. *Break down barriers between departments.* People in research, design, material procurement, sales, and production must work as a team. They must understand the requirements and specifications. Teamwork leads to improvements in quality and productivity.

10. *Eliminate slogans, exhortations, and targets for the work force.* Exhortations such as asking for zero defects and new levels of productivity only create adversarial relationships. The bulk of the causes of low quality and low productivity belong to the system and thus lie beyond the power of the work force.

11. *Eliminate numerical quotas for the work force and eliminate numerical goals for people in management.* Quotas lead to the deterioration of quality. Learn the capabilities of processes and methods to improve them.

12. *Remove barriers that rob people of pride of workmanship.* Quality is achieved in the company when all employees are satisfied and motivated. Management must create an environment where the workers take pride in their job.

13. *Encourage education and self-improvement for everyone.* An organization needs people that are improving with education.

14. *Take action to accomplish the transformation.* The transformation is everybody's job.

Joseph M. Juran

Joseph M. Juran is the founder of the Juran Institute, which offers consulting and management training in quality. Juran obtained his degree in electrical engineering from the University of Minnesota in 1924 and then worked with Shewhart at Western Electric. He worked with a team from Bell Laboratories in 1926 to set up the first statistical process control techniques for factories. He published his *Quality Control Handbook* in 1951. In 1954 he was invited to Japan, where he conducted training courses in quality management. Juran contributed to quality through his original ideas and the vast amount of literature he developed on quality. He defined quality as *fitness for use*. Juran proposed the *quality trilogy*: quality planning, quality control, and quality improvement to develop a universal thought process for quality. Quality planning is the process for preparing to meet the company's goals. Both internal and external customers are identified and their needs are determined. Products and services are developed to fulfill these needs. Quality control is the process for meeting company goals during operations, and statistical process control techniques are the primary tools of control. Quality improvement is the process for breaking through to superior, unprecedented levels of performance. Juran stated categorically that waste must be identified and eliminated. Juran conceptualized the

Pareto principle, which helps in identifying the *vital few* out of the *trivial many*. This is commonly referred to as the 80–20 principle – 80% of the problems are created by 20% of the causes.

Philip B. Crosby

Philip B. Crosby is a businessman and author who influenced quality improvement through his writings and lectures. He started the Crosby Associates, which provides consulting and training in quality management. In his book *Quality Is Free*, Crosby provides a detailed quality management grid which provides various stages of management understanding and attitude, quality organization status, problem handling, cost of quality in relation to sales, quality improvement actions, and company quality posture. Crosby's response to the quality crisis was the principle of "doing it right the first time." He also included four major principles: (1) quality is "conformance to requirements," (2) the management system is prevention, (3) the performance standard is *zero defects*, and (4) the measurement system is the cost of nonconformance. The concept of zero defects was ahead of its time. More recently the concept of zero defects has led to the creation and development of Six Sigma, pioneered by Motorola, which has since been adopted worldwide by many other organizations.

Armand V. Feigenbaum

Armand V. Feigenbaum is a pivotal figure in the history of quality. He received his Ph.D. from MIT. In 1951 he published his book *Quality Control: Principles, Practice and Administration*, which was later published under the title *Total Quality Control* in 1961. Feigenbaum broadened a discipline that had relied primarily on production employees to a new stage in which everyone in an organization participates in the process of quality improvement. His book influenced much of the early philosophy of quality management in Japan in the early 1950s. He proposed a three-step process for quality improvement: quality leadership, quality technology, and organizational commitment. Total quality control is an effective system for integrating the quality development, quality maintenance, and quality improvement efforts of the various groups in an organization, enabling production and service to operate at the most economical level to achieve full customer satisfaction.

We have briefly described the philosophies of Deming, Juran, Crosby, and Feigenbaum. Each of them stressed the importance of quality and the pivotal role management must play in the implementation of quality improvement.

1.5 Conclusion

We have presented a brief description of quality history and the philosophies that influenced the quality movement. There are many others who contributed to quality improvement. Readers are encouraged to study current trends such as the Six Sigma approach and ISO 9000 certification. Quality improvement is an ongoing process, and the implementation of quality principles is not limited to industry – these principles are for all businesses, offices, services, education, healthcare, and other organizations.

Questions for Discussion

1. The chapter discussed four quality gurus. Who are the other major contributors to quality improvement?
2. What is the Malcolm Baldridge National Quality Award? What are the criteria for this award? Which organizations are the past recipients?
3. What is the underlying philosophy of Six Sigma? Which companies spearheaded this movement?
4. Who proposed the "zero defects" concept? Why did this philosophy lose its appeal?
5. How does improvement in quality benefit a manufacturing company?
6. Why do quality and reliability go hand in hand?

2

Tolerances and Fits

2.1 Introduction

Quality in a product starts at the specification stage. A number of choices are made in a design with respect to dimensions, sizes, and tolerances. *Preferred numbers* play an important role in simplifying these choices. The concept of preferred numbers is discussed and then the topics of tolerances and fits are developed, followed by a discussion of how tolerances are related to manufacturing processes.

2.2 Preferred Numbers

To reduce variety, it is a standard practice to select the sizes of objects from a preferred series of numbers, which follow a geometric progression. The size recommendations in worldwide metric standards use these numbers. Charles Renard, a French army captain, developed a set of these numbers in 1877 and reduced the number of different sizes of rope used in military balloons from 425 to 17. The series of numbers are designated R5, R10, R20, and R40 (where R stands for Renard). Each is a geometric series of numbers that are rounded. The R5 series has a progression ratio of $\sqrt[5]{10} = 1.584893192\ldots$ (~ 1.6), the R10 series has a progression ratio of $\sqrt[10]{10} = 1.2589254117\ldots$ (~ 1.25), and so on. The numbers are calculated using these exact ratios and are then rounded. The preferred numbers are included in the ANSI Z17.1-1973 and ISO 3-1973 standards. Four basic series of preferred numbers are given here:

R5: 10, 16, 25, 40, 63, 100.
R10: 10, 12.5, 16, 20, 25, 31.5, 40, 50, 63, 80, 100.
R20: 10, 11.2, 12.5, 14, 16, 18, 20, 22.4, 25, 28, 31.5, 35.5, 40, 45, 50, 56, 63, 71, 80, 90, 100.

R40: 10, 10.6, 11.2, 11.8, 12.5, 13.2, 14, 15, 16, 17, 18, 19, 20, 21.2, 22.4, 23.6, 25, 26.5, 28, 30, 31.5, 33.5, 35.5, 37.5, 40, 42.5, 45, 47.5, 50, 53, 56, 60, 63, 67, 71, 75, 80, 85, 90, 95, 100.

We note that multiples of 10, 100, ... and fractions 1/10, 1/100, ... are also included within each series.

Other rounded numbers are suggested for preferred metric sizes. These may be used for choosing dimensions, capacities, and so on:

First choice	10		16		25		40		60		100
Second choice		12		20		30		50		80		
Third choice	11	14	18	22	28	35	45	55	70	90		

As an example, the automotive engine capacity designations 1 liter, 1.6 liter, 2.5 liter, 4 liter, etc., are numbers that originate from this series. The rounded off preferred numbers here may be used as preliminary values for design. For more detailed use, we recommend the use of values from ISO or ANSI standards.

The use of preferred sizes is a step toward standardization, which reduces variety.

Example 2.1. An automotive engine manufacturer produces engines with 1.6-liter, 2.5-liter, and 4-liter capacities from the R5 series. Based on the demand, it is determined that a new engine is to be introduced with a capacity between 1.6 and 2.5 liters. What capacity do you suggest?

Solution. We note that 1.6 and 2.5 are also included in the R10 series and 2 is the value between these numbers. The suggested capacity is 2 liters. We also observe that 2 is the geometric mean ($2 \sim \sqrt{(1.6)(2.5)}$) of 1.6 and 2.5. Thus, the geometric mean is included in one of the R series.

Another example of preferred sizes is the spindle speed in machine tools. Geometric speed steps are used in multiple-speed gear boxes.

Part sizes and tolerance values are chosen from the preferred numbers. The tolerance of a part is a measure of quality, and the specification of tolerance plays an important role in making economic decisions.

2.3 Tolerances and Fits

No engineering component can be manufactured repeatedly to an exact size. If a number of components of specified dimensions are made by the same process, their sizes will vary and these variations will be random in nature. The variations may be reduced by controlling the variables of the process. The designer must make a

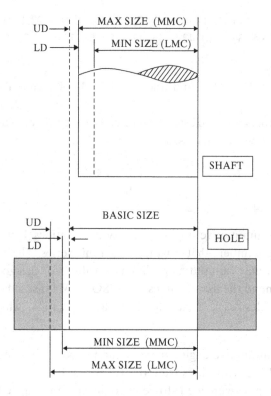

Figure 2.1. Definitions of tolerances for shafts and holes (LD, lower deviation; UD, upper deviation; MMC, maximum material condition; LMC, least material condition)

decision about these variations based on the function of the part. The smaller the specified variation, the higher the cost. This leads to the concept of tolerance.

Tolerance is the difference between the maximum and minimum size limits on a part. The tolerance is specified by the international tolerance (IT) grade. International tolerance grades range from IT01 to IT16. The tolerance values for various grades are discussed after introducing some important definitions.

Engineering products generally consist of several parts that are assembled together. An assembly may be viewed as a shaft fitting in a hole, as shown in Fig. 2.1. The various definitions related to tolerances, limits, and fits become clear from this figure. The *basic size* is the size to which tolerance and deviations are assigned; it is the same for the hole and its mating shaft. The *deviation* is the algebraic difference between a size and the corresponding basic size. The *upper deviation* is the algebraic difference between the maximum size limit and the basic size, whereas the *lower deviation* refers to the algebraic difference between the minimum size limit and the basic size. The *fundamental deviation* is the deviation of the size closer to the basic size and is specified by a letter: Upper case letters designate holes and lower case letters designate shafts. The tolerance on a part is completely specified by specifying the basic size, fundamental deviation, and the tolerance grade; for example, 50H7 specifies the tolerance on a hole of basic size 50 mm, and 50g6 specifies the tolerance on a shaft of basic size 50 mm. The maximum size of a shaft and the minimum size of the hole defining a fit indicate the *maximum material condition* (MMC). It is clear that the part has the most material in it at MMC. The MMC represents the most severe condition for a fit. The maximum size of a hole and the minimum size of the shaft may be referred to as the *least material condition* (LMC). A loose fit has its maximum clearance at LMC.

Tolerance Grades

In defining the tolerance grades IT5 through IT16, we make use of the tolerance unit i μm (1 μm = 10^{-6} m or 0.001 mm) defined by

$$i = 0.45 D^{\frac{1}{3}} + 0.001 D \ \mu\text{m}, \tag{2.1}$$

where D is the dimension in millimeters.

The tolerance values for various grades are given as follows:

Grade   IT5 IT6 IT7 IT8 IT9 IT10 IT11 IT12 IT13 IT14 IT15 IT16
Tolerance 7i 10i 16i 25i 40i 64i 100i 160i 250i 400i 640i 1000i.   (2.2)

The tolerance value increases tenfold after five steps; thus, the values follow the R5 series of preferred numbers.

The tolerances for IT01, IT0, and IT1 are taken as $0.3 + 0.008D$, $0.5 + 0.012D$, and $0.8 + 0.020D$, respectively. Tolerance grades IT2 and IT3 are placed between IT1 and IT5 to form a geometric progression.

Example 2.2. Calculate tolerance grades IT01 through IT16 for the nominal size of 40 mm.

Solution. The unit tolerance is calculated using $D = 40$ mm:

$$i = 0.45 D^{\frac{1}{3}} + 0.001 D = 1.578978 \ \mu\text{m}.$$

For grade IT5, $7i = 11.1 \ \mu$m; for grade IT01, $0.3 + 0.008D = 0.6 \ \mu$m; for grade IT0, $0.5 + 0.012D = 1.0 \ \mu$m; and for grade IT1, $0.8 + 0.020D = 1.6 \ \mu$m. Grades IT6 to IT16 are calculated using Eq. (2.2). Grades IT2, IT3, and IT4 are placed to form a geometric series. The final result is as follows:

IT Grade   01   0   1   2   3   4   5   6   7   8   9   10 11 12 13 14 15   16
Tolerance (μm) 0.6 1 1.6 2.6 4.2 6.8 11.1 15.8 25.3 39.5 63.2 101 158 253 395 632 1011 1579

Table 2.1. International tolerance system

D1 < D ≤ D2		Grade																	
D1	D2	IT01	IT0	IT1	IT2	IT3	IT4	IT5	IT6	IT7	IT8	IT9	IT10	IT11	IT12	IT13	IT14	IT15	IT16
0	3	0.3	0.5	0.8	1.2	2	3	4	6	10	14	25	40	60	100	140	250	400	600
3	6	0.4	0.6	1	1.5	2.5	4	5	8	12	18	30	48	75	120	180	300	480	750
6	10	0.4	0.6	1	1.5	2.5	4	6	9	15	22	36	58	90	150	220	360	580	900
10	18	0.5	0.8	1.2	2	3	5	8	11	18	27	43	70	110	180	270	430	700	1100
18	30	0.6	1	1.5	2.5	4	6	9	13	21	33	52	84	130	210	330	520	840	1300
30	50	0.6	1	1.5	2.5	4	7	11	16	25	39	62	100	160	250	390	620	1000	1600
50	80	0.8	1.2	2	3	5	8	13	19	30	46	74	120	190	300	460	740	1200	1900
80	120	1	1.5	2.5	4	6	10	15	22	35	54	87	140	220	350	540	870	1400	2200
120	180	1.2	2	3.5	5	8	12	18	25	40	63	100	160	250	400	630	1000	1600	2500
180	250	2	3	4.5	7	10	14	20	29	46	72	115	185	290	460	720	1150	1850	2900
250	315	2.5	4	6	8	12	16	23	32	52	81	130	210	320	520	810	1300	2100	3200
315	400	3	5	7	9	13	18	25	36	57	89	140	230	360	570	890	1400	2300	3600
400	500	4	6	8	10	15	20	27	40	63	97	155	250	400	630	970	1550	2500	4000
500	630	–	–	–	–	–	–	–	44	70	110	175	280	440	700	1100	1750	2800	4400
630	800	–	–	–	–	–	–	–	50	80	125	200	320	500	800	1250	2000	3200	5000
800	1000	–	–	–	–	–	–	–	56	90	140	230	360	560	900	1400	2300	3600	5600
1000	1250	–	–	–	–	–	–	–	66	105	165	260	420	660	1050	1650	2600	4200	6600
1250	1600	–	–	–	–	–	–	–	78	125	195	310	500	780	1250	1950	3100	5000	7800
1600	2000	–	–	–	–	–	–	–	92	150	230	370	600	920	1500	2300	3700	6000	9200
2000	2500	–	–	–	–	–	–	–	110	175	280	440	700	1100	1750	2800	4400	7000	11000
2500	3150	–	–	–	–	–	–	–	135	210	330	540	860	1350	2100	3300	5400	8600	13500

Unit = 0.001 mm (1 μm).

2.3 Tolerances and Fits

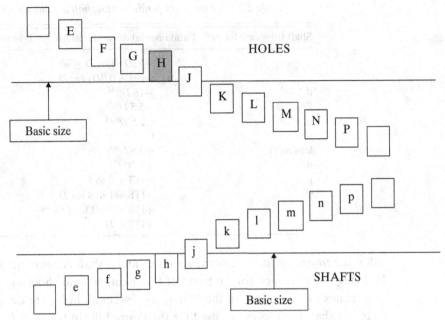

Figure 2.2. Position of tolerance zones for holes and shafts

The preceding calculations are given to explain the rationale of tolerance grades. These calculated values have been rounded and provided in Table 2.1.

Letter designations are given to define the placement of the tolerance zone with respect to the basic size. The placement is based on the fundamental deviation. The relative placement of the tolerance zones for various letter designations is shown schematically in Fig. 2.2.

We observe that H for holes and h for shafts have a fundamental deviation of zero. Holes with the H designation are of basic size and larger and shafts with the h designation are of basic size and smaller. The fundamental deviation formulas for some typical shaft designations are given in Table 2.2.

The fundamental deviation for 50f7 is $-5.5\,(50)^{0.41} = -27.35\,\mu m$. The rounded value from the standard tables is $-25\,\mu m$. The IT7 tolerance of $25\,\mu m$ is now placed in line with the zone suggested in Fig. 2.2. Thus, the tolerance is -25 to $-50\,\mu m$. The standard tables for the tolerance values are given in ISO 286–2:1988, ANSI B4.2–1978, and other national standards. We provide typical recommended fits in Table 2.3.

The tolerance conditions of the hole and the shaft result in a *fit* in an assembly. If clearance occurs under all tolerance conditions we call it the *clearance fit*. If interference occurs under all tolerance conditions we call it the *interference fit*. If clearance or interference is determined by the actual size within the permissible tolerances, we

Table 2.2. *Formulas for fundamental shaft deviations*

Shaft tolerance letter	Fundamental deviation in μm for D in mm
c	$-52D^{0.2}$ for $D \leq 40$
	$-(95 + 0.8D)$ for $D > 40$
d	$-16 D^{0.44}$
f	$-5.5 D^{0.41}$
g	$-2.5 D^{0.34}$
h	0
k (4 to 7)	$+0.6 D^{0.34}$
n	$+5 D^{0.34}$
p	$+\text{IT7} + 0$ to 5
s	$+\text{IT8} + 1$ to 4 for $D \leq 50$
	$+\text{IT7} + 0.4D$ for $D > 50$
u	$+\text{IT7} + D$

call it the *transition fit*. A fit between a hole and a shaft is conveniently determined by fixing the tolerance zone for one of the members and choosing different tolerance zones for the other. In the *hole basis* system all holes are designated H, and different shaft tolerances are used for the desired fit. In the *shaft basis* system, the

Table 2.3. *Description of preferred fits (from ANSI B4.2)*

Hole basis	Shaft basis	Description
H11/c11	C11/h11	*Loose running* fit for wide commercial tolerances
H9/d9	D9/h9	*Free running* fit for use where accuracy is essential, but good for large temperature variations, high running speeds, or heavy journal pressures
H8/f7	F8/h7	*Close running* fit for running on accurate machines and for accurate location at moderate speeds and journal pressures
H7/g6	G7/h6	*Sliding* fit not intended to run freely but to move and turn freely and locate accurately
H7/h6	H7/h6	*Locational clearance* fit provides snug fit for locating stationary parts but can be freely assembled and disassembled
H7/k6	K7/h6	*Locational transition* fit for accurate location, a compromise between clearance and interference
H7/n6	N7/h6	*Locational transition* fit for more accurate location where greater interference is permissible
H7/p6	P7/h6	*Locational interference* fit for parts requiring rigidity and alignment with prime accuracy of location but without special bore pressure requirements
H7/s6	S7/h6	*Medium drive* fit for ordinary steel parts or shrink fits on light sections; the tightest fit usable with cast iron
H7/u6	U7/h6	*Force* fit suitable for parts that can be highly stressed or for shrink fits where the heavy pressing forces required are impractical

shafts are fixed as h and the holes of various designations are used for different fits. As an example, the outer dimension of a ball bearing is h-type (shaft basis), and the inner diameter is H-type (hole basis). The housing and shaft tolerances are chosen to obtain the desired fit. Suggested fits from the standards are given in Table 2.3. The production of a hole is more difficult than the production of a shaft of the same size. In choice of a fit, the tolerance of a hole is generally designated one grade higher than that of its mating shaft. In Table 2.3, common hole-basis fits are H7/g6, H7/h6, H7/k6, H7/p6, H7/s6, and H7/u6. The tolerances for these recommended fits are given in Tables 2.4 and 2.5.

Example 2.3. Make your choice for a sliding fit for a basic size of 50 mm and provide the tolerances for the hole-basis and shaft-basis approaches.

Solution. The choice is 50 H7/g6 for the hole basis and 50 G7/h6 for the shaft basis.

Hole basis (50 H7/g6 tolerances): from Table 2.5, the tolerance for 50H7 is $+0$ to $+25$ μm; from Table 2.4, the tolerance for 50g6 is -9 to -25 μm.

Shaft basis (50 G7/h6 tolerances): from Table 2.5, the tolerance for 50G7 is $+9$ to $+34$ μm; from Table 2.4, the tolerance for 50h6 is 0 to -16 μm.

The part tolerances are related to the parts' manufacturing processes.

2.4 Manufacturing Processes and Tolerances

For the most common parts used in engineering assemblies, the tolerance grades range from IT4 through IT11. The range of grades IT01 to IT7 is generally used for components in measuring tools and optics. Grades IT8 though IT16 are used for material stock specifications. The tolerance grades obtainable with various manufacturing processes are given in Table 2.6.

Table 2.6 shows that there is a close relationship between the process and the tolerance grade. In each process, the achieved tolerance is related to a number of variables, such as precision of the machine used, the skill of the operator, the processing variables that are selected, and so on.

2.5 Tolerance Selection in Assemblies

The tolerances of components are chosen at the design stage. An assembly has several components. If the components are stacked as shown in Fig. 2.3, the resulting

Table 2.4. Tolerance zones for shafts (for hole-basis fits)

Over	To	c11	d9	f7	g6	h6	k6	n6	p6	s6	u6
0	3	−120/−60	−45/−20	−16/−6	−8/−2	−6/0	0/+6	+4/+10	+6/+12	+14/+20	+18/+24
3	6	−145/−70	−60/−30	−18/−10	−12/−4	−8/0	+1/+9	+8/+16	+12/+20	+19/+27	+23/+31
6	10	−170/−80	−76/−40	−28/−13	−14/−5	−9/0	+1/+10	+10/+19	+15/+24	+23/+32	+28/+37
10	14	−205/−95	−93/−50	−34/−16	−17/−6	−11/0	+1/+12	+12/+23	+18/+29	+28/+39	+33/+44
14	18	−205/−95	−93/−50	−34/−16	−17/−6	−11/0	+1/+12	+12/+23	+18/+29	+28/+39	+33/+44
18	24	−240/−110	−117/−65	−41/−20	−20/−7	−13/0	+2/+15	+15/+28	+22/+35	+35/+48	+41/+54
24	30	−240/−110	−117/−65	−41/−20	−20/−7	−13/0	+2/+15	+15/+28	+22/+35	+35/+48	+48/+61
30	40	−280/−120	−142/−80	−50/−25	−25/−9	−16/0	+2/+18	+17/+33	+26/+42	+43/+59	+60/+76
40	50	−290/−130	−142/−80	−50/−25	−25/−9	−16/0	+2/+18	+17/+33	+26/+42	+43/+59	+70/+86
50	65	−330/−140	−174/−100	−60/−30	−29/−10	−19/0	+2/+21	+20/+39	+32/+51	+53/+72	+87/+106
65	80	−340/−150	−174/−100	−60/−30	−29/−10	−19/0	+2/+21	+20/+39	+32/+51	+59/+78	+102/+121
80	100	−390/−170	−207/−120	−71/−36	−34/−12	−22/0	+3/+25	+23/+45	+37/+59	+71/+93	+124/+146
100	120	−400/−180	−207/−120	−71/−36	−34/−12	−22/0	+3/+25	+23/+45	+37/+59	+79/+101	+144/+166
120	140	−450/−200	−245/−145	−83/−43	−39/−14	−25/0	+3/+28	+27/+52	+43/+68	+92/+117	+170/+195
140	160	−460/−210	−245/−145	−83/−43	−39/−14	−25/0	+3/+28	+27/+52	+43/+68	+100/+125	+190/+215
160	180	−480/−230	−245/−145	−83/−43	−39/−14	−25/0	+3/+28	+27/+52	+43/+68	+108/+133	+210/+235
180	200	−530/−240	−285/−170	−96/−50	−44/−15	−29/0	+4/+33	+31/+60	+50/+79	+122/+151	+236/+265
200	225	−550/−260	−285/−170	−96/−50	−44/−15	−29/0	+4/+33	+31/+60	+50/+79	+130/+159	+258/+287
225	250	−570/−280	−285/−170	−96/−50	−44/−15	−29/0	+4/+33	+31/+60	+50/+79	+140/+169	+284/+313
250	280	−620/−300	−325/−190	−108/−56	−49/−17	−32/0	+4/+36	+34/+66	+56/+88	+158/+190	+315/+347
280	315	−650/−330	−325/−190	−108/−56	−49/−17	−32/0	+4/+36	+34/+66	+56/+88	+170/+202	+350/+382
315	355	−720/−360	−350/−210	−119/−62	−54/−18	−36/0	+4/+40	+37/+73	+62/+98	+190/+226	+390/+426
355	400	−760/−400	−350/−210	−119/−62	−54/−18	−36/0	+4/+40	+37/+73	+62/+98	+208/+244	+435/+471
400	450	−840/−440	−385/−230	−131/−68	−60/−20	−40/0	+5/+45	+40/+80	+68/+108	+232/+272	+490/+530
450	500	−880/−480	−385/−230	−131/−68	−60/−20	−40/0	+5/+45	+40/+80	+68/+108	+252/+292	+540/+580

All units in μm; $1 \mu m = 0.001$ mm.
(Ref. ANSI B4. 2−1978)

Table 2.5. Tolerance zones for holes (for shaft-basis fits)

Over	To	C11	D9	F8	G7	H7	K7	N7	P7	S7	U7
0	3	+60/+120	+20/+45	+6/+20	+2/+12	0/+10	−10/0	−14/−4	−16/−6	−24/−14	−28/−18
3	6	+70/+145	+30/+60	+10/+28	+4/+16	0/+12	−9/+3	−16/−4	−20/−8	−27/−15	−31/−19
6	10	+80/+170	+40/+76	+13/+35	+5/+20	0/+15	−10/+5	−19/−4	−24/−9	−32/−17	−37/−22
10	14	+95/+205	+50/+93	+16/+43	+6/+24	0/+18	−12/+6	−23/−5	−29/−11	−39/−21	−44/−26
14	18	+95/+205	+50/+93	+16/+43	+6/+24	0/+18	−12/+6	−23/−5	−29/−11	−39/−21	−44/−26
18	24	+110/+240	+65/+117	+20/+53	+7/+28	0/+21	−15/+6	−28/−7	−35/−14	−48/−27	−54/−33
24	30	+110/+240	+65/+117	+20/+53	+7/+28	0/+21	−15/+6	−28/−7	−35/−14	−48/−27	−61/−40
30	40	+120/+280	+80/+142	+25/+64	+9/+34	0/+25	−18/+7	−33/−8	−42/−17	−59/−34	−76/−51
40	50	+130/+290	+80/+142	+25/+64	+9/+34	0/+25	−18/+7	−33/−8	−42/−17	−59/−34	−86/−61
50	65	+140/+330	+100/+174	+30/+76	+10/+40	0/+30	−21/+9	−39/−9	−51/−21	−72/−42	−106/−76
65	80	+150/+340	+100/+174	+30/+76	+10/+40	0/+30	−21/+9	−39/−9	−51/−21	−78/−48	−121/−91
80	100	+170/+390	+120/+207	+36/+90	+12/+47	0/+35	−25/+10	−45/−10	−59/−24	−93/−58	−146/−111
100	120	+180/+400	+120/+207	+36/+90	+12/+47	0/+35	−25/+10	−45/−10	−59/−24	−101/−66	−166/−131
120	140	+200/+450	+145/+245	+43/+106	+14/+54	0/+40	−28/+12	−52/−12	−68/−28	−117/−77	−195/−155
140	160	+210/+460	+145/+245	+43/+106	+14/+54	0/+40	−28/+12	−52/−12	−68/−28	−125/−85	−215/−175
160	180	+230/+480	+145/+245	+43/+106	+14/+54	0/+40	−28/+12	−52/−12	−68/−28	−133/−93	−235/−195
180	200	+240/+530	+170/+285	+50/+122	+15/+61	0/+46	−33/+13	−60/−14	−79/−33	−151/−105	−265/−219
200	225	+260/+550	+170/+285	+50/+122	+15/+61	0/+46	−33/+13	−60/−14	−79/−33	−159/−113	−287/−241
225	250	+280/+570	+170/+285	+50/+122	+15/+61	0/+46	−33/+13	−60/−14	−79/−33	−169/−123	−313/−267
250	280	+300/+620	+190/+320	+56/+137	+17/+69	0/+52	−36/+16	−66/−14	−88/−36	−190/−138	−347/−295
280	315	+330/+650	+190/+320	+56/+137	+17/+69	0/+52	−36/+16	−66/−14	−88/−36	−202/−150	−382/−330
315	355	+360/+720	+210/+350	+62/+151	+18/+75	0/+57	−40/+17	−73/−16	−98/−41	−226/−169	−426/−369
355	400	+400/+760	+210/+350	+62/+151	+18/+75	0/+57	−40/+17	−73/−16	−98/−41	−244/−187	−471/−414
400	450	+440/+840	+230/+385	+68/+165	+20/+83	0/+63	−45/+18	−80/−17	−108/−45	−272/−209	−530/−467
450	500	+480/+880	+230/+385	+68/+165	+20/+83	0/+63	−45/+18	−80/−17	−108/−45	−292/−229	−580/−517

All units in μm; 1 μm = 0.001 mm.
(Ref: ANSI B4.2-1978)

Table 2.6. *Manufacturing processes and IT grades*

Process	IT grade range
Lapping and honing	4–5
Grinding	5–8
Precision turning and boring	5–7
Broaching	5–8
Reaming	6–10
Powder metallurgy (sintered)	7–10
Turning and boring	7–11
Milling, planing, shaping	10–11
Drilling	10–11
Punching, blanking	10–11
Die casting	10–12
Casting	14–16

(Ref: ANSI B4.2)

dimension Y is the sum of the component dimensions:

$$Y = X_1 + X_2 + X_3. \tag{2.3}$$

Fig. 2.4 shows Y as a gap formed by the difference between dimensions X_1 and X_2:

$$Y = X_1 - X_2. \tag{2.4}$$

We now denote t_Y as the tolerance of Y and t_1, t_2, and t_3 as tolerances of X_1, X_2, and X_3, respectively; t_Y does not have a linear relationship to the individual tolerances. Each dimension has a higher probability of occurrence in an interval closer to the mean value. For a normal distribution of each dimension, which is generally the case, the tolerances follow the rule of variances. (Statistics fundamentals will be presented in detail in Chapter 4.) Under the assumption that the process capabilities for the components are equal, the tolerance relationship for the assembly in Fig. 2.3 is

$$t_Y^2 = t_1^2 + t_2^2 + t_3^2.$$

For the assembly shown in Fig. 2.4, the tolerance relationship is

$$t_Y^2 = t_1^2 + t_2^2.$$

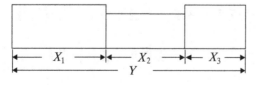

Figure 2.3. Assembly – addition of dimensions

2.5 Tolerance Selection in Assemblies

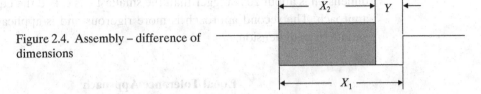

Figure 2.4. Assembly – difference of dimensions

We observe that *the square of the resulting tolerance is the sum of squares of the component tolerances*. Whether a dimension is added or subtracted when obtaining Y, the square of its tolerance is added to the right-hand side of t_Y^2. We now state this for a general case.

Let $X_1, X_2, \ldots, X_m$ be the m member dimensions and let Y be the resulting dimension obtained by a linear combination of the member dimensions with coefficients $+1$ or -1. The resulting tolerance is

$$t_Y^2 = t_1^2 + t_2^2 + \cdots + t_m^2. \tag{2.5}$$

Tolerance-related problems may be classified into two categories:

1. If component tolerances $t_1, t_2, \ldots$ are known, what is the resulting tolerance t_Y?
2. If the final tolerance, t_Y, is known, how do we allocate it to the component tolerances?

The solution of the problem of the first type is straightforward and follows from Eq. (2.5). We illustrate it by means of an example.

Example 2.4. Dimensions Y, X_1, X_2, X_3, and X_4 are related by the relationship $Y = X_1 + X_2 - X_3 + X_4$. If X_1, X_2, X_3, and X_4 are specified as 20 ± 0.025 mm, 40 ± 0.045 mm, 30 ± 0.035 mm, and 16 ± 0.015 mm, respectively, find the mean dimension of Y and specify its tolerance.

Solution. Since the mean dimensions of X_1, X_2, X_3, and X_4 are given, we obtain the mean dimension of Y as follows:

$$Y = 20 + 40 - 30 + 16 = 46 \text{ mm}.$$

The tolerances of the four dimensions are $t_1 = 0.05$ mm, $t_2 = 0.09$ mm, $t_3 = 0.07$ mm, and $t_4 = 0.03$ mm. Using Eq. (2.5),

$$t_Y = \sqrt{t_1^2 + t_2^2 + t_3^2 + t_4^2} = \sqrt{0.05^2 + 0.09^2 + 0.07^2 + 0.03^2} = 0.1281.$$

The tolerance on Y may be specified as $Y \pm t_Y/2$, which is 46 ± 0.064 mm.

Problems of the second type, where t_Y is known and the component tolerances have to be found, need special considerations. An approach that is recommended when the component dimensions are nearly equal – say, the largest dimension is about 20% bigger than the smallest – is called the equal tolerance approach. The second approach is more rigorous and is applicable for wider variations in dimension.

Equal Tolerance Approach

In this approach we set all component tolerances equal:

$$t = t_1 = t_2 = \cdots = t_m. \tag{2.6}$$

Then from Eq. (2.5) we get

$$t = \frac{t_Y}{\sqrt{m}}. \tag{2.7}$$

Example 2.5. A gap of 0.2 mm is to be maintained in an assembly with a tolerance of ± 16 μm. The gap is arrived at as a combination of two dimensions as follows:

$$0.2 = 30.5 - 30.3.$$

Determine the tolerances of each dimension.

Solution. The component dimensions are nearly equal in this problem. We use the equal allocation strategy. We have $t_Y = 32$ μm. The number of component dimensions is $m = 2$. From Eq. (2.7), we get

$$t = \frac{t_Y}{\sqrt{m}} = \frac{32}{\sqrt{2}} = 22.63 \,\mu\text{m}.$$

The tolerances may be set as 30.5 ± 11.3 μm and 30.3 ± 11.3 μm.

Unit Tolerance Approach

Equation (2.1) defines the tolerance unit $i = 0.45 D^{\frac{1}{3}} + 0.001 D$ μm used for a dimension D mm. We use the following notation:

$$i_k = 0.45 X_k^{\frac{1}{3}} + 0.001 X_k \quad X_k, k = 1, 2, \ldots, m. \tag{2.8}$$

We assume that all members have the same IT grades. We may then assume that the tolerance of a dimension is proportional to the unit tolerance. Let C be the constant of proportionality such that

$$t_k = C i_k \quad k = 1, 2, \ldots, m. \tag{2.9}$$

We evaluate C by substituting in Eq. (2.5) and rearranging the terms:

$$C = \frac{t_Y}{\sqrt{i_1^2 + i_2^2 + \cdots + i_m^2}} \tag{2.10}$$

Tolerances are then placed appropriately with respect to the basic dimensions. Note that C is a constant of proportionality so that the tolerance will have the same units as t_Y; however, the dimensions must be in millimeters for the applicability of Eq. (2.8). This weighted approach gives a good allocation that is consistent with the IT system.

Example 2.6. Dimension Y is obtained as the composite of dimensions X_1, $X_2, \ldots, X_4$ as follows:

$$Y = X_1 + X_2 - X_3 + X_4.$$

The desired tolerance of Y is ± 0.4 mm and $X_1 = 30$ mm, $X_2 = 40$ mm, $X_3 = 20$ mm, $X_4 = 10$ mm. Determine the tolerances of each dimension.

Solution. The smallest dimension is 10 mm and the largest is 40 mm. Since the variation is large, we use the unit tolerance approach for the tolerance allocation. The unit tolerance values for the dimensions are as follows:

$$i_1 = 0.45\sqrt[3]{30} + 0.001 \times 30 = 1.4283$$
$$i_2 = 0.45\sqrt[3]{40} + 0.001 \times 40 = 1.579$$
$$i_3 = 0.45\sqrt[3]{20} + 0.001 \times 20 = 1.2415$$
$$i_4 = 0.45\sqrt[3]{10} + 0.001 \times 10 = 0.9795.$$

From Eq. (2.10), we get

$$C = \frac{t_Y}{\sqrt{i_1^2 + i_2^2 + \cdots + i_m^2}} = \frac{0.8}{\sqrt{1.4283^2 + 1.579^2 + 1.2415^2 + 0.9795^2}} = 0.3031.$$

The tolerances are obtained using Eq. (2.9):

$$t_1 = Ci_1 = 0.432 \text{ mm}$$
$$t_2 = Ci_2 = 0.4785 \text{ mm}$$
$$t_3 = Ci_3 = 0.3763 \text{ mm}$$
$$t_4 = Ci_4 = 0.297 \text{ mm}.$$

We use symmetric placement to arrive at the dimensions 30 ± 0.216 mm, 40 ± 0.24 mm, 20 ± 0.19 mm, and 10 ± 0.15 mm.

2.6 Summary

The chapter explains the basis for choosing dimensions from the preferred number series. The IT system and the specification of tolerances for holes and shafts have been covered, and recommended fits have been given in complete detail. The theory of tolerance allocation has been presented. The material in this chapter will be of use in tolerance specification for quality improvement at the design stage.

EXERCISE PROBLEMS

2.1. What are hole-basis and shaft-basis systems for specification of fits? Where do you recommend each of these systems?

2.2. What is the basis for preferred numbers? How do they help in standardization?

2.3. A close running fit is desired between a shaft and a hole. The nominal diameter is 20 mm.

  a. Give your choice of the fit.
  b. Provide the tolerance values.

2.4. A hole–shaft assembly is assigned the tolerance $\phi 40$ H7/p6. Answer the following:

  a. Is the tolerance hole basis or shaft basis?
  b. Give the MMC and LMC sizes of the shaft and the hole.
  c. What type of fit does it result in?
  d. Give the maximum and minimum values of interference or clearance as appropriate. Also determine the corresponding values using the sum of squares rule.

2.5. A gap of 0.2 mm is to be maintained in an assembly with a tolerance of ± 16 μm. The gap is arrived at as a combination of two dimensions as follows:

$$0.2 = 30.5 - 30.3.$$

Determine the tolerances on each dimension, using (a) the equal tolerance approach and (b) the unit tolerance approach. Draw your conclusions.

2.6. In an assembly, the gap Y is obtained as $Y = X_1 - X_2 - X_3$. The tolerances on X_1, X_2, and X_3 are specified as 100 ± 0.5 mm, 70 ± 0.3 mm, and 28 ± 0.5 mm, respectively. Determine the tolerance of the gap.

2.7. Dimension Y is obtained as the composite of dimensions X_1, X_2, and X_3 as

$$Y = X_1 + X_2 - X_3.$$

The tolerance desired for Y is ± 0.05 mm and $X_1 = 12$ mm, $X_2 = 35$ mm, and $X_3 = 45$ mm. Determine the tolerances of each dimension using the unit tolerance approach.

2.8. A cluster gear is assembled using three spur gears which are randomly selected from the widths specified as 16 ± 0.3, 14 ± 0.15, and 15 ± 0.18 (mm). Determine the tolerance of the composite width of 45 mm. State any assumptions made.

2.9. The gap of 0.1 mm in an assembly is to be maintained with a tolerance of ±0.012 mm. The gap is arrived at as a combination of several dimensions as follows:

$$0.1 = 30.5 + 20.1 + 15.2 - 65.7.$$

Determine the tolerances of each dimension. State any assumptions made.

2.10. The fit between a shaft and a hole is specified as $\phi45$ H7/g6. What type of fit does it represent? Give the mean clearance or interference as appropriate and its minimum and maximum values using the sum-of-squares rule.

3

Geometric Tolerances

3.1 Introduction

The tolerances presented in Chapter 2 dealt with variations in the size of a part. The form of the part may vary within the tolerance zone of the size of a part. If the dimensional tolerances are specified for a cylinder, the boundary of the cross section of the cylinder may be of any form within the bounds defined by the upper and lower limits. However, such a form may not be acceptable from a functional point of view. There are other geometric features such as the position of a hole or the relationship of one surface with respect to another, such as parallelism or perpendicularity, that need to be specified. Tolerances on such features are called *geometric tolerances*. We present the basic ideas of geometric tolerances in this chapter.

3.2 Geometric Tolerances – Some Basic Ideas

Geometric tolerances are covered in ASME Y14.5M-1994, ISO 10303-519:2000, and other national standards. We follow the ASME Y14.5M notation in our discussion. Geometric tolerances are classified as shown in Fig. 3.1. The various symbols used for the tolerance specifications are also shown in the figure. Tolerances of form – straightness, flatness, circularity, and cylindricity – refer to the variations from a true straight line, a plane, a circle, and a cylinder, respectively. Thus, they are not defined relative to any reference feature or *datum*. The orientation, location, and runout are always relative to another feature such as an axis or a surface, referred to as a datum. Tolerances of line or surface profiles fall into either category.

Geometric tolerances are specified on a drawing in a feature control frame. An example of a feature control frame for position tolerance is shown in Fig. 3.2. It has rectangular areas designated for the tolerance symbol, tolerance, and reference datums. Other modifying symbols for specifying the feature, such as the diameter

3.2 Geometric Tolerances – Some Basic Ideas

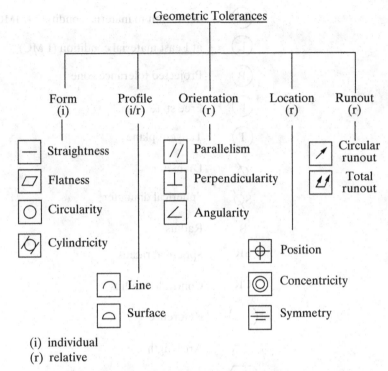

Figure 3.1. Geometric tolerances

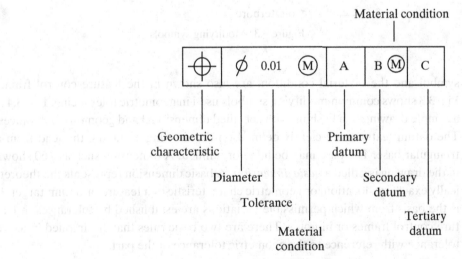

Figure 3.2. Feature control frame

Ⓜ	at Maximum material condition (MMC)
Ⓛ	at Least material condition (LMC)
Ⓟ	Projected tolerance zone
Ⓕ	Free state
Ⓣ	Tangent plane
⌀	Diameter
S⌀	Spherical diameter
R	Radius
SR	Spherical radius
CR	Controlled radius
( )	Reference
⌒	Arc length
⟨ST⟩	Statistical tolerance
↔	Between
⊔	Counterbore

Figure 3.3. Modifying symbols

symbol and the material condition, are also shown in the feature control frame. Fig. 3.3 shows common modifying symbols used in geometric tolerancing. Fig. 3.4 is a sample drawing of a bushing with specified dimensional and geometric tolerances. The datum features are clearly defined on the drawing by taking the lead from a triangular base. The base may be filled or unfilled. A dimension such as $\phi 60$ shown in the drawing is called a *basic dimension*. A basic dimension represents the theoretically exact size, location, or geometric characteristic of a feature or datum target. It is the basis from which permissible variations are established by tolerances in feature control frames or in notes. There are two basic rules that are implied in a size tolerance with reference to the geometric tolerance of the part.

Rule 1. When only a size tolerance is specified, the limits of size of an individual feature control the form as well as the size of the feature.

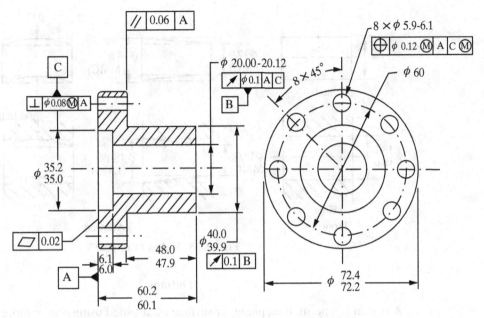

Figure 3.4. Symbols shown on a drawing

Rule 1 is also referred to as *Taylor's principle* after Frederick W. Taylor (1856–1915). The rule implies that no element of the feature may extend beyond the envelope of the perfect form at maximum material condition (MMC). The local size at any section must be within the least material condition (LMC) size limit. A part produced at LMC size may have the maximum form variation. Rule 1 does not apply for stock such as bars, sheets, tubing, and so on, or for parts subjected to free-state variation. If it is desired that a surface of a feature should exceed the boundary, a note to that effect must be specified. An interpretation of Rule 1 is shown in Fig. 3.5.

Rule 2. The condition *regardless of feature size* (RFS) applies with respect to the individual tolerance, datum reference, or both, as applicable. MMC or LMC must be specified on the drawing where it is required.

If MMC is specified for an individual tolerance, the precise tolerance applies at MMC and any deviation from MMC is added to the tolerance. This additional tolerance is referred to as the *bonus tolerance*. This bonus tolerance enables more parts to be acceptable in comparison with conventional tolerancing; we present some examples later.

We stated earlier that the geometric tolerances of orientation, location, and runout are relative to some reference feature called a datum. We now develop the concepts for defining datums.

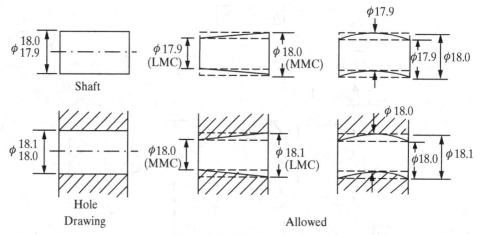

Figure 3.5. Rule 1 example

Datums

A datum is a point, line, plane, or surface established using one or more features of a part. Surfaces are generally flat or cylindrical. Datums are needed at the design stage for specifying tolerances, at the manufacturing stage for locating the parts to produce the surfaces, and at the measurement stage for monitoring quality. In a box-shaped part, the datum features are used to establish a frame of reference, consisting of three mutually perpendicular planes as shown in Fig. 3.6. The largest surface on the part defines the *primary datum*. Theoretically, three points on the surface define the plane. The *secondary datum* is defined by the line contact established by two points. The *tertiary datum* is theoretically a one-point contact. The part is first brought into contact with the primary datum (three-point contact) and then slid to touch the secondary datum (two-point contact). While keeping contact

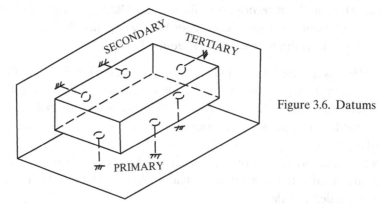

Figure 3.6. Datums

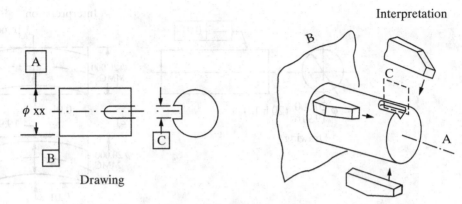

Figure 3.7. Cylindrical part

with the primary and secondary datums, the part is moved to contact the tertiary datum (one-point contact). The part is now completely located. This is the *3-2-1 principle of location*. A rigid part in space has six degrees of freedom (df) – three translational df along the axes and three rotational df about the axes. A fully located part requires all six df suppressed. If it is not possible to use an entire surface to establish a datum, points, lines, or areas may be defined as *datum targets* consistent with the 3-2-1 principle. The ASME Y14.5M standard provides symbols for these targets. Fig. 3.7 shows a cylindrical part with its axis defined as the primary datum. The axis is established by holding the part using a self-centering arrangement (four points) and one end face is the secondary datum (one point) and the keyway defines the tertiary datum (one point). One may observe that the three-jaw self-centering arrangement arrests four degrees of freedom, leaving the part free to translate along and rotate about the axis. We now discuss the various geometric tolerances.

3.3 Tolerances of Form

Tolerances of form consist of straightness, flatness, circularity, and cylindricity. These geometric tolerances represent variations from a true form and do not need any datum reference.

Straightness

Straightness is a condition where an element of a surface or axis is a straight line. A straightness tolerance specifies a tolerance zone within which the considered element or derived median line must lie. For cylindrical parts, the straightness of a surface element or the straightness of its axis may be specified.

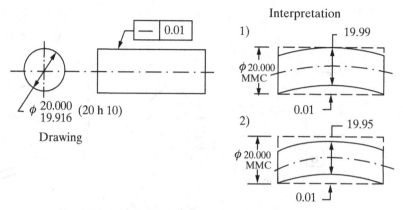

Figure 3.8. Straightness of surface elements

A straightness representation of surface elements is shown in Fig. 3.8. Every surface element must lie between two parallel lines separated by the straightness specification of 0.01 mm in a plane common with the nominal axis. *The entire object must lie within the boundary defined by the MMC.* If the cylindrical part has a uniform size of ϕ19.995 mm, the straightness error allowed is 20.000 mm (MMC) − 19.995 mm = 0.005 mm.

If the axis is controlled by the straightness tolerance as shown in Fig. 3.9, the applicable tolerance is valid at every size. The part itself can go outside the MMC

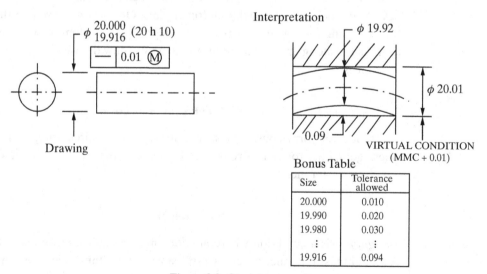

Figure 3.9. Straightness axis

3.3 Tolerances of Form

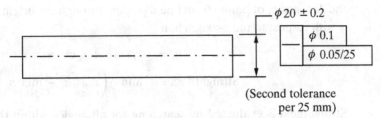

(Second tolerance per 25 mm)

Figure 3.10. Unit straightness

boundary. If the MMC symbol is not given in the tolerance control frame, then for every size from 19.916 to 20.000 mm, the straightness tolerance is 0.01 mm (RFS). The MMC specification defines the *virtual condition* (MMC + form tolerance = 20.01). The straightness tolerance allowed is "virtual condition − size of the part," as shown in the bonus table in Fig. 3.9. The MMC specification allows bonus tolerance. In some situations, if a tighter tolerance is to be specified on a per unit basis, a double-height feature control frame is used as shown in Fig. 3.10.

We provide the mathematical definition of straightness in the mid-plane x-y. This definition is useful for measuring straightness using coordinate measuring machines. Let (x_i, y_i), $i = 1, 2, \ldots, n$, shown in Fig. 3.11 be n points measured using the coordinate measuring machine (CMM). As seen in the figure, we define

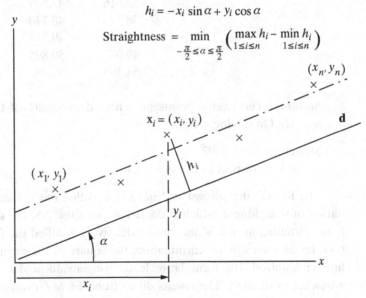

$$h_i = -x_i \sin \alpha + y_i \cos \alpha$$

$$\text{Straightness} = \min_{-\frac{\pi}{2} \leq \alpha \leq \frac{\pi}{2}} \left(\max_{1 \leq i \leq n} h_i - \min_{1 \leq i \leq n} h_i \right)$$

Figure 3.11. Straightness

the distance h_i of point i from line **d** passing through the origin, and inclined at angle α with respect to the x-axis. Then

$$h_i = -x_i \sin\alpha + y_i \cos\alpha$$

$$\text{Straightness} = \min_{-\frac{\pi}{2} \leq \alpha \leq \frac{\pi}{2}} \left(\max_{1 \leq i \leq n} h_i - \min_{1 \leq i \leq n} h_i \right). \quad (3.1)$$

Straightness is evaluated by searching for all angles within the range shown. This has been implemented in the program *Straightness.xls*.

Example 3.1. The following data were generated to test for the straightness of a surface element. Find the straightness and orientation of the surface element. (Data were generated using http://www.npl.co.uk/ssfm/theme2/rds/chebyshev_best_fit_line.html.)

i	x_i	y_i
1	−4.392	39.993
2	0.945	41.119
3	6.282	42.243
4	11.644	43.247
5	16.984	44.363
6	22.331	45.440
7	27.665	46.583
8	33.016	47.639
9	38.358	48.740
10	43.720	49.745
11	49.052	50.895
12	54.395	51.993

Solution. The twelve points are entered in sheet1 of the program *Straightness.xls*. On solving, we get

Straightness = 0.0902
Angle with respect to x-axis = 11.46°.

The formulation shown in Eq. (3.1) is called the *minimum zone* (MZ) evaluation of straightness, which is consistent with the ASME Y14.5M standard. The other formulation that is also used extensively is called the *least squares formulation*. In the least squares formulation, the square of the error from the best-fitting line is minimized. This formulation leads to eigenvalue and eigenvector analysis and a quicker evaluation. The results differ from the MZ evaluation but are close for small errors. The direction established from the least squares formulation may be used for searching for the MZ error.

3.3 Tolerances of Form

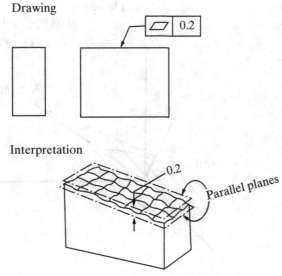

Figure 3.12. Flatness specification

The straightness of an axis is the smallest diameter of the cylinder inside which it is enclosed. Thus, the straightness of an axis is closely related to cylindricity, which we discuss in a later section.

Flatness

Flatness is the condition of a surface where all elements lie in one plane. A flatness tolerance zone is defined by two parallel planes between which the surface must lie. The surface must lie within the specified limits of the size (Rule 1). An example of the flatness specification and its interpretation is shown in Fig. 3.12. Flatness may also be specified on a unit basis, in which case the flatness on a unit area measuring 25×25 may be added in a double-height feature control frame. The unit tolerance is smaller than the overall flatness tolerance on the surface.

If $\mathbf{x}_i = (x_i, y_i, z_i), i = 1$ to n, are coordinate measurements of n points on a surface for which flatness is to be evaluated, the MZ flatness is given by

$$\min \left(\max_{1 \leq i \leq n} \mathbf{x}_i \cdot \mathbf{d} - \min_{1 \leq i \leq n} \mathbf{x}_i \cdot \mathbf{d} \right) \quad (3.2)$$
$$\text{subject to} \quad \mathbf{d} \cdot \mathbf{d} = 1$$

A geometric interpretation of Eq. (3.2) is shown in Fig. 3.13. The constraint in the preceding formulation leads to the search on a unit hemisphere. The program *Flatness.xls* uses this formulation for flatness evaluation of data. The Nelder–Mead

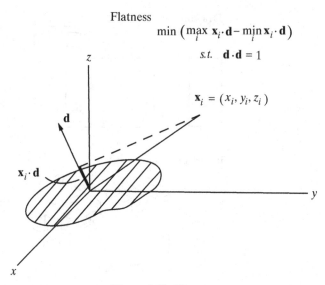

Figure 3.13. Flatness

simplex method is used in the program, and the starting direction is established by the least squares formulation.

Example 3.2. The following data represent coordinates of points on a surface. Find the flatness and orientation of the surface.

i	x_i	y_i	z_i
1	27.154	65.776	61.291
2	12.809	79.713	61.286
3	−1.550	93.635	61.311
4	−15.880	107.586	61.276
5	38.663	77.627	72.565
6	24.313	91.557	72.573
7	9.954	105.480	72.596
8	−4.383	119.424	72.576
9	50.156	89.460	83.874
10	35.815	103.400	83.862
11	21.460	117.326	83.879
12	7.118	131.266	83.867

Solution. The twelve points are entered on sheet1 of the program *Flatness.xls*. On solving, we get

Flatness $= 0.0405$
Unit normal direction for the surface $= (-0.3932, -0.4047, 0.8256)$.

3.3 Tolerances of Form

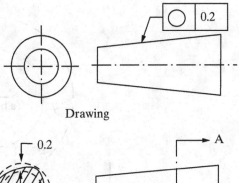

Drawing

Figure 3.14. Circularity specification

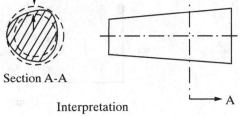

Section A-A

Interpretation

Circularity (Roundness)

Circularity is the condition of a section where all points lie on a circle. For a cylinder or a cone, the cross section is taken perpendicular to its axis, and for a sphere, the cross section is taken through the center of the sphere. Fig. 3.14 shows the circularity specification on a conical surface. The circularity tolerance zone of 0.2 is established at each section normal to the axis. The circularity of a perfect ellipse is half of the difference between its major and minor axes.

If $(x_i, y_i), i = 1$ to n, are coordinate measurements of n points expected to be on a circle as shown in Fig. 3.15, the circularity is given by

$$\text{Circularity} = \min_{a,b} \left(\max_{1 \leq i \leq n} r_i - \min_{1 \leq i \leq n} r_i \right) \\ r_i = \sqrt{(x_i - a)^2 + (y_i - b)^2} \quad (3.3)$$

The formulation in Eq. (3.3), which involves a two-variable search, has been implemented in the program *CircleSphere.xls*. The program also finds the *sphericity*, which is defined by

$$\text{Sphericity} = \min_{a,b,c} \left(\max_{1 \leq i \leq n} r_i - \min_{1 \leq i \leq n} r_i \right) \\ r_i = \sqrt{(x_i - a)^2 + (y_i - b)^2 + (z_i - c)^2} \quad (3.4)$$

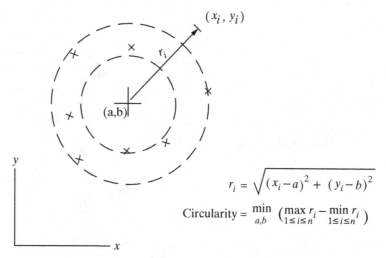

Figure 3.15. Circularity measurement

If the data include three-dimensional measurements (x, y, z), the program finds the sphericity. A three-variable search is used to find the sphericity.

Example 3.3. The following data represent points on a circle at a cross section. Determine the circularity of the data set and the location of the center. (Data were generated using http://www.npl.co.uk/ssfm/theme2/rds/chebyshev_best_fit_circle.html.)

i	x_i	y_i
1	7.473	56.946
2	41.703	15.765
3	11.912	25.817
4	22.164	17.909
5	6.481	54.367
6	46.561	17.293
7	52.313	69.484
8	21.179	18.349
9	64.970	43.509
10	16.266	68.418
11	64.656	49.443
12	41.923	74.168
13	11.499	26.330
14	17.019	20.958
15	9.361	60.620

3.3 Tolerances of Form

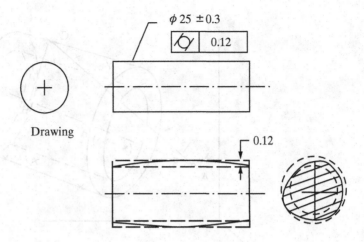

Drawing

Interpretation

Figure 3.16. Cylindricity interpretation

Solution. The fifteen points are entered on sheet1 of the program *Circle-Sphere.xls*. On solving, we get

Circularity = 0.044
Center location = (35, 45)
Maximum radius = 30.022
Minimum radius = 29.978.

Cylindricity

Cylindricity is a condition of a surface of revolution where all the points are equidistant from a common axis. Cylindricity is identified by the minimum radial distance between two coaxial cylinders which enclose the surface. Cylindricity is a composite form tolerance that includes circularity, straightness, and the taper of a cylindrical feature.

An example of the cylindricity specification is shown in Fig. 3.16. To determine the cylindricity, the coordinates $\mathbf{x}_i = (x_i, y_i, z_i)$, $i = 1, 2, \ldots, n$, of a set of n points are measured using a CMM. To evaluate the cylindricity from the data, we need to determine a point $\mathbf{x}_0$ and the direction of a unit vector, $\mathbf{d}$, as shown in Fig. 3.17 so that

$$\text{Cylindricity} = \min_{\mathbf{x}_0} \left(\max_{1 \leq i \leq n} r_i - \min_{1 \leq i \leq n} r_i \right)$$

$$r_i = \sqrt{(\mathbf{x}_i - \mathbf{x}_0) \cdot (\mathbf{x}_i - \mathbf{x}_0) - [(\mathbf{x}_i - \mathbf{x}_0) \cdot \mathbf{d}]^2} \qquad (3.5)$$

subject to

$$\mathbf{d} \cdot \mathbf{d} = 1.$$

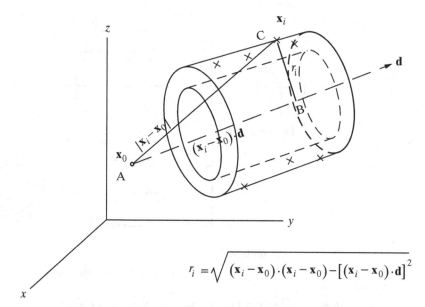

$$r_i = \sqrt{(\mathbf{x}_i - \mathbf{x}_0) \cdot (\mathbf{x}_i - \mathbf{x}_0) - [(\mathbf{x}_i - \mathbf{x}_0) \cdot \mathbf{d}]^2}$$

Figure 3.17. Cylindricity determination

The cylindricity evaluation formulation, Eq. (3.5), has been implemented in the program *Cylindricity.xls*, which uses the Nelder–Mead simplex method. Several data sets from published literature have been included on sheet2 of the program. The program gives the fixed point $\mathbf{x}_0$, the direction $\mathbf{d}$, and the cylindricity value. Starting values are randomly chosen by the program.

We note that, for a given direction $\mathbf{d}$, the cylindricity is the same as the circularity on a plane normal to $\mathbf{d}$. This observation is used to improve the efficiency of calculations in the computer program.

The straightness of an axis in three dimensions is similar to the cylindricity defined in Eq. (3.5) except that min r_i is not to be included:

$$\text{Straightness} = \min_{\mathbf{x}_0} \left(\max_{1 \leq i \leq n} r_i \right)$$

$$r_i = \sqrt{(\mathbf{x}_i - \mathbf{x}_0) \cdot (\mathbf{x}_i - \mathbf{x}_0) - [(\mathbf{x}_i - \mathbf{x}_0) \cdot \mathbf{d}]^2} \qquad (3.6)$$

subject to

$$\mathbf{d} \cdot \mathbf{d} = 1.$$

The program *Straightness3D.xls* solves formulation (3.6). Several example data sets from the published literature are included on the program spreadsheet.

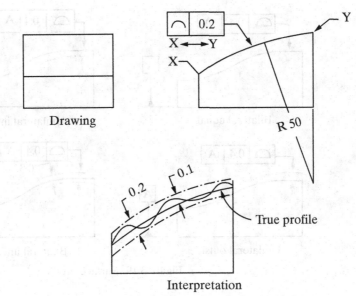

Figure 3.18. Line profile

3.4 Profile Tolerances

A *profile* is the outline of an object or the cross section through the object. Profile tolerances are assumed to be bilateral unless specified otherwise. The true profile is defined by basic radii, basic angular dimensions, basic coordinate dimensions, basic size dimensions, or formulas. The profile tolerance is a uniform boundary along the true profile within which the line or surface must lie.

A *line profile* specification is shown in Fig. 3.18. The tolerance of 0.2 is equally distributed on either side of the true profile to establish the boundary limits. Unequal or unilateral distribution must be shown as illustrated in Fig. 3.19. The surface profile in Fig. 3.19 is established with reference datum A.

A surface definition that uses basic (exact) dimensions and profile control through tolerance specification enables precise control of fits of complex forms. A door panel in an automobile is a good example of a profile form.

3.5 Orientation Tolerances

Parallelism, perpendicularity, and angularity are orientation tolerances that control how features are oriented to one another.

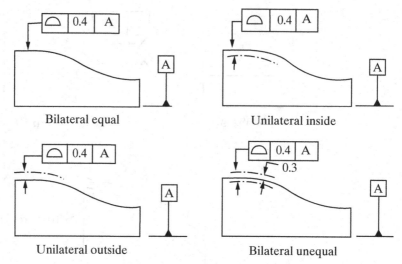

Figure 3.19. Surface profile

Parallelism

Parallelism is the condition of a surface or mid-plane equidistant from a datum plane or datum axis. Parallelism tolerance establishes the distance between two planes parallel to a datum plane within which the surface must be enclosed (see Fig. 3.20). If the parallelism occurs with respect to an axis, the tolerance zone is defined by the diameter of a cylinder whose axis is parallel to the datum axis (see Fig. 3.21). The MMC provides the bonus tolerance for deviation from MMC.

Perpendicularity

Perpendicularity is the condition of a surface, mid-plane, or axis perpendicular to a datum plane or axis. The tolerance zone is established as the distance between two

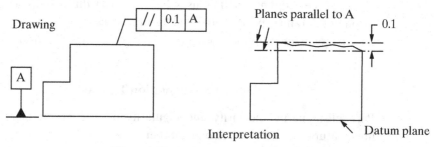

Figure 3.20. Parallelism of a plane surface

3.5 Orientation Tolerances

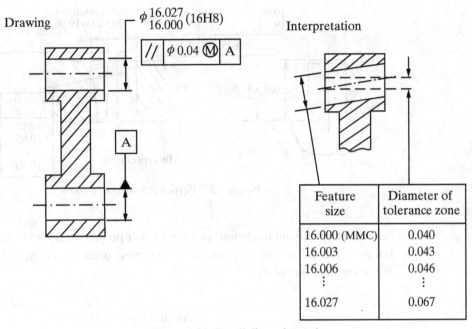

Figure 3.21. Parallelism of an axis

parallel planes at a right angle to the datum plane. To determine the perpendicularity of an axis, the tolerance zone is established as the diameter of a cylinder with its axis perpendicular to the datum axis. Examples of these two cases are shown in Figs. 3.22 and 3.23. Fig. 3.23 also shows the effect of the MMC specification, which defines the virtual condition. The perpendicularity of a hole, such as a bolt hole, is

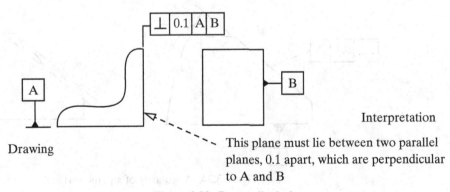

Figure 3.22. Perpendicularity

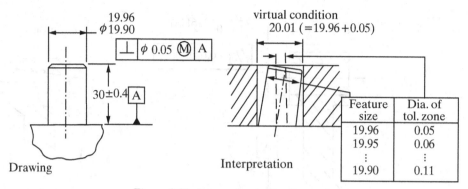

Figure 3.23. Perpendicularity of an axis

defined by specifying the height of a pin or bolt projecting from the hole. The symbol from Fig. 3.3 for the projected tolerance zone precedes the height specification in the feature control frame.

Angularity

Angularity is the condition of a surface, mid-plane, or axis at a specified angle with respect to a datum plane or axis. The angle dimension is basic. The tolerance zone is established as the distance between two planes parallel to the datum plane at the true specified angle as shown in Fig. 3.24. The tolerance zone for an axis is established as the diameter of a cylinder whose axis is at the true orientation with respect to the datum. Once the orientation is established, the specification is similar to that of perpendicularity and parallelism.

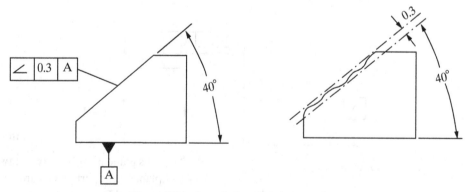

Figure 3.24. Angularity of a plane surface

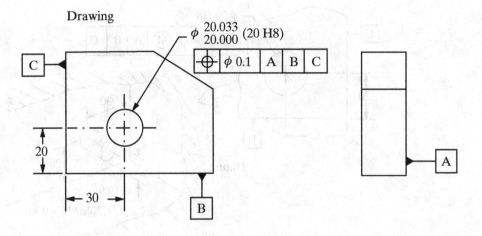

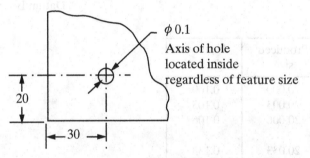

Figure 3.25. Position tolerance

3.6 Tolerances of Location

Location tolerances included in this section are tolerances of position, concentricity, and symmetry.

Position

A *position tolerance* specifies a zone within which the center, axis, or mid-plane of a feature of size is permitted to vary from a true (theoretically exact) position. Based on the MMC or LMC, the *virtual condition* boundary is defined, which may not be violated by the surface or surfaces of the feature under consideration.

Position tolerance, shown in Fig. 3.25, implies that the center must lie in a cylinder of diameter 0.1 located at (30, 20) regardless of the size of the hole. If the MMC

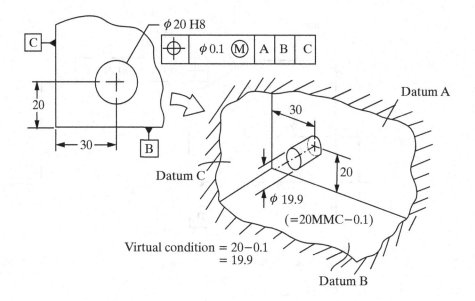

Produced size	Position tolerance
20.000	0.100
20.003	0.103
20.006	0.106
⋮	⋮
20.033	0.133

Figure 3.26. Position tolerance with MMC

is specified as shown in Fig. 3.26, the position tolerance of 0.1 is allowed at the MMC size of 20.000. Any variation from this size will be added to the tolerance. The bonus table shows the tolerance.

Virtual condition = 20.0 (MMC size) − 0.1 (position tolerance) = 19.9.

A pin with size equal to the virtual condition that is placed at the true position defines a go gage. A part that fits into this gage is an acceptable part (see Fig. 3.26).

If the LMC is specified in the drawing, the position tolerance applies at LMC. Deviation from LMC provides the bonus tolerance. Fig. 3.27 shows the position tolerance at LMC. The LMC specification guarantees a minimum wall thickness.

3.6 Tolerances of Location

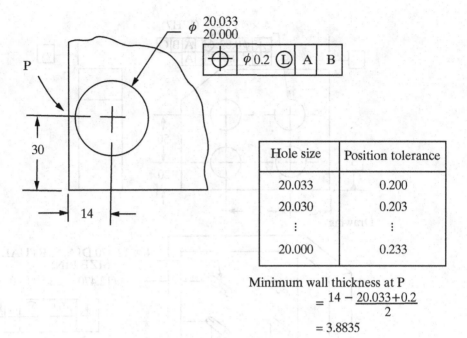

Figure 3.27. Position tolerance at LMC

In some specifications, the relationship between the features and the datums and the interrelationships of the features (pattern location) may have to be distinguished. An example is illustrated in Fig. 3.28. The feature control frame has two parts. The upper tolerance is the conventional one already discussed. The virtual size pins for this are shown in gage 1. The lower tolerance is in relation to datum A only; it refers to the tolerance zone for the relative position of the holes. Gage 2 shows the virtual size pins and their locations for this lower tolerance zone. Position tolerance is covered in great detail in the ASME Y14.5M-1994 standard.

Concentricity

Concentricity (or *coaxiality*) is the condition where the median points of all diametrically opposed elements of a surface of revolution are congruent with the axis of a datum feature. Fig. 3.29 shows an example of concentricity specification. The median points of all diametrically opposed surface lines must lie within a cylinder of diameter 0.3 whose axis coincides with the datum axis.

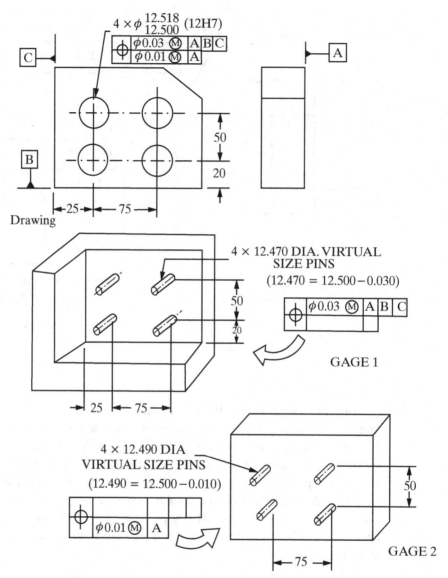

Figure 3.28. Feature relating and pattern locating tolerances

Symmetry

Symmetry is the condition where the median points of all similarly located elements of two or more feature surfaces are congruent with the centerline/centerplane of the datum feature. Fig. 3.30 shows the symmetry specification.

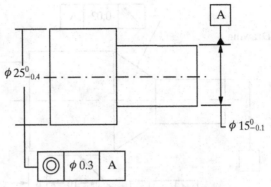

Figure 3.29. Concentricity

3.7 Tolerances of Runout

Runout is a composite tolerance used to control a combination of circularity, cylindricity, and concentricity relationships of one or more features of a part to a datum axis.

Circular Runout

Circular runout is used to control the coaxiality and circularity of a feature. Circular runout is measured using a dial indicator set at right angles to the surface; the part is rotated about the datum axis. The full indicator movement (FIM) measures the circular runout. Fig. 3.31 shows the runout specification for three surfaces and its interpretation.

Total Runout

Total runout is used to control the cumulative variations of circularity, straightness, concentricity, angularity, taper, and profile of a surface. As illustrated in Fig. 3.32,

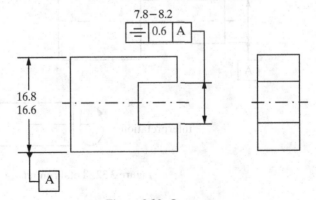

Figure 3.30. Symmetry

Geometric Tolerances

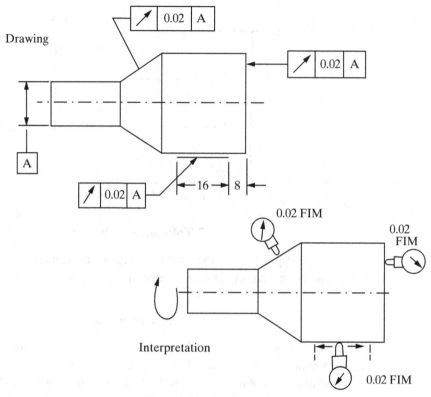

Figure 3.31. Circular runout

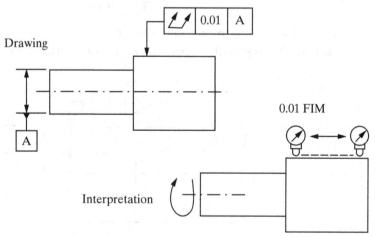

Figure 3.32. Total runout

the part is turned about the datum axis and the dial indicator is moved over the surface to record the FIM.

3.8 Summary

In this chapter we have presented the main ideas of geometric dimensioning and tolerancing. Various form tolerances have been presented with considerable detail. The minimum zone evaluations of form tolerances have been implemented in Excel-based computer programs. The tolerances of profile, orientation, position, and runout have been discussed. Readers are encouraged to refer to American, ISO, and other national standards that present the various tolerances in greater detail.

Computer programs discussed in this chapter include the following:

Straightness.xls
Flatness.xls
CircleSphere.xls
Cylindricity.xls
Straightness3D.xls

EXERCISE PROBLEMS

3.1. The straightness of a pin is given as shown in Fig. P3.1.
  a. What is the virtual size?
  b. If the diameter of a pin produced is 19.96 mm, what is the permissible straightness error?

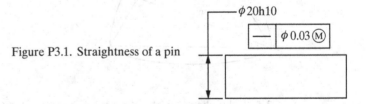

Figure P3.1. Straightness of a pin

3.2. Calculate the allowable true position values for the hole shown in Fig. P3.2. (Hint: Show the tolerance zone for the MMC in the sketch and develop a table for the hole sizes that deviate from the MMC.)

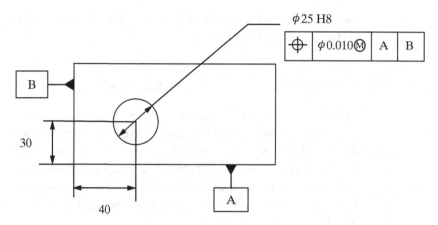

Figure P3.2. Position tolerance

3.3. A rectangular block 40 mm × 60 mm has a hole ϕ25 H7 at the center. The thickness of the block is 25 mm. Draw a sketch showing a position tolerance of ϕ0.016 mm at MMC. Provide the bonus tolerance chart.

3.4. The Reuleux triangle is a constant-width curve used for piston shape in rotary engines. The distance between two opposing points is constant. If the equilateral triangle shown in Fig. P3.4 has a side equal to 45 mm, determine its circularity. (Hint: Explore the Reuleux triangle by searching the web.)

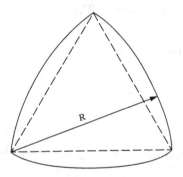

Figure P3.4. Reuleux triangle

3.5. Lobe shapes with an odd number of sides ($n = 3, 5, 7, \ldots$) are produced during centerless grinding due to support conditions. If the smaller radius is r and the larger radius is R (see Fig. P3.5), show that the circularity is $(R-r)(\frac{1}{\cos \varphi} - 1)$, where $\varphi = \pi/(2n)$. For $R = 40$ mm, $r = 5$ mm, plot n vs. circularity. (Note: The case $r = 0$ represents the Reuleux polygon. Check if the formula applies to Exercise Problem 3.4.)

Exercises 3.6–3.8

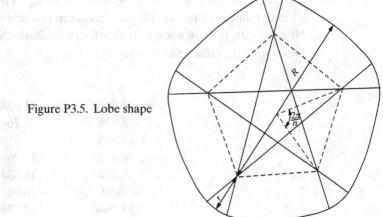

Figure P3.5. Lobe shape

3.6. The parallelism of an edge is specified as 0.08 with reference to datum A. The part was supported on the surface plate on datum A, and the y coordinates of the edge were measured on a CMM at regular 10-mm intervals along the entire x-coordinate edge. The data are

x	0	10	20	30	40	50	60 mm
y	−5	1	3	5	9	12	17 µm

Determine the parallelism error.

3.7. For the data measured in Problem 3.6, what is the straightness of the edge?

3.8. The following measurements were made on a surface for flatness evaluation. Determine the flatness of the surface. What is the orientation of the minimum zone plane?

Point #	x	y	z
1	−8.01	35.75	12.28
2	−29.54	56.64	12.29
3	−51.07	77.54	12.31
4	−72.59	98.44	12.31
5	3.47	47.57	23.6
6	−18.03	68.49	23.57
7	−39.57	89.38	23.6
8	−61.08	110.29	23.58
9	14.98	59.42	34.88
10	−6.53	80.33	34.86
11	−28.05	101.24	34.86
12	−49.57	122.14	34.86

3.9. The following data were generated using center location (12, 5); radius 25 mm; number of points, 12; and maximum residual deviation, 0.05 mm (from the NPL website, http://www.npl.co.uk/ssfm/theme2/rds/chebyshev_best_fit_circle.html). Determine the circularity for the data.

x	y
30.642	−11.628
8.976	29.766
−13.016	4.518
30.937	−11.269
−10.224	16.384
−8.016	−10.062
35.364	−3.938
22.700	−17.619
27.002	−14.936
15.497	−19.779
−6.709	21.605
32.016	20.062

(Note: The data have been rounded to three decimal places.)

3.10. The following data were generated using centroid location (3, 4); angle, 0.3 radians; line length, 85 mm; number of points, 10; and maximum residual deviation 0.044 (from the NPL website, http://www.npl.co.uk/ssfm/theme2/rds/chebyshev_best_fit_line.html). Determine the straightness for the data.

x	y
−37.596	−8.577
−28.586	−5.745
−19.547	−3.009
−10.525	−0.216
−1.522	2.639
7.495	5.448
16.534	8.186
25.561	10.965
34.575	13.782
43.612	16.528

(Note: The data have been rounded to three decimal places.)

3.11. Determine the sphericity for the following data. Also find the radius of the sphere.

x	y	z
1	0	0
2	1	0
1	2	0
0	1	0
0	0	1
2	0	1
2	2	1
0	2	1
1	0	2
2	1	2
1	2	2
0	1	2

(Note: The twelve points are the mid-points of the edges of a $2 \times 2 \times 2$ cube.)

3.12. Determine the cylindricity for the following data.

x	y	z
10	0	0
0	10	0
−10	0	0
0	−10	0
10.1	0	25
0	10.1	25
−10.1	0	25
0	−10.1	25
10.15	0	50
0	10.15	50
−10.15	0	50
0	−10.15	50

4

Elements of Probability and Statistics

4.1 Introduction

In this chapter we develop probability and statistics concepts, which will lay the foundation for the study of quality and reliability. *Probability* is used to describe the chance of occurrence of an event. *Statistics* deals with the collection, tabulation, analysis, interpretation, and presentation of quantitative data. There is an inherent relationship between probability and statistics: events occur in a random manner, and the variables for which data are collected are random in nature. We now attempt to develop the concepts: first elements of probability, followed by statistical terms and probability distributions. Sampling and sampling theory are introduced in Chapter 5.

4.2 Probability

The key words used in developing probability concepts are *experiment*, *sample space*, and *event*. An experiment is any procedure or action that generates a set of possible observations or outcomes. Examples of experiments include tossing a coin a number of times, measuring the diameter of a shaft, and finding the tensile strength of a bar. The *sample space, S*, is the set of all outcomes of an experiment.

Example 4.1. A coin is flipped twice. What is the sample space? Which outcomes define the event A, where at least one head shows up?

Solution. If the heads outcome is designated H and tails is designated T, then the sample space S is given by

$$S = \{HH, HT, TH, TT\}.$$

We observe that there are three sample outcomes in S that constitute event A, where at least one head shows up:

$$A = \{HH, HT, TH\}.$$

The complement of A is the space of occurrence of everything but A and is designated A^c. Thus,

$$A^c = S - A.$$

In Example 4.1, $A^c = \{TT\}$.

Let A and B be two events defined over the same sample space, S. Then the *intersection* of A and B, $A \cap B$, is the event whose outcomes belong to both A and B. The *union* $A \cup B$ is the event whose outcomes belong to either A or B or both.

Example 4.2. In the coin flip experiment described in Example 4.1, we define the event B, where at least one tail shows up; find $A \cap B$ and $A \cup B$.

Solution.

$$A = \{HH, HT, TH\} \quad B = \{HT, TH, TT\}$$
$$A \cap B = \{HT, TH\}$$
$$A \cup B = \{HH, HT, TH, TT\} = S.$$

When two events have no outcome in common, they are said to be *mutually exclusive*. If A and B are mutually exclusive, $A \cap B = \emptyset$ (null).

Relationships among two or more events may be represented graphically using *Venn diagrams*. Fig. 4.1 shows union, intersection, complement, and mutually exclusive representations of events using Venn diagrams.

From the Venn diagram, it is easy to see that

$$A \cup B = A + B - A \cap B. \tag{4.1}$$

The probability of an event A, $P(A)$, is the likelihood of occurrence of A when the experiment is performed. $P(A)$ is a number between 0 and 1 and can be approximated by dividing the number of occurrences of event A by the size of the sample space.

In an experiment, if there are N equally likely mutually exclusive events and if a number of these events, n_A, result in event A, then the probability of event A is

$$P(A) = \frac{n_A}{N}. \tag{4.2}$$

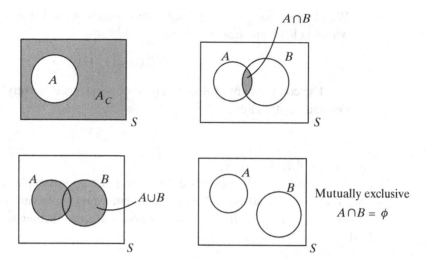

Figure 4.1. Venn diagrams

Example 4.3. A die has six faces marked by dots: 1, 2, ..., 6. If two dice are thrown, determine the probability that the sum of two tosses is 6.

Solution.

$$S = \{(1,1), (1,2), \ldots, (1,6), (2,1), (2,2), \ldots, (2,6), \ldots, (6,1), (6,2), \ldots (6,6)\}$$
$$N = 36$$
$$A = \{(1,5), (2,4), (3,3), (4,2), (5,1)\}$$
$$n_A = 5$$
$$P(A) = \frac{5}{36}.$$

If a coin is tossed once, the size of the sample space $\{H, T\}$ is 2. The probability of a head or tail for a single toss is $\frac{1}{2}$, and the following relationships are easily established:

$$P(S) = 1$$
$$P(A^c) = 1 - P(A). \tag{4.3}$$

For any two events A and B,

$$P(A \cup B) = P(A) + P(B) - P(A \cap B). \tag{4.4}$$

If A and B are mutually exclusive, then $A \cap B = \emptyset$ and $P(A \cap B) = 0$.

4.2 Probability

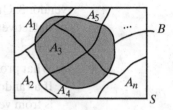

Figure 4.2. Partition of S

Conditional probability $P(A|B)$ is the probability of A given that event B has occurred. If $P(B) > 0$, the conditional probability $P(A|B)$ is given by

$$P(A|B) = \frac{P(A \cap B)}{P(B)}.$$

The *multiplication rule* is

$$P(A \cap B) = P(A|B)P(B). \tag{4.5}$$

We also note that $P(A \cap B) = P(B|A)P(A)$. Two events, A and B, are *independent* if the outcome of one has no influence on the outcome of the other. For example, A and B are independent if $P(A|B) = P(A)$ (or $P(B|A) = P(B)$). If A and B are mutually exclusive events with finite probabilities, then $P(A|B) = 0 \neq P(A)$ and they are not independent; A and B are independent if and only if $P(A \cap B) = P(A)P(B)$.

In modern manufacturing, companies often receive components from several vendors. If each vendor supplies a certain percentage that are defective, the total probability of defective components should be determined. If a defective component is found, the probability that it is from a certain vendor may need to be ascertained. Such problems also arise in drug testing performed by drug manufacturers. Reverend Thomas Bayes (1702–1761) gave formulas for such problems.

We consider a set of n mutually exclusive events, $A_1, A_2, \ldots, A_n$, such that their union is S. The set is illustrated in Fig. 4.2. For any event B in S, the total probability is given by

$$P(B) = \sum_{i=1}^{n} P(B|A_i)P(A_i) \tag{4.6}$$

and Bayes' theorem states

$$P(A_j|B) = \frac{P(B|A_j)P(A_j)}{\sum_{i=1}^{n} P(B|A_i)P(A_i)}. \tag{4.7}$$

Example 4.4. An automobile manufacturer receives steering subassemblies from three vendors – 40% from vendor 1, 35% from vendor 2, and 25% from vendor 3. Quality control standards for the three vendors differ. The percentage of defective subassemblies from vendors 1, 2, and 3 are 2, 2.6, and 3%, respectively.

a. What proportion of the steering assemblies received is defective?
b. If a randomly picked assembly is defective, determine the probability that it is from vendor 1.

Solution. Let A_i be the event "Steering assembly came from supplier i" ($i = 1, 2, 3$). Let B be the event "Steering assembly received is defective." We have

$$P(A_1) = 0.4, \quad P(A_2) = 0.35, \quad P(A_3) = 0.25$$

and

$$P(B|A_1) = 0.02, \quad P(B|A_2) = 0.026, \quad P(B|A_3) = 0.03.$$

a. From the total probability relationship, Eq. (4.6),

$$P(B) = (0.02)(0.4) + (0.026)(0.35) + (0.03)(0.25) = 0.0246.$$

Thus, the manufacturer can expect 2.46% of the subassemblies to be defective.

b. From Bayes' theorem, Eq. (4.7), the probability that the defective unit is from vendor 1 is

$$P(A_1|B) = \frac{P(B|A_1)P(A_1)}{\sum_{i=1}^{n} P(B|A_i)P(A_i)} = \frac{(0.02)(0.4)}{0.0246} = 0.3252.$$

4.3 Statistics

Statistics is the branch of mathematics that deals with the collection, tabulation, analysis, interpretation, and presentation of quantitative data. Statistics has two branches – descriptive statistics and inferential statistics.

Descriptive statistics describes the basic features of data collected in a study; it provides summary statistics, statistical measures, and graphical presentation of the collected data, which forms the basis for all quantitative analyses.

The data collection is performed on a randomly chosen *sample* – a subset of the *population*, which is the set of all items of a certain characteristic. Consider a batch of 5,000 shafts produced by a manufacturer using the same process settings. This forms the population of interest. If the diameter is the *parameter* of interest, we may select a sample of fifty shafts and measure the diameter. The collection and tabulation of data are part of descriptive statistics. Then conclusions may be drawn about the population.

Inferential statistics draws conclusions about the population based on information contained in a *random sample*. Every element of the population has an equal chance of being chosen in a random sample.

A characteristic in a random sample is described by a *random variable*. These variables may be *continuous* or *discrete*. Examples of continuous variables are the diameter of a shaft, the temperature of a quench tank, and the tensile strength of a specimen. Discrete variables are countable using integers. Examples of discrete variables include the number of customers at a bank counter during a ten-minute interval or the number of successes in a given number of trials.

Statistical Measures

We now discuss some numerical measures used to prepare summary information about the gathered data. The measures of central tendency are the mean, median, and mode. Measures of spread include the range, standard deviation, skewness, kurtosis, and correlation.

We provide a sample data set which is given in two columns.

Data Set

2.34	4.6
4.23	3.7
1.36	3.3
2.5	2
3.6	2

All related calculations of statistical measures are available in the file *Worksheet-Func.xls*. The file also gives other Excel functions useful in probability and statistics calculations.

A random sample of size n drawn from a population of size N is denoted $X_1, X_2, \ldots, X_n$. If the aforementioned data are put into the Excel spreadsheet cells B1:C5 then we refer to B1:C5 as the cellRange. As an example, we obtain the sum by using the formula =SUM(cellRange).

Mean

The *mean* is the average of the observations. The sample mean is denoted by $\overline{X}$ and the population mean is denoted by μ. The sample mean is calculated using the observations in the sample:

$$\overline{X} = \frac{\sum_{i=1}^{n} X_i}{n}. \tag{4.8}$$

The population mean is calculated using every member of the population:

$$\mu = \frac{\sum_{i=1}^{N} X_i}{N}. \qquad (4.9)$$

In the example data set, we have $n = 10$ and the mean is 2.963. The Excel function to calculate the mean is AVERAGE(cellRange). To exclude stray data at the low and high ends, a *trimmed mean* is used. The trimmed mean excludes a proportion p of the data points from the low and high ends together. The Excel function TRIMMEAN(cellRange, 0.2) for the data set excludes two data points (1.36 and 4.8) and calculates the mean of the remaining eight entries as 2.95875.

Median

The *median* is the value that falls in the middle of a series of values. If there is an even number of data entries, the average of the two middle numbers is used. The value of MEDIAN(range) for the example data set is the average of 2.5 and 3.3, which is 2.9.

Mode

The *mode* is the most frequently occurring value in the data set. The MODE(cellRange) for the example data set is 2.

Range

The *range* is the difference between the largest and smallest values in the data set. The range for the example data set is 3.24, which is found by using the Excel function combination MAX(cellRange) − MIN(cellRange). The range is often used as a measure of dispersion in quality control studies.

Variance

The *variance* is a measure of the fluctuations of data about the mean. The *population variance* σ^2 is calculated using all N members of the population and the mean μ:

$$\sigma^2 = \frac{\sum_{i=1}^{N}(X_i - \mu)^2}{N}. \qquad (4.10)$$

4.3 Statistics

The *sample variance* s^2 is calculated using n members of the sample and the mean $\overline{X}$:

$$s^2 = \frac{\sum_{i=1}^{n}(X_i - \overline{X})^2}{n-1}. \tag{4.11}$$

Note that the sum of the squared deviation is divided by $n-1$. Since $\sum_{i=1}^{n}(X_i - \overline{X}) = 0$ from Eq. (4.8), we have $n-1$ independent terms. Thus, it is customary to say that s^2 is based on $n-1$ *degrees of freedom* (df). The sample variance is what is calculated whenever a sample is taken.

The numerator of Eq. (4.11) is calculated using

$$\sum_{i=1}^{n}(X_i - \overline{X})^2 = \sum_{i=1}^{n}(X_i^2 - 2X_i\overline{X} + \overline{X}^2)$$

$$= \sum_{i=1}^{n} X_i^2 - 2\overline{X}\sum_{i=1}^{n} X_i + \sum_{i=1}^{n} \overline{X}^2 = \sum_{i=1}^{n} X_i^2 - 2n\overline{X}^2 + n\overline{X}^2$$

$$= \sum_{i=1}^{n} X_i^2 - \frac{\left(\sum_{i=1}^{n} X_i\right)^2}{n}. \tag{4.12}$$

In this calculation, prior evaluation of the mean is not necessary. A similar expression may be written for the numerator of Eq. (4.10).

The variance of the population is given by VARP(cellRange) and the variance of the sample by VAR(cellRange) in the Excel spreadsheet. For the example data set, VARP( ) = 1.0374 and VAR( ) = 1.1527.

Standard Deviation

The *standard deviation* is the square root of the variance and has the same units as the variable. The *population standard deviation* σ is given by

$$\sigma = \sqrt{\frac{\sum_{i=1}^{N}(X_i - \mu)^2}{N}} \tag{4.13}$$

and the *sample standard deviation* s is given by

$$s = \sqrt{\frac{\sum_{i=1}^{n}(X_i - \overline{X})^2}{n-1}}. \tag{4.14}$$

The numerator under the square root is conveniently calculated as stated in Eq. (4.12). For the example data set, $\sum_{i=1}^{n} X_i = \text{SUM(cellRange)} = 29.31$, and $\sum_{i=1}^{n} X_i^2 = \text{SUMSQ(cellRange)} = 98.1681$. The sample standard deviation is

$$s = \sqrt{\frac{98.1681 - \frac{29.31^2}{10}}{10 - 1}} = 1.07364,$$

which may be directly obtained using $\text{STDEV(cellRange)} = 1.07364$; the population standard deviation is given by $\text{STDEVP(cellRange)} = 1.01855$. The population standard deviation is meaningful if the data represent the entire population.

Interquartile Range

The lower quartile, Q_1, such that one-quarter of the number of observations fall below it (and three-quarters above it); Q_2 is the second quartile, which corresponds to the median; and Q_3 is the third quartile such that three-quarters of the observations fall below it (and one-quarter above it). Then the interquartile range (IQR) is given by

$$\text{IQR} = Q_3 - Q_1. \tag{4.15}$$

For the example data set, the Excel calculations are $Q_1 = \text{QUARTILE(cellRange,1)} = 2.085$, $Q_3 = \text{QUARTILE(cellRange,3)} = 3.675$, and $Q_3 - Q_1 = 1.59$. The IQR provides the spread corresponding to the middle 50% of the data.

Skewness

The *skewness* of the data, γ_1, is determined by a parameter that represents the lack of symmetry:

$$\gamma_1 = \frac{\sum_{i=1}^{n} (X_i - \overline{X})^3}{ns^3}.$$

The skewness formula that is attributed to Pearson and is used in the Excel spreadsheet is given by

$$\gamma_1 = \frac{n}{(n-1)(n-2)} \frac{\sum_{i=1}^{n} (X_i - \overline{X})^3}{s^3}. \tag{4.16}$$

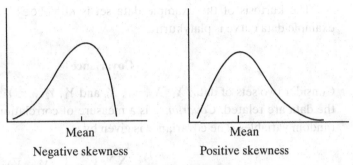

Figure 4.3. Skewness

The skewness of the example data set is given by SKEW(cellRange) = 0.0978. Fig. 4.3 shows the skewness shape. Negative skewness is indicated by a tail on the left of the curve.

Kurtosis

The *kurtosis* of the data, γ_2, is a parameter that represents the steepness/flatness of the data curve:

$$\gamma_2 = \frac{\sum_{i=1}^{n}(X_i - \overline{X})^4}{ns^4} - 3.$$

The skewness formula attributed to Pearson and used in the Excel spreadsheet is given by

$$\gamma_2 = \frac{n(n+1)}{(n-1)(n-2)(n-3)} \frac{\sum_{i=1}^{n}(X_i - \overline{X})^4}{s^4} - \frac{3(n-1)^2}{(n-2)(n-3)}. \qquad (4.17)$$

The kurtosis parameter is normalized by subtracting the second term so that 0 represents peakiness of a standard normal curve, or *mesokurtic distribution*. A negative value represents a *platykurtic distribution* and a positive value represents a *leptokurtic distribution*. These are illustrated in Fig. 4.4.

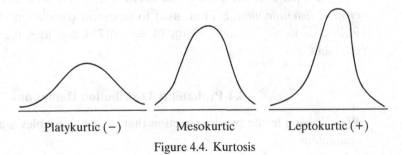

Figure 4.4. Kurtosis

The kurtosis of the example data set is KURT(cellRange) = −1.259. The example data curve is platykurtic.

Covariance

Consider two sets of data: $X_1, X_2, \ldots, X_n$ and $Y_1, Y_2, \ldots, Y_n$. We wish to know how the data are related. *Covariance* is a measure of correlation between two or more random variables. The covariance is given by

$$\text{Cov}(X, Y) = \frac{\sum_{i=1}^{n}(X_i - \overline{X})(Y_i - \overline{Y})}{n}. \tag{4.18}$$

Note that $\text{Cov}(X, X) = \sigma_X^2$. In the example problem, treating the two columns as X and Y variables, respectively, COVAR(cellRangeX, cellRangeY) = −0.13412.

Correlation Coefficient

The *correlation coefficient*, r, is a measure of how two random variables change with respect to each other; it is also called Pearson's correlation. The coefficient is related to covariance and has a value between −1 and +1. A positive value shows that the two variables vary together and a negative value indicates that they change in opposite directions. A value of zero indicates no correlation.

$$r = \frac{\text{Cov}(X, Y)}{\sigma_X \sigma_Y} = \frac{\sum_{i=1}^{n}(X_i - \overline{X})(Y_i - \overline{Y})}{\sqrt{\sum_{i=1}^{n}(X_i - \overline{X})^2 \sum_{i=1}^{n}(Y_i - \overline{Y})^2}}. \tag{4.19}$$

The correlation coefficient shows the slope of the best fitting line of the (X, Y) scatter plot. For the example data set, CORREL(cellRangeX, cellRangeY) = −0.13243. Fig. 4.5 shows a scatter plot of the example data set.

The square of the correlation coefficient, r^2 (or R^2), also called the *coefficient of determination*, is often used to represent correlation. The Excel function RSQ(cellRangeY, cellRangeX) = 0.01754 provides the R^2 value for the data set.

4.4 Probability Distribution Definitions

We discussed in the previous section that random variables are either discrete or continuous.

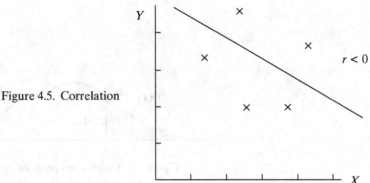

Figure 4.5. Correlation

Discrete Distribution

A random variable X defined on a countable set S is said to have a *discrete probability distribution*. The *probability distribution function* (pdf) $p(x)$ is defined by

$$p(x) = P(X = x) \text{ for every } x \text{ in } S. \tag{4.20}$$

The pdf of a discrete distribution satisfies the following properties:

1. $p(x) \geq 0$ for every x in S
2. $\sum_{x \text{ in } S} p(x) = 1.$

The *cumulative distribution function* (cdf) is denoted by $F(x)$ and is defined as

$$F(x) = \sum_{y \leq x \text{ in } S} p(y). \tag{4.21}$$

Continuous Distribution

Let X be a continuous random variable. The associated pdf $f(x)$ is such that

$$P(a \leq X \leq b) = \int_a^b f(x)dx \tag{4.22}$$

and $f(x)$ satisfies the following conditions:

1. $f(x) \geq 0$ for all x
2. $\int_{-\infty}^{\infty} f(x)dx = 1.$

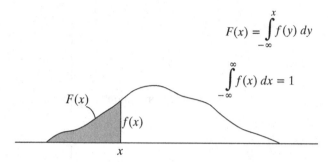

Figure 4.6. Continuous probability distribution

The cdf $F(x)$ is defined as

$$F(x) = \int_{-\infty}^{x} f(y)dy. \qquad (4.23)$$

A pdf for a continuous probability distribution is illustrated in Fig. 4.6.

Expected Value

The *mean* or *expected value*, $E(X)$, of a discrete distribution is given by

$$\mu = E(X) = \sum_{x \text{ in } S} xp(x), \qquad (4.24)$$

and the mean or expected value, $E(X)$, of a continuous probability distribution is given by

$$\mu = E(X) = \int_{-\infty}^{\infty} xf(x)dx. \qquad (4.25)$$

This is the first moment of the probability distribution.

The *variance*, $\text{Var}(X)$, is the expected value of $(X-\mu)^2$; thus, for discrete distribution,

$$E[(X-\mu)^2] = \sum_{x \text{ in } S} (x-\mu)^2 p(x), \qquad (4.26)$$

and for continuous distribution,

$$E[(X-\mu)^2] = \int_{-\infty}^{\infty} (x-\mu)^2 f(x)dx. \qquad (4.27)$$

The variance is calculated using the relation $E[(X-\mu)^2] = E(X^2) - [E(X)]^2$.

4.4 Probability Distribution Definitions

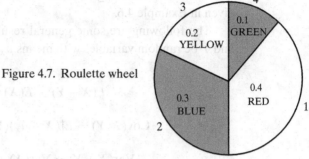

Figure 4.7. Roulette wheel

Example 4.5. A simple roulette wheel is painted in four colors: 40% of the total area in red, 30% of the area in blue, 20% of the area in yellow, and 10% of the area in green, as shown in Fig. 4.7. If red, blue, yellow, and green are assigned the numbers 1, 2, 3, and 4, respectively, find the expected value and variance.

Solution. We have $p(1) = 0.4$, $p(2) = 0.3$, $p(3) = 0.2$, and $p(4) = 0.1$.

$$E(X) = (1)(0.4) + (2)(0.3) + (3)(0.2) + (4)(0.1)$$
$$= 0.4 + 0.6 + 0.6 + 0.4$$
$$= 2.0$$
$$E(X^2) = (1)^2(0.4) + (2)^2(0.3) + (2)^2(0.4) + (1)^2(0.4)$$
$$= 0.4 + 1.6 + 3.6 + 6.4 = 12$$
$$\text{Var}(X) = E(X^2) - [E(X)]^2 = 12 - (1.9)^2 = 8.39.$$

Example 4.6. A *uniform distribution* is defined by the pdf $f(x) = 1 (0 \leq x \leq 1)$ and $f(x) = 0$ elsewhere. Find the expected value and the variance.

Solution.

$$E(X) = \int_0^1 x f(x) dx = \int_0^1 x \, dx = \frac{1}{2}$$

$$E(X^2) = \int_0^1 x^2 f(x) dx = \int_0^1 x^2 \, dx = \frac{1}{3}$$

$$\text{Var}(X) = E(X^2) - [E(X)]^2 = \frac{1}{3} - \left(\frac{1}{2}\right)^2 = \frac{1}{12}.$$

The random numbers generated on a computer are from a uniform distribution given in Example 4.6.

The following are some general results for a pair of random variables. Let X and Y be random variables with means μ_x and μ_y, respectively. Then

$$E(X+Y) = E(X) + E(Y) = \mu_x + \mu_y \qquad (4.28)$$

$$\text{Cov}(X, Y) = E[(X - \mu_x)(Y - \mu_y)] = E(XY) - \mu_x \mu_y \qquad (4.29)$$

$$\text{Var}(X+Y) = \text{Var}(X) + \text{Var}(Y) + 2\text{Cov}(X, Y). \qquad (4.30)$$

The last statement follows by expanding the right-hand side of $\text{Var}(X+Y) = E[(X - \mu_x + Y - \mu_y)^2]$. If X and Y are independent, then $\text{Cov}(X, Y) = 0$.

4.5 Discrete Probability Distributions

A few key discrete distributions, such as hypergeometric, binomial, multinomial, and Poisson distributions, find applications in quality studies.

Hypergeometric Distribution

Let D be the number of defective items in a population of size N. A sample of n items is chosen from the population as shown in Fig. 4.8. The probability of finding

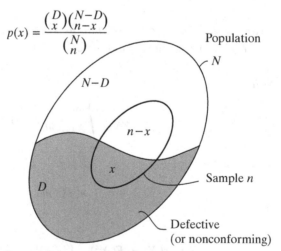

Figure 4.8. Hypergeometric distribution

x defective items in the sample, $p(x)$, is given by the hypergeometric distribution:

$$p(x) = \frac{\binom{D}{x}\binom{N-D}{n-x}}{\binom{N}{n}}, \quad (4.31)$$

where $\binom{a}{b}$ is the number of combinations of a items taken b at a time and is given by $\binom{a}{b} = \frac{a!}{b!(a-b)!}$; $a!$ is factorial of a given by $a! = a(a-1)(a-2)\cdots(3)(2)(1)$ and $0! = 1$.

The mean and variance of the distribution are given by

$$\mu = \frac{nD}{N} \quad (4.32)$$

and

$$\sigma^2 = \frac{nD}{N}\left(1 - \frac{D}{N}\right)\left(\frac{N-n}{N-1}\right). \quad (4.33)$$

The Excel function for the hypergeometric distribution is HYPGEOMDIST (x, D, n, N).

Example 4.7. There are 4 nonconforming capacitors in a lot of 25. A sample of 6 capacitors is taken. What is the probability of finding 2 nonconforming capacitors in the sample? Also find the mean and the standard deviation.

Solution. We have $N = 25$, $D = 4$, $n = 6$, and $x = 2$. The probability, mean, and standard deviation, respectively, are

$$P(X=2) = \frac{\binom{4}{2}\binom{21}{4}}{\binom{25}{6}} = 0.20277 \quad [= \text{HYPGEOMDIST}(2, 6, 4, 25)]$$

$$\mu = \frac{(6)(4)}{25} = 0.96$$

$$\sigma = \sqrt{\frac{(6)(4)}{25}\left(1 - \frac{4}{25}\right)\left(\frac{25-6}{25-1}\right)} = \sqrt{0.6384} = 0.799.$$

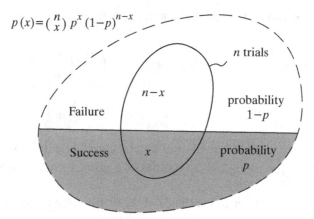

Figure 4.9. Binomial distribution

Binomial Distribution

We consider a sequence of n independent trials, each with one of two outcomes – success or failure, nonconforming or conforming, and defective or good. Trials that have binary outcomes are called *Bernoulli trials*. For such trials, if the probability of success is p then the probability of failure is $1 - p$. The experiment is illustrated in Fig. 4.9. The probability of x successes in n trials is given by

$$p(x) = \binom{n}{x} p^x (1-p)^{n-x} \quad (x = 0, 1, 2, \ldots, n). \tag{4.34}$$

Success in a quality study may be defined as finding a nonconforming or defective item. If the experiment involves tossing a coin and we define success as coming up with a head and failure otherwise (coming up with a tail), then $p = 0.5$ and $(1 - p) = 0.5$. If success in tossing a die is defined as coming up with a 6 then $p = 1/6$ and $(1 - p) = 5/6$.

The mean and variance for a binomial distribution are given by

$$\mu = np \tag{4.35}$$

$$\sigma^2 = np(1 - p). \tag{4.36}$$

If the population is of finite size N and is approximated using the binomial distribution, the variance may be corrected in line with Eq. (4.33) as $\sigma^2 = np(1-p)(\frac{N-n}{N-1})$.

The binomial probabilities $\binom{n}{x} p^x (1-p)^{n-x}$ $(x = 0, 1, 2, \ldots, n)$ are the successive terms in the binomial expansion of $[(1-p) + p]^n$ and their sum is 1. This shows

4.5 Discrete Probability Distributions

that, in evaluating the cumulative binomial probabilities, we may use the following identity:

$$\sum_{i=0}^{k}\binom{n}{i}p^i(1-p)^{n-i} = 1 - \sum_{j=k+1}^{n}\binom{n}{j}p^j(1-p)^{n-j}. \qquad (4.37)$$

For term-by-term calculation of the cumulative distribution, it is convenient to work with the lower number of terms depending on the value of k. Appendix Table A.1 provides a list of values for cumulative binomial distribution. The Excel function for the binomial term is BINOMDIST(x, n, p, False) and for the cumulative distribution it is BINOMDIST(x, n, p, True).

Example 4.8. A manufacturer knows that 10% of the items produced in a certain large batch are nonconforming. What is the probability of 2 nonconforming items occurring in a sample of 5? What is the probability that there are 1 or fewer nonconforming items in the sample?

Solution. We have $p = 0.1, x = 2$, and $n = 5$:

$$P(X=2) = \binom{5}{2}p^2(1-p)^{5-2}$$

$$= \frac{5!}{2!3!}0.1^2 0.9^3$$

$$= 0.0729.$$

The probability of exactly 2 nonconforming items in a sample of 5 is 7.29%. To find the probability of 1 or fewer nonconforming items,

$$P(X \le 1) = P(X=0) + P(X=1)$$

$$= \binom{5}{0}p^0(1-p)^5 + \binom{5}{1}p^1(1-p)^{5-1}$$

$$= 0.5905 + 0.3280$$

$$= 0.9185.$$

Using Table A.1,

$$P(X=2) = P(X \le 2) - P(X \le 1) = 0.991 - 0.919 = 0.072$$

and $P(X \le 1) = 0.919$. Note that the table values are rounded to three decimal places.

Multinomial Distribution

A *multinomial distribution* is a generalization of the binomial distribution. Consider, for example, a die with six faces, one face marked with one dot, two faces marked with two dots each, and three faces marked with three dots each. The probability of hitting 1, 2, and 3 each in a random throw are 1/6, 2/6, and 3/6, respectively. The multinomial distribution answers the following question: if we throw the die 10 times, what is the probability of hitting 1 twice, 2 three times, and 3 five times?

Consider n independent and identical trials with each trial having k possible outcomes. Let $p_i = P(\text{outcome } i \text{ on any trial})$. Let the random variable X_i equal the number of trials resulting in outcome i ($i = 1, 2, \ldots, k$); then the joint probability of $X_1, X_2, \ldots, X_k$, is given by

$$p(x_1, x_2, \ldots, x_k) = \frac{n!}{x_1! x_2! \ldots x_k!} p_1^{x_1} p_2^{x_2} \cdots p_k^{x_k} \tag{4.38}$$
$$x_i = 0, 1, 2, \ldots, n; \quad x_1 + x_2 + \cdots + x_k = n; \quad p_1 + p_2 + \cdots + p_k = 1.$$

Using this, the answer to the probability question posed earlier is

$$\frac{10!}{2!3!5!} \left(\frac{1}{6}\right)^2 \left(\frac{2}{6}\right)^3 \left(\frac{3}{6}\right)^5 = 0.08102.$$

Poisson Distribution

Random occurrences, such as the number of surface defects per square meter of a fabric, the number of vehicles traveling on a given road in a ten-minute interval, the number of customers arriving at a bank window in one hour, the number of machine breakdowns per month, are modeled via *Poisson distribution*. Let x represent the observed value of the random variable X. Let λ be the average rate per unit time or unit surface area or volume. Then the probability of x events in the unit entity, $p(x)$, is given by

$$p(x) = \frac{e^{-\lambda} \lambda^x}{x!} \quad (x = 0, 1, 2, \ldots), \tag{4.39}$$

where e is the base of the natural logarithm, equal to approximately 2.71828.

The mean and variance for the binomial distribution are given by

$$\mu = \lambda \tag{4.40}$$

$$\sigma^2 = \lambda. \tag{4.41}$$

Using the Maclauren expansion of e^λ, we get the identity

$$1 = e^{-\lambda}(e^\lambda) = e^{-\lambda}\left(1 + \lambda + \frac{\lambda^2}{2!} + \cdots + \frac{\lambda^x}{x!} + \cdots\right).$$

This shows that the sum of probabilities of all occurrences is 1, which fulfills the condition for the pdf.

Appendix Table A.2 gives cumulative Poisson probabilities for various λ values. The Excel function for the Poisson term is POISSON(x, λ, False) and for the cumulative distribution it is POISSON(x, λ, True).

Example 4.9. An interval of 20 seconds is needed to take a right turn at an intersection. The average number of vehicles moving in the direction of interest is 6 per minute. What is the probability of there being no vehicle on the road (for a successful turn)?

Solution. The average number of vehicles during the interval of interest, 20 seconds (or $1/3$ minute), is $\lambda = 2$. The probability of no vehicle during the interval is

$$P(X=0) = \frac{e^{-\lambda}\lambda^0}{0!} = e^{-2} = 0.1353.$$

Example 4.10. The average number of paint defects in the bumper area of an automobile is 3. In a randomly selected vehicle, determine the probability of finding 2 or fewer paint defects.

Solution.

$$P(X \leq 2) = \frac{e^{-3}3^0}{0!} + \frac{e^{-3}3^1}{1!} + \frac{e^{-3}3^2}{2!}$$
$$= 0.0498 + 0.1494 + 0.2240$$
$$= 0.4232.$$

Using Table A.2, for $\lambda = 3$ and $x = 2$, $P(X \leq 2) = 0.4232$.

If the sample size n is large and p is small in a binomial distribution, probability can be approximated by Poisson distribution by setting $\lambda = np$.

4.6 Continuous Distributions

Several continuous distributions are of particular interest in statistical quality control. The distributions discussed here are the normal distribution, the chi-squared distribution, the t-distribution, the F-distribution, and some noncentral distributions. Exponential and Weibull distributions will be presented in a later chapter on reliability.

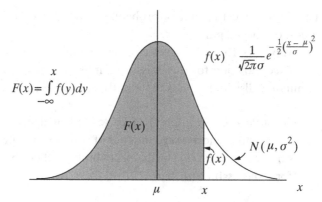

Figure 4.10. Normal distribution

Normal Distribution

Normal distribution, or *Gaussian distribution*, is of great importance in many fields. This distribution is most widely used in statistical quality control. Measurement errors in engineering products, scores on various tests, and numerous economic measures and indicators have a normal distribution.

The pdf for the normal distribution is given by

$$f(x) = \frac{1}{\sqrt{2\pi}\sigma} e^{-\frac{1}{2}\left(\frac{x-\mu}{\sigma}\right)^2} \quad (-\infty < x < \infty), \qquad (4.42)$$

where μ $(-\infty < \mu < \infty)$ is the population mean and σ is the standard deviation.

The statement that the random variable X is normally distributed with parameters μ and σ^2 is abbreviated $X \sim N(\mu, \sigma^2)$. The curve representing Eq. (4.42) is shown in Fig. 4.10. The graph is symmetric about μ and is bell shaped. Small values of σ show a high peak and large values of σ result in a larger spread about the mean. $F(x)$ is the area under the curve from negative infinity to x. Fig. 4.11 gives the area under the curve from $\mu - k\sigma$ to $\mu + k\sigma$ for various values of k; 99.73% of the area is contained in the interval $\mu - 3\sigma$ to $\mu + 3\sigma$. This interval plays an important role in statistical quality control. The $F(x)$ value can be obtained using the Excel function NORMDIST(x, μ, σ, true), and $f(x)$ can be obtained using NORMDIST (x, μ, σ, false). The value of x can be evaluated for a given $F(x)$ by using the inverse function NORMINV(F(x), μ, σ). Examples of using these functions are given in the file *WorksheetFunc.xls*.

The shape of the normal curve changes as the values of parameters μ and σ change. A change of variable will allow us to make the curve invariant, enabling us to use standard tables. We define a standard normal curve by introducing the standardized normal random variable, Z:

$$Z = \frac{X - \mu}{\sigma}. \qquad (4.43)$$

4.6 Continuous Distributions

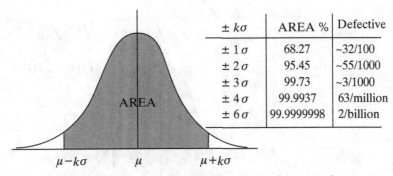

Figure 4.11. Area under normal curve $\mu - k\sigma$ to $\mu + k\sigma$

With this change of variable, Z has a mean value of 0 and standard deviation of 1; thus, $Z \sim N(0, 1)$. The pdf of the *standard normal distribution* is given by

$$\phi(z) = \frac{1}{\sqrt{2\pi}} e^{-\frac{z^2}{2}} \quad (-\infty < z < \infty). \tag{4.44}$$

The cdf $\Phi(z)$ is given by

$$\Phi(z) = \int_{-\infty}^{z} f(t) dt \quad (-\infty < z < \infty) \tag{4.45}$$

and X can be obtained from Z using the inverse relation corresponding to Eq. (4.43):

$$X = \mu + \sigma Z. \tag{4.46}$$

The transformation from X to Z and Z to X is shown in Fig. 4.12.

Values of normal cdf $\Phi(z)$ for various values of z are given in Table A.3 in the Appendix. The inverse of this is to find the z-value for a given tail area. It is general practice to denote z_α as the z-value for a given right-tail area of α as shown in Fig. 4.13. Table A.4 gives z_α for a given right-tail area of α. These values are provided for α in the range 0 to 0.5. For $\alpha > 0.5$, the z-value is $-z_{1-\alpha}$. For a left-tail area of α, the z-value is $-z_\alpha$. The value of $\Phi(z)$ can be obtained using the Excel function NORMSDIST(z), and its inverse z by using NORMSINV($\Phi(z)$). The Excel function uses the left-tail area; thus, z_α = NORMSINV(1 − α). We now apply these relations to some examples.

Example 4.11. The yield strength of an alloy steel is normally distributed with a mean of 220 MPa and a standard deviation of 20 MPa.

a. If the specimen is acceptable only if its yield strength is more than 192 MPa, what is the probability that a randomly chosen sample has an acceptable yield strength?

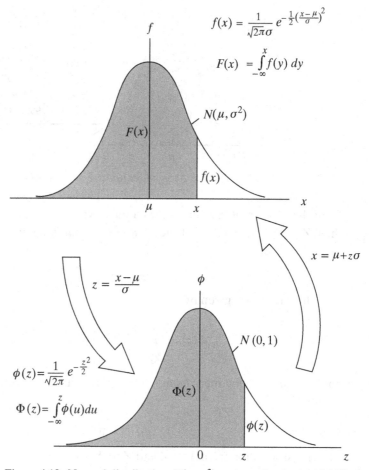

Figure 4.12. Normal distribution $N(\mu, \sigma^2)$ and standard normal $N(0, 1)$

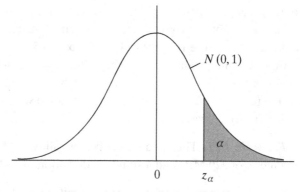

Figure 4.13. z_α for right tail area α

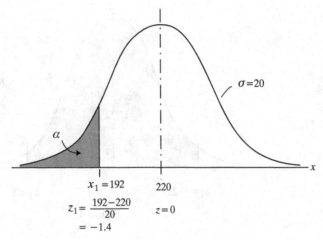

Figure 4.14.

b. If the acceptable yield strength can be lowered, determine the yield stress at which 95% of the specimens are acceptable.

Solution. For this problem, we have $\mu = 220$ MPa and $\sigma = 20$ MPa.

a. The lower acceptance limit is set as $x_1 = 192$ MPa. The corresponding z-value for the standard normal distribution is

$$z_1 = \frac{x_1 - \mu}{\sigma} = \frac{192 - 220}{20} = -1.4,$$

which is illustrated in Fig. 4.14. The area to the left of x_1 under the normal curve is the same as the area to the left of z_1 under the standard normal curve. Using the values in Table A.3, we get

$$\alpha = F(x_1) = \Phi(z_1) = 0.0808.$$

This shows that $1 - 0.0808 = 0.9192$ or 91.92% of the parts have an acceptable yield strength.

b. For the second case, the area to the left is given as $1 - 0.95 = 0.05$, or $\alpha = 0.05$. We now use the inverse of the standard normal distribution from Table A.4 to get $z_1 = -z_{0.05} = -1.6449$. Therefore, $x_1 = \mu + z_1\sigma = 220 - 1.6449 \times 20 = 187.1$ MPa.

Example 4.12. Shafts produced in a manufacturing plant have a normal distribution with an average diameter of 50.05 mm; the standard deviation of the process is 0.02 mm. Determine the proportion of shafts in the acceptable range of 50 to 50.11 mm.

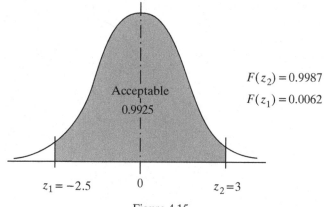

Figure 4.15.

Solution. We have $\mu = 50.05$ mm and $\sigma = 0.02$ mm. The acceptable range is set as $x_1 = 50$ mm and $x_2 = 50.11$ mm. The standardized z limits are

$$z_1 = \frac{x_1 - \mu}{\sigma} = \frac{50 - 50.05}{0.02} = -2.5$$

$$z_2 = \frac{x_2 - \mu}{\sigma} = \frac{50.11 - 50.05}{0.02} = 3.$$

The standard normal curve with z_1 and z_2 is shown in Fig. 4.15.

From Table A.3, $\Phi(z_1) = 0.0062$ and $\Phi(z_2) = 0.9987$. Thus, the proportion of acceptable parts is $\Phi(z_2) - \Phi(z_1) = 0.9987 - 0.0062 = 0.9925$ or 99.25% of the parts produced.

Gaussian Random Numbers

In a number of simulation studies related to quality, we need random numbers from a normal or Gaussian distribution. The Excel function =rand() and similar functions give random numbers from a uniform distribution on [0, 1]. The Box–Muller[1] transformation provides an elegant approach for generating Gaussian random numbers. Hitting Alt-F11 from an Excel screen and then choosing "Insert" and

---

[1] See G.E.P. Box and M.E. Muller, A Note on the Generation of Random Normal Deviates, *Annals of Mathematical Statistics*, v29, (1958) pp. 610–611.

"Module" from the Menu bar and entering the following Visual Basic code provides the function =randn(), which provides a random number from $N(0, 1)$:

```
Function randn()
    Randomize Timer
    '--- Box-Muller Approach
    Do
        u1 = 2 * Rnd - 1
        u2 = 2 * Rnd - 1
        u = u1 * u1 + u2 * u2
    Loop While (u >= 1)
    w = Sqr(-2 * Log(u) / u)
    randn = u1 * w
End Function
```

Gaussian random numbers can be used to generate simulated experimental data from a normal distribution. See programs *BoxMuller.xls* and *NormalData.xls*.

Chi-Squared Distribution

The *chi-squared* or χ^2 *distribution* is closely associated with the normal distribution. If $Z_1, Z_2, \ldots, Z_\nu$ are ν standard normal variables [$\sim N(0, 1)$], then the random variable X given by

$$X = Z_1^2 + Z_2^2 + \cdots + Z_\nu^2 \tag{4.47}$$

has a chi-squared distribution with ν df. We use the notation $X \sim \chi_\nu^2$.

The probability distribution function is given by

$$f(x) = \frac{1}{2^{\nu/2}\Gamma(\nu/2)} x^{(\nu/2)-1} e^{-(x/2)} \quad x > 0, \tag{4.48}$$

where $\Gamma(\alpha) = \int_0^\infty t^{\alpha-1} e^{-t} dt$ is the gamma function of α. For integer n, $\Gamma(n+1) = n!$ and $\Gamma(1/2) = \sqrt{\pi}$.

A typical chi-squared distribution is shown in Fig. 4.16. The chi-squared value for a right-tail area of α is denoted $\chi_{\alpha,\nu}^2$. The mean and variance for the chi-squared distribution are given by

$$\mu = \nu \tag{4.49}$$

$$\sigma^2 = 2\nu. \tag{4.50}$$

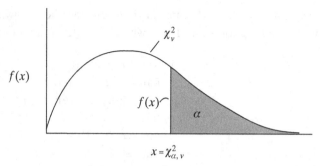

Figure 4.16. χ^2-distribution

Critical chi-squared values for a given right-tail area α near left and right tails are given in Table A.5 in the Appendix. The Excel function CHIDIST(x,df) gives the area to the right of x. CHIINV(α,df) gives the x-value for right-tail area α.

If s is the standard deviation of a sample of size n, then $\frac{(n-1)s^2}{\sigma^2}$ has a chi-squared distribution with $(n-1)$ df. This is used in Chapter 5 to evaluate the confidence intervals on σ^2.

If $\delta_1, \delta_2, \ldots, \delta_\nu$ are constants such that $\delta^2 = \delta_1^2 + \delta_2^2 + \cdots + \delta_\nu^2$, then the random variable $X = (Z_1 + \delta_1)^2 + (Z_2 + \delta_2)^2 + \cdots + (Z_\nu + \delta_\nu)^2$ has a *noncentral chi-squared distribution* with ν df and noncentrality parameter δ^2, which is denoted $\chi_\nu^2(\delta^2)$. The functions chinc(x, df, delta) and chincinv(p, df, delta) are given in *NoncentralDistributions.xls*. Here *p* is the left-tail area. These functions may be used if the need arises for a noncentral chi-squared distribution.

t-Distribution

The *t-distribution* or *Student's t-distribution* is widely used in sampling studies. If a random variable X has a standard normal distribution and an independent random variable Y has a chi-squared distribution with ν df, then the random variable

$$T = \frac{X}{\sqrt{Y/\nu}} \qquad (4.51)$$

has a *t*-distribution with ν df, or $T \sim t_\nu$.

The pdf of a *t*-distribution is given by

$$f(t) = \frac{\Gamma\left(\frac{\nu+1}{2}\right)}{\Gamma(\nu/2)\sqrt{\nu\pi}} \left(1 + \frac{t^2}{\nu}\right)^{-\frac{\nu+1}{2}} \qquad -\infty < t < \infty, \qquad (4.52)$$

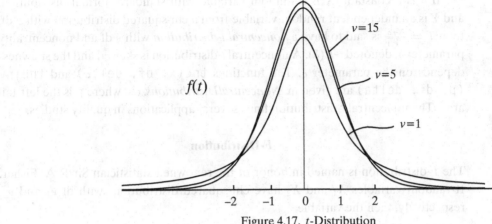
Figure 4.17. *t*-Distribution

and the *t*-distribution is symmetric about $t = 0$ and has a different curve for each value of v, as shown in Fig. 4.17. The variance of the *t*-distribution is

$$\sigma^2 = \frac{v}{v-2}. \tag{4.53}$$

As the number of df approaches infinity, the *t*-distribution approaches the standard normal distribution. The *t*-value for v df and a given right-tail area α is denoted $t_{\alpha,v}$, as shown in Fig. 4.18. Table A.6 in the Appendix gives *t*-values for standard cumulative probabilities for various df values. The applications of *t*-distributions are discussed in Chapter 5. The Excel function `TDIST(t,df,1)` provides the right-tail area and `TDIST(t,df,2)` provides the two-tail area. The inverse function `TINV(`α`,df)` gives the positive *t*-value for the area α included in the two tails.

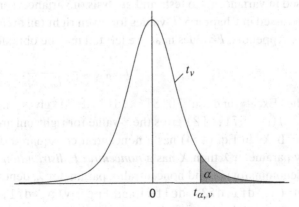
Figure 4.18. $t_{\alpha,v}$ definition

If δ is a constant, X is a random variable with standard normal distribution, and Y is an independent random variable from a chi-squared distribution with v df, then $T = \frac{X+\delta}{\sqrt{Y/v}}$ is said to have a *noncentral t-distribution* with v df and noncentrality parameter δ, denoted $\sim t_v(\delta)$. A noncentral t-distribution is skewed and the skewness depends on the parameter δ. The functions tnc(x, df, delta) and tncinv(p, df, delta) are given in *NoncentralDistributions.xls*, where p is the left-tail area. The noncentral t-distribution finds several applications in quality studies.

F-Distribution

The *F*-distribution is named in honor of the renowned statistician Sir R.A. Fisher. If random variables X_1 and X_2 have chi-squared distributions with df v_1 and v_2, respectively, then the variable

$$X = \frac{X_1/v_1}{X_2/v_2} \tag{4.54}$$

has an *F*-distribution with v_1 (numerator) and v_2 (denominator) df, or $X \sim F_{v_1, v_2}$.

The pdf of the *F*-distribution is given by

$$f(x) = \frac{(v_1/v_2)^{\frac{v_1}{2}}}{B\left(\frac{v_1}{2}, \frac{v_2}{2}\right)} \frac{x^{\frac{v_1}{2}-1}}{\left(1 + \frac{v_1}{v_2}x\right)^{(v_1+v_2)/2}} \quad 0 < x < \infty \tag{4.55}$$

where $B(m, n) = \Gamma(m)\Gamma(n)/\Gamma(m+n)$ is the beta function.

The mean value is given by

$$E(X) = \frac{v_2}{v_2 - 2} \quad \text{for } v_2 > 2. \tag{4.56}$$

The *F*-distribution is illustrated in Fig. 4.19. The *F*-value for numerator df (v_1), denominator df (v_2), and right-tail area (α) is denoted F_{α, v_1, v_2}. *F*-distributions are used in variance ratio tests and analysis of variance, some applications of which are discussed in Chapter 5. *F*-values for given right tail areas α are listed in Table A.7 in the Appendix. *F*-values near the left tail may be obtained using the relationship

$$F_{1-\alpha, v_1, v_2} = \frac{1}{F_{\alpha, v_2, v_1}}. \tag{4.57}$$

The Excel function FDIST(x,df1,df2) gives the area to the right of x; FINV(α, df1,df2) gives the x-value for right-tail area α.

If X_1 in Eq. (4.54) has a noncentral chi-squared distribution with noncentrality parameter λ then X has a *noncentral F-distribution* with v_1 (numerator) and v_2 (denominator) df and noncentrality parameter λ, denoted $X \sim F_{v_1, v_2}(\lambda)$. Functions Fnc(x, df1,df2, delta) and Fncinv(p, df1,df2, delta) are given in *NoncentralDistributions.xls*, where p is the left-tail area.

4.6 Continuous Distributions

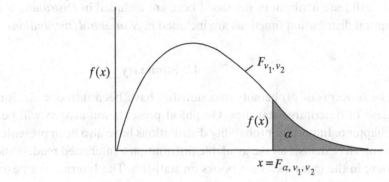

Figure 4.19. *F*-distribution

Studentized Range Distribution

Multiple mean comparison methods use the studentized range distribution. Let Y_1, $Y_2, \ldots, Y_k$ be independent normally distributed random variables $\sim N(\mu, \sigma^2)$, and let S^2 be an estimator of σ^2 based on a chi-squared random variable with v df, which is independent of the value of Y_i. Let $R = \max_i Y_i - \min_i Y_i$ denote the range. The studentized range is the ratio

$$Q_{k,v} = \frac{R}{S}. \qquad (4.58)$$

The studentized range distribution is shown in Fig. 4.20. The studentized range value for a right-tail area of α is denoted $Q_{\alpha,k,v}$. Studentized range values are given in Table A.10 in the Appendix. Two functions have been created in the spreadsheet

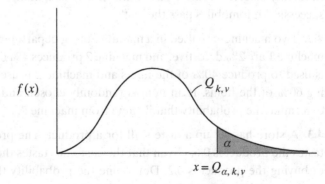

Figure 4.20. Studentized range distribution

StudentRange.xls: the function `strng(q,k,v)` evaluates the probability (left-tail area) for given studentized range q, and the function `strnginv(p,k,v)` evaluates the studentized range for a given value of confidence, $p = 1 - \alpha$.

All the distributions discussed here are included in *Distributions.xls*. The noncentral distribution functions are included in *NoncentralDistributions.xls*.

4.7 Summary

The concepts of probability and statistics have been introduced, along with some ideas of descriptive statistics. Graphical presentation aspects will be discussed in Chapter 6. Important probability distributions have also been presented, but this by no means exhausts all the available distributions. Interested readers should look for these in the many excellent books on statistics. The Internet is an excellent source to pursue interest in statistics. In the next chapter we enter the inferential statistics arena, where we intend to draw conclusions about a population based on data collected from samples.

The computer programs discussed in this chapter include the following:

WorksheetFunc.xls (gives Excel worksheet functions with examples)
Distributions.xls
BoxMuller.xls
NormalData.xls (generates normal data for given values of mean and variance)
NoncentralDistributions.xls

All tables in the Appendix are active Excel worksheets.

EXERCISE PROBLEMS

4.1. Automobiles are inspected for emissions every year. If the probability that a randomly selected vehicle passes the test is 0.8, what is the probability that two successive automobiles pass the test?

4.2. Two machines are used in a manufacturing department. The items produced by machine 1 are 2% defective, and machine 2 produces 4% defective items. Machine 1 is used to produce 40% of the items and machine 2 is used to produce the remaining 60% of the items. If an item is randomly chosen and is found to be defective, determine the probability that it came from machine 2.

4.3. A store has set up a taste stall for a product. The probability that a customer tastes the product is 0.6. Given that the customer tastes the product, the probability of buying the product is 0.2. Determine the probability that a customer tastes the product and buys it.

4.4. For events A, B, and C, show that

$$P(A \cup B \cup C) = P(A) + P(B) + P(C) - P(A \cap B)$$
$$- P(B \cap C) - P(C \cap A) + P(A \cap B \cap C).$$

4.5. A container has 5 good parts and 2 defective parts, and another container has 6 good parts and 4 defective parts. If one part is chosen from each container, find the probability that

   a. both parts are good,
   b. one part is good and one part is defective, and
   c. both parts are defective.

4.6. Determine the mean, median, and mode for the following 10 data points:

$$5, 6, 3, 8, 9, 6, 4, 2, 7, 11.$$

4.7. Calculate the standard deviation for the data in Problem 4.6.

4.8. The following forty observations represent kilometer-per-liter data in evaluating the performance of an engine using a certain brand of gasoline:

9.4	10.7	9.8	8.1
9.5	10.3	10.7	10.2
9.9	8.9	9.9	8.9
10.3	9.6	9.5	9
10.1	8.6	8.8	10.2
9.7	9.6	10	9
9.1	9.8	9.9	10.3
9.9	9.2	10.6	8.4
8.9	9	8.2	8.9
8.6	9.6	9.4	9.9

   a. Find the mean, median, mode, and standard deviation.
   b. Find the skewness, kurtosis, and the IQR.

Comment about the distribution of data based on the preceding calculations.

4.9. A pdf is defined as $f(x) = 2x$ when $0 \leq x \leq 1$ and $f(x) = 0$ elsewhere. Find the expected value and the variance.

4.10. Bin 1 contains 7 circuit boards, of which 2 are nonconforming. Bin 2 contains 8 circuit boards, of which 3 are nonconforming. Two circuit boards randomly picked from bin 1 were transferred to bin 2. If one circuit board is picked from bin 2, what is the probability that it is nonconforming?

4.11. A batch of 25 units produced contains 2 defective units. If 12 units are randomly picked, what is the probability that the number of defective units is 1 or fewer?

4.12. The proportion of defective parts in a large lot is 0.10. These defects are observed to occur randomly and independently. A sample of 20 parts is chosen. Answer the following:

   a. What probability distribution do you use for this case?
   b. What is the probability that there are exactly 3 defects in the sample?
   c. What is the probability that there are 3 or more defects in the sample?
   d. What are the expected values of the mean and standard deviation?

4.13. The number of automobiles of a certain model and year that sometime in the future will suffer catastrophic failure of a suspension mechanism is denoted by X. If X has a Poisson distribution with the average number of failures given as 10, determine the following:

   a. What is the probability that at most 12 cars suffer such a failure?
   b. What is the probability that between 10 and 15 (inclusive) cars will suffer such a failure?

4.14. The fraction of nonconforming parts in a production process is 0.03. A sample of 60 parts is drawn. What is the probability of observing a fraction of 0.04 nonconformities in the sample?

4.15. In the tires produced by a company, a tire is said to be defective if it fails during the first 32,000 km of its life of 64,000 km. The probability of the occurrence of defective tires in manufacture is 0.02. If four new tires made by this company are installed on a car, what is the probability that one tire will fail during the first 32,000 km?

4.16. The average number of paint defects on a four-door sedan produced by an automobile manufacturer is 6. These defects are observed to occur randomly and independently.

   a. What probability distribution do you use to predict the number of defects?
   b. Determine the probability that there are exactly 4 defects in an automobile.
   c. Determine the probability that there are 4 or more defects in an automobile.

4.17. In a selective assembly process, shafts with tolerances of 0 to 30 μm on the diameter are divided into three equal classes: class I, $0 < x \leq 10$; class II, $10 < x \leq 20$; and class III, $20 < x \leq 30$ μm. The probability of shafts falling in class I is 0.2, class II is 0.6, and class III is 0.2. If a random sample of 4 shafts is chosen, what is the probability that it consists of 1 shaft from class I, 1 shaft from class II, and 2 shafts from class III?

4.18. The average number of patients admitted daily to the emergency room of a hospital is 5. What is the probability that

    a. no patients will be admitted on a given day?

    b. 5 or more patients will be admitted on a given day?

4.19. The Rockwell hardness of a hardened alloy steel is normally distributed with a mean of 48 and a standard deviation of 2. Assume that the Rockwell hardness is measured on a continuous scale. If the acceptable tolerance range is $(48 - c, 48 + c)$, what value of c would guarantee 95% acceptable parts?

4.20. The mean diameter and standard deviation of shafts produced in a manufacturing facility are 45 and 2 mm, respectively. If the lower specification limit is 40 mm and the upper specification limit is 51 mm, determine the proportion of acceptable parts produced. What strategy do you suggest to improve the process?

4.21. A manufacturer produces 10,000 shafts. The specification limits are set as 55 ± 0.03 mm. The standard deviation of the process is 0.01 mm. Each undersized part that is scrapped is valued at $40 and the cost of reworking an oversized part is $0.50. Determine the cost to the manufacturer if the process mean is maintained at

    a. 55 mm.

    b. 55.01 mm.

Make your recommendation.

4.22. If a random variable z has a standard normal distribution, how is z^2 distributed?

5

Sampling Concepts

5.1 Introduction

Inspecting every part that is produced is counterproductive in batch and mass production. W. Edwards Deming states "cease dependence on mass inspection." Gathering sample data and using statistical methods is the way to ensure quality in a product. This approach falls into the area of inferential statistics, where we draw conclusions about population characteristics using sample data. In this chapter we present the central limit theorem, which forms the basis for practically all statistical approaches. We then develop the concepts of confidence limits, hypothesis testing, and Type I and Type II errors.

5.2 The Central Limit Theorem

Let $X_1, X_2, \ldots, X_n$ be a random sample taken from a population with mean μ and variance σ^2. We may then state that $X_1, X_2, \ldots, X_n$ are independent random variables from an identical distribution, or $\{X_n\}$ are i.i.d. (independent identically distributed) random variables. The *central limit theorem* addresses the question, "how is the sum $X_1 + X_2 + \cdots + X_n$ distributed?"

The central limit theorem states that $X_1 + X_2 + \cdots + X_n$ has an approximate normal distribution with mean $n\mu$ and variance $n\sigma^2$. It approaches the normal distribution as n approaches infinity. This means that the mean of the sample, $\overline{X}$, has an approximate normal distribution with mean μ and variance σ^2/n. The following is the widely used form of the central limit theorem.

5.2 The Central Limit Theorem

Central Limit Theorem

If $\overline{X}$ is the mean of a random sample of size n from a population with mean μ and standard deviation σ, then

- $\overline{X}$ is approximately normally distributed $[\overline{X} \sim N(\mu, \sigma^2/n)]$.
- The mean of the sampling distribution $\mu_{\overline{X}}$ is

$$\mu_{\overline{X}} = \mu. \tag{5.1}$$

- The standard deviation $\sigma_{\overline{X}}$ of the sampling distribution is

$$\sigma_{\overline{X}} = \frac{\sigma}{\sqrt{n}}. \tag{5.2}$$

Alternatively, we may state the theorem using a standard normal variable in the following form.

Central Limit Theorem – Alternative Statement

If $\overline{X}$ is the mean of a random sample of size n from a population with mean μ and standard deviation σ, then

$$Z = \frac{\overline{X} - \mu}{\sigma/\sqrt{n}} \text{ is approximately } N(0, 1). \tag{5.3}$$

The sample mean tends toward a normal distribution irrespective of the distribution of the population. Even when the population has a uniform distribution, the sample mean can be approximated by a normal distribution for a sample size of five or more. If the population has a normal distribution, the sample mean has a normal distribution that satisfies the central limit theorem statements.

Example 5.1. A random sample of 5 shafts is chosen from a population with a mean diameter of 45 mm and standard deviation 0.2 mm. What is the probability that the mean diameter of a sample is less than 44.92 mm?

Solution. We need to evaluate the probability $P(\overline{X} \leq 44.92)$. Denoting $z = \frac{\bar{x} - \mu}{\sigma/\sqrt{n}}$, the required probability in the standard form is

$$P\left(z \leq \frac{44.92 - 45}{0.2/\sqrt{5}}\right)$$

$$P(z \leq -0.8944).$$

From Table A.3, we have $P(\overline{X} \leq 44.92) = \Phi(-0.8944) = 0.1856$.

5.3 Confidence Interval Estimation

The central limit theorem leads to many important steps in statistical inference. Newspapers often publish the results of surveys that were conducted. For example, based on a survey conducted on a random sample of 400 people the support for a certain political party is 55% and the margin of error is ±5% with 95% confidence; this means the interval of 50 to 60% is correct 95% of the time and extends to the whole population. In quality studies, we would like to predict intervals for the mean or some other parameter of the population with a certain degree of confidence using the sample data.

In confidence interval estimation, we first need to choose (1) a *parameter* of interest, θ (mean, variance, etc.), and (2) a *confidence level*, $1 - \alpha$. We then seek to establish an interval for the parameter so that it falls in that interval with a probability of $1 - \alpha$; α is the probability that the parameter is outside the limit and is referred to as the *level of significance*. For some parameters we may seek both a lower and an upper limit. An example is the mean diameter of a shaft. The *two-sided confidence limit* is of the form

$$P(\theta_L \leq \theta \leq \theta_U) = 1 - \alpha. \tag{5.4}$$

For some parameters such as yield strength in material testing, we may seek only a lower limit. We may seek an upper limit only for finding the percentage level of an impurity in a material. The *one-sided confidence limit* is of the following form:

$$\text{lower } P(\theta_L \leq \theta) = 1 - \alpha$$
$$\text{or upper } P(\theta \leq \theta_U) = 1 - \alpha. \tag{5.5}$$

5.4 Confidence Interval for the Mean of a Normal Population

The confidence interval for the mean is established when the variance of the population is known or when the variance is unknown. In both cases, the central limit theorem helps in developing the relations.

Variance σ^2 Known

Choose a random sample of size n and calculate the sample mean $\overline{X}$. From the central limit theorem, we know that $Z = \frac{\overline{X}-\mu}{\sigma/\sqrt{n}} \sim N(0, 1)$. By setting the confidence level,

5.4 Confidence Interval for the Mean of a Normal Population

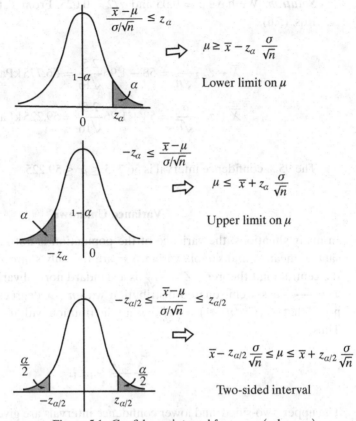

Figure 5.1. Confidence interval for mean (σ known)

$1 - \alpha$, on the standard normal curve as shown in Fig. 5.1, we get the confidence interval as follows:

$$-z_\alpha \leq \frac{\overline{X} - \mu}{\sigma/\sqrt{n}} \Rightarrow \mu \leq \overline{X} + z_\alpha \frac{\sigma}{\sqrt{n}} \quad \text{(upper)}$$

$$-z_{\alpha/2} \leq \frac{\overline{X} - \mu}{\sigma/\sqrt{n}} \leq z_{\alpha/2} \Rightarrow \overline{X} - z_{\alpha/2}\frac{\sigma}{\sqrt{n}} \leq \mu \leq \overline{X} + z_{\alpha/2}\frac{\sigma}{\sqrt{n}} \quad \text{(two-sided)} \quad (5.6)$$

$$\frac{\overline{X} - \mu}{\sigma/\sqrt{n}} \leq z_\alpha \Rightarrow \mu \geq \overline{X} - z_\alpha \frac{\sigma}{\sqrt{n}} \quad \text{(lower)}.$$

The lower tail limit on z gives the upper bound on μ and vice versa. If we fix $\overline{X}$ and treat μ as a random variable, μ appears to have a normal distribution.

Example 5.2. The pressure in a pneumatic conveying system has a standard deviation of 2.5 kPa. The mean value of 16 pressure readings is obtained as 58 kPa. Find a two-sided 95% confidence interval for the pressure.

Solution. We have $\alpha = 0.05$ and $\alpha/2 = 0.025$. From Table A.4, $z_{\alpha/2} = 1.96$. In Eqs. (5.6),

$$\overline{X} - z_{\alpha/2}\frac{\sigma}{\sqrt{n}} = 58 - 1.96\frac{2.5}{\sqrt{16}} = 56.775 \text{ kPa, and}$$

$$\overline{X} + z_{\alpha/2}\frac{\sigma}{\sqrt{n}} = 58 + 1.96\frac{2.5}{\sqrt{16}} = 59.225 \text{ kPa.}$$

The 95% confidence interval is $56.775 \leq \mu \leq 59.225$.

Variance Unknown

In many situations, the variance of the population is not known. In that case the sample mean $\overline{X}$ and sample variance s^2 are the only sources of information. From the central limit theorem, $Z = \frac{\overline{X}-\mu}{\sigma/\sqrt{n}}$ is a standard normal variable. From Chapter 4, $Y = \frac{(n-1)s^2}{\sigma^2}$ has a chi-squared distribution with $n-1$ degrees of freedom. We also noted that $Z/\sqrt{Y/(n-1)} = \frac{\overline{X}-\mu}{s/\sqrt{n}}$ has a t-distribution with $n-1$ degrees of freedom. Thus,

$$\frac{\overline{X} - \mu}{s/\sqrt{n}} \sim t_{n-1}. \tag{5.7}$$

The upper, two-sided, and lower confidence intervals are given by

$$-t_{\alpha,n-1} \leq \frac{\overline{X}-\mu}{s/\sqrt{n}} \Rightarrow \mu \leq \overline{X} + t_{\alpha,n-1}\frac{s}{\sqrt{n}} \quad \text{(upper)}$$

$$-t_{\alpha/2,n-1} \leq \frac{\overline{X}-\mu}{s/\sqrt{n}} \leq t_{\alpha/2,n-1} \Rightarrow \overline{X} - t_{\alpha/2,n-1}\frac{s}{\sqrt{n}} \leq \mu \leq \overline{X} + t_{\alpha/2,n-1}\frac{s}{\sqrt{n}} \quad \text{(two-sided)}$$

$$\frac{\overline{X}-\mu}{s/\sqrt{n}} \leq t_{\alpha,n-1} \Rightarrow \mu \geq \overline{X} - t_{\alpha,n-1}\frac{s}{\sqrt{n}} \quad \text{(lower).}$$
$$\tag{5.8}$$

Note that Table A.6 gives $t_{\alpha,\nu}$ values for some standard values of α.

Example 5.3. A random sample of 25 is selected and a tensile test is performed. The sample mean and sample standard deviation are 118 and 10.5 MPa, respectively. Find the one-sided 90% confidence interval if tensile strength at the higher level is desirable.

Solution. $\overline{X} = 118$ MPa and $s = 10.5$ MPa. The lower limit on μ is obtained by placing the entire $\alpha = 0.1$ at the upper end of the t-variable: $\frac{\overline{X}-\mu}{s/\sqrt{n}}$. We determine $t_{\alpha,n-1} = t_{0.1,24} = 1.3178$ (from Table A.6). From Eqs. (5.8), we get

$$\mu \geq \overline{X} - t_{\alpha,n-1}\frac{s}{\sqrt{n}} = 118 - 1.3178\frac{10.5}{\sqrt{25}} = 115.23.$$

The 90% confidence interval is $\mu \geq 115.23$.

5.5 Confidence Interval for the Difference between Two Means

Sample Variances Known

If $\overline{X}_1$ is the mean of a sample of size n_1 from population $N(\mu_1, \sigma_1^2)$ and $\overline{X}_2$ is the mean of a sample of size n_2 from population $N(\mu_2, \sigma_2^2)$, then we wish to find a $1 - \alpha$ confidence interval for $\mu_1 - \mu_2$. From the central limit theorem, note that $\overline{X}_1 - \overline{X}_2 \sim N(\mu_1 - \mu_2, \frac{\sigma_1^2}{n_1} + \frac{\sigma_2^2}{n_2})$. We may obtain the two-sided confidence limits as

$$\overline{X}_1 - \overline{X}_2 - z_{\alpha/2}\sqrt{\frac{\sigma_1^2}{n_1} + \frac{\sigma_2^2}{n_2}} \leq \mu_1 - \mu_2 \leq \overline{X}_1 - \overline{X}_2 + z_{\alpha/2}\sqrt{\frac{\sigma_1^2}{n_1} + \frac{\sigma_2^2}{n_2}}. \quad (5.9)$$

Sample Variances Not Known

If population variances are not known, we make use of the calculated values of sample variances, s_1^2 and s_2^2. The confidence interval is defined using the t-distribution as

$$\overline{X}_1 - \overline{X}_2 - t_{\alpha/2,\nu} s_D \leq \mu_1 - \mu_2 \leq \overline{X}_1 - \overline{X}_2 + t_{\alpha/2,\nu} s_D. \quad (5.10)$$

The degrees of freedom, ν and s_D are defined for each of two cases, when the population variances are assumed equal and when they are not equal.

$$\text{Case 1: } \sigma_1^2 = \sigma_2^2$$

If s_1^2 and s_2^2 are the calculated sample variances, then the weighted average, s_p^2, called the *pooled estimator of the variance of the samples*, is given by

$$s_p^2 = \frac{(n_1 - 1)s_1^2 + (n_2 - 1)s_2^2}{(n_1 - 1) + (n_2 - 1)} \quad (5.11)$$

$$s_D = s_p\sqrt{\frac{1}{n_1} + \frac{1}{n_2}} \quad (5.12)$$

$$\nu = n_1 + n_2 - 2.$$

Case 2: $\sigma_1^2 \neq \sigma_2^2$

If s_1^2 and s_2^2 are the calculated sample variances, then s_D is given by

$$s_D = \sqrt{\frac{s_1^2}{n_1} + \frac{s_2^2}{n_2}}$$

$$\nu = \frac{\left(\frac{s_1^2}{n_1} + \frac{s_2^2}{n_2}\right)^2}{\left(s_1^2/n_1\right)^2/(n_1-1) + \left(s_2^2/n_2\right)^2/(n_2-1)}. \tag{5.13}$$

Example 5.4. The assembly time for a product from two vendors is under consideration. The average assembly time for 10 units from vendor 1 is 24.5 minutes with a sample standard deviation of 2.5 minutes. The average assembly time for 12 units from vendor 2 is 21.2 minutes with a standard deviation of 3.1 minutes. Find a 90% confidence interval for the difference in assembly time between vendor 1 and vendor 2, assuming that the population variabilities may not be equal.

Solution. We have $n_1 = 10$, $\bar{x}_1 = 24.5$, and $s_1 = 2.5$, and $n_2 = 12$, $\bar{x}_2 = 21.2$, and $s_2 = 3.1$ for this problem:

$$s_D = \sqrt{\frac{s_1^2}{n_1} + \frac{s_2^2}{n_2}} = \sqrt{\frac{2.5^2}{10} + \frac{3.1^2}{12}} = 1.1941$$

$$\nu = \frac{\left(\frac{2.5^2}{10} + \frac{3.1^2}{12}\right)^2}{(2.5^2/10)^2/9 + (3.1^2/12)^2/11} = 19.99 \approx 20.$$

For the 90% confidence interval, $\alpha/2 = 0.05$; from Table A.6, $t_{0.05,20} = 1.7247$. Substituting this in Eq. (5.10), we get

$$24.5 - 21.2 - (1.7247)(1.1941) \leq \mu_1 - \mu_2 \leq 24.5 - 21.2 + (1.7247)(1.1941)$$
$$1.2405 \leq \mu_1 - \mu_2 \leq 5.3594.$$

Since lower and upper limits are both positive, we may conclude that $\mu_1 > \mu_2$ at a level of significance of 0.10.

5.6 Confidence Interval for a Proportion

We have the following result from statistics. If X is a binomial random variable defined on n independent trials where p represents the probability of success, then the random variable $\frac{X-np}{\sqrt{np(1-p)}}$ has an approximate standard normal distribution. This implies that $\frac{X/n-p}{\sqrt{p(1-p)/n}}$ has a standard normal distribution.

If $\hat{p}$ is the proportion of successes in the sample with p as the probability of success then $\hat{p} \sim N(p, p(1-p)/n)$. When n is large enough to satisfy $n\hat{p} \geq 5$ and $n(1-\hat{p}) \geq 5$, the $1-\alpha$ confidence interval for p is given by

$$\hat{p} - z_{\alpha/2}\sqrt{\frac{\hat{p}(1-\hat{p})}{n}} \leq p \leq \hat{p} + z_{\alpha/2}\sqrt{\frac{\hat{p}(1-\hat{p})}{n}}. \tag{5.14}$$

For small samples, the Clopper–Pearson[1] confidence limits are recommended. These limits are based on the binomial probability. We work directly with the number of successes, k, and obtain p_L and p_U such that

$$\sum_{j=k}^{n} \binom{n}{j} p_L^j (1-p_L)^{n-j} = \alpha/2; \quad \sum_{j=k+1}^{n} \binom{n}{j} p_U^j (1-p_U)^{n-j} = 1 - \alpha/2. \tag{5.15}$$

Then p_L and p_U are the lower and upper confidence limits for $p(p_L \leq p \leq p_U)$. Equation (5.15) is solved using the F-distribution transformation, which is introduced in the program *ConfidenceInt.xls*. As an example, p_L is solved using $p_L = 1/(1 + \frac{m}{u} F_{\alpha/2,m,u})$, where $m = 2(n-k+1)$, $u = 2k$.

We now turn our attention to the example posed at the beginning of Section 5.3. If 55% of a random sample of 400 people support a party, let us find the 95% confidence interval. We have $\hat{p} = 0.55$ and $\alpha = 0.05$; $z_{\alpha/2} = z_{0.025} = 1.96$ (from Table A.4) and $1.96\sqrt{\frac{(0.55)(0.45)}{400}} = 0.049$. Using this, the confidence interval is $0.501 \leq p \leq 0.599$. This is a more precise calculation of what was stated earlier.

Sometimes it may be necessary to compare the difference between population proportions. If $\hat{p}_1$ is the proportion of successes in a sample of size n_1 from a population with probability of success p_1, and $\hat{p}_2$ is the proportion of successes in a sample of size n_2 from a population with probability of success p_2, then the $1 - \alpha$ confidence interval for $p_1 - p_2$ is given by

$$\left(\hat{p}_1 - \hat{p}_2 - z_{\alpha/2}\sqrt{\frac{\hat{p}_1(1-\hat{p}_1)}{n_1} + \frac{\hat{p}_2(1-\hat{p}_2)}{n_2}}, \hat{p}_1 - \hat{p}_2 + z_{\alpha/2}\sqrt{\frac{\hat{p}_1(1-\hat{p}_1)}{n_1} + \frac{\hat{p}_2(1-\hat{p}_2)}{n_2}}\right).$$

(5.16)

Example 5.5. The difference in the proportion of nonconforming components produced by two processes is to be established. A random sample of 96 parts

---

[1] C.J. Clopper and E.S. Pearson, The Use of Confidence or Fiducial Limits Illustrated in the Case of the Binomial, *Biometrica*, v26, n4 (1934) pp. 404–413.

taken from one process produced 6 nonconforming parts and a sample of 120 parts taken from the second process had 8 nonconforming parts. Find the 95% confidence interval for the difference in proportion of nonconforming parts.

Solution. We have $\hat{p}_1 = 6/96 = 0.0625$, $n_1 = 96$, $\hat{p}_2 = 8/120 = 0.0667$, $n_2 = 120$, and $\alpha = 0.05$; $z_{\alpha/2} = z_{0.025} = 1.96$ is given from Table A.4. Substituting in Eq. (5.15) we get the confidence interval on the difference of proportions:

$$z_{1-\alpha/2}\sqrt{\frac{\hat{p}_1(1-\hat{p}_1)}{n_1} + \frac{\hat{p}_2(1-\hat{p}_2)}{n_2}}$$

$$= 1.96\sqrt{\frac{0.0625(1-0.0625)}{96} + \frac{0.0667(1-0.0667)}{120}} = 0.0659$$

$$0.0625 - 0.0667 - 0.0659 \leq p_1 - p_2 \leq 0.0625 - 0.0667 + 0.0659$$
$$-0.07 \leq p_1 - p_2 \leq 0.0617.$$

Since 0 is included in the interval, the proportions may be equal at a confidence level of 0.95.

5.7 Confidence Interval for the Variance

A random sample of size n is taken from a normal population and its standard deviation, s, is calculated. We wish to establish a $1 - \alpha$ confidence interval on the population standard deviation using the sample data. In Chapter 4 we discussed that $\frac{(n-1)s^2}{\sigma^2}$ has a chi-squared distribution with $n - 1$ degrees of freedom:

$$\frac{(n-1)s^2}{\sigma^2} \sim \chi^2_{n-1} \tag{5.17}$$

and

$$\chi^2_{1-\alpha/2,n-1} \leq \frac{(n-1)s^2}{\sigma^2} \leq \chi^2_{\alpha/2,n-1}$$
$$\frac{(n-1)s^2}{\chi^2_{\alpha/2,n-1}} \leq \sigma^2 \leq \frac{(n-1)s^2}{\chi^2_{1-\alpha/2,n-1}}. \tag{5.18}$$

The values of $\chi^2_{\alpha/2,n-1}$ and $\chi^2_{1-\alpha/2,n-1}$ are obtained from Table A.5, which gives important critical values for the chi-squared distribution.

Example 5.6. Cylindrical pins are produced on a centerless grinding machine. A random sample of 10 cylindrical pins has been collected and the standard deviation of the sample is obtained as 0.025 mm (25 μm). Determine the 95% confidence interval for the population standard deviation, σ.

Solution. We have $\alpha = 0.05$ and $n = 10$. From Table A.5, $\chi^2_{1-\alpha/2,n-1} = \chi^2_{0.975,9} = 2.7$, and $\chi^2_{\alpha/2,n-1} = \chi^2_{0.025,9} = 19.023$. From Eq. (5.17), the confidence interval is

$$\frac{(9)(0.025)^2}{19.023} \le \sigma^2 \le \frac{(9)(0.025)^2}{2.7}$$

$$0.000296 \le \sigma^2 \le 0.002083$$

$$0.0172 \le \sigma \le 0.0456.$$

5.8 Confidence Interval for the Ratio of Two Variances

In Chapter 4 we learned that if random variables X_1 and X_2 have chi-squared distributions with degrees of freedom ν_1 and ν_2, respectively, then the variable $X = \frac{X_1/\nu_1}{X_2/\nu_2}$ has an F-distribution with ν_1 (numerator) and ν_2 (denominator) degrees of freedom: $X \sim F(\nu_1, \nu_2)$. If a sample variance of s_1 is the sample standard deviation of a random sample of size n_1 from a population $N(\mu_1, \sigma_1^2)$ and a sample variance of s_2 is the sample standard deviation of a random sample of size n_2 from a population $N(\mu_2, \sigma_2^2)$, then consider the ratios $\frac{(n_1-1)s_1^2}{\sigma_1^2}/(n_1 - 1)$ and $\frac{(n_2-1)s_2^2}{\sigma_2^2}/(n_2 - 1)$. These ratios have an F-distribution with $n_1 - 1$ and $n_2 - 1$ degrees of freedom:

$$\frac{s_1^2/\sigma_1^2}{s_2^2/\sigma_2^2} \sim F_{n_1-1, n_2-1}. \qquad (5.19)$$

A confidence interval of $1 - \alpha$ for the population variance ratio is then given by

$$\frac{s_1^2/s_2^2}{F_{\alpha/2,n_1-1,n_2-1}} \le \frac{\sigma_1^2}{\sigma_2^2} \le \frac{s_1^2/s_2^2}{F_{1-\alpha/2,n_1-1,n_2-1}}. \qquad (5.20)$$

Table A.7 only has values given for $F_{\alpha/2,\nu_1,\nu_2}$; $F_{1-\alpha/2,n_1-1,n_2-1}$ can be obtained by reading $F_{\alpha/2,n_2-1,n_1-1}$ and using Eq. (4.52):

$$F_{1-\alpha/2,n_1-1,n_2-1} = \frac{1}{F_{\alpha/2,n_2-1,n_1-1}}. \qquad (5.21)$$

Example 5.7. Determine a 90% confidence interval for the ratio of population standard deviations from the two vendors in Example 5.4.

Solution. In Example 5.4, we considered the assembly times from two vendors: a standard deviation $s_1 = 2.5$ minutes on 10 units ($n_1 = 10$) from vendor 1 and $s_2 = 3.1$ minutes on 12 units ($n_2 = 12$) from vendor 2. For 90% confidence, $\alpha = 0.1$. From Table A.7, for F-distribution, $F_{\alpha/2,n_1-1,n_2-1} = F_{0.05,9,11} = 2.90$. We do not have $F_{9,11,0.95}$ available in the table, but $F_{0.05,11,9}$ can be approximately read by interpolation as 3.12. Then, using Eq. (5.20), we have $F_{0.95,9,11} = 1/F_{0.05,11,9} = 0.321$. The value can be directly obtained in the spreadsheet file that is included in the disk.

The confidence interval is

$$\frac{2.5^2/3.1^2}{2.90} \leq \frac{\sigma_1^2}{\sigma_2^2} \leq \frac{2.5^2/3.1^2}{0.321}$$

$$0.224 \leq \frac{\sigma_1^2}{\sigma_2^2} \leq 2.026$$

$$0.4733 \leq \frac{\sigma_1}{\sigma_2} \leq 1.4234.$$

More precise values can be obtained by using the Excel function. The calculations are provided in *ConfidenceInt.xls*.

5.9 Hypothesis Testing

Hypothesis testing is an integral part of scientific investigation. An investigator comes up with a theory or hypothesis and this hypothesis continues to hold unless proven otherwise. Setting up and testing hypotheses is at the heart of inferential statistics. Consider an example, where a company uses a certain alloy steel and the average tensile strength has been established at 390 MPa with a standard deviation of 6.5 MPa. Recently the supplier stated that they made some process improvements in steel production. A recent test conducted on 12 specimens gave an average value of tensile strength as 394 MPa. With this data in hand, do we conclude that the process has changed the strength? With the mean value of $\mu_0 = 390$ MPa, we state the hypotheses:

Null hypothesis $\quad H_0: \mu = \mu_0$
Alternative hypothesis $\quad H_1: \mu \neq \mu_0.$

The null hypothesis (H_0) implies no change in the status quo. It continues to hold unless it is proved that the alternative hypothesis (H_1) holds. The alternative hypothesis, H_1, is all that is not H_0. The aforementioned hypotheses define the two-sided rejection region. To test a hypothesis, we need to select a *level of significance*, α. In our example, let us set the level of significance as $\alpha = 0.05$. From the central limit theorem, $\frac{\bar{X}-\mu_0}{\sigma/\sqrt{n}}$ has a normal distribution, $N(0, 1)$. From Table A.4, $z_{\alpha/2} = 1.96$, and we evaluate z_0 for the sample:

$$z_0 = \frac{\bar{x}-\mu_0}{\sigma/\sqrt{n}} = \frac{392-390}{6.5/\sqrt{12}} = \frac{2}{1.876} = 1.066.$$

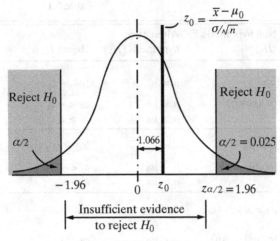

Figure 5.2. Two-sided test

Since $1.066 < 1.96$, the evidence is not sufficient to reject H_0. This is illustrated in Fig. 5.2.

Hypothesis testing is closely associated with the confidence interval study presented in the previous section. In our example, if we calculate the confidence interval for the mean, we get the interval $(392 - 1.96 \times 1.876, 392 + 1.96 \times 1.876) = (388.32, 395.68)$; $\mu_0 = 390$ MPa is in this interval, which implies that the evidence is not sufficient to reject the null hypothesis.

Another approach to making the decision about rejecting the null hypothesis is called the *p-method*. For the example considered, we evaluate the area $1 - \Phi(z)$ to the right of $z_0 = 1.066$ from Table A.3 as $1 - 0.8568 = 0.1432$. Since the rejection region is two-sided, we multiply this by 2 to get $p = 0.2864$. If $p \geq \alpha$, the evidence is not sufficient to reject the null hypothesis and it is rejected otherwise. The *p*-evaluation for a two-sided test is illustrated in Fig. 5.3.

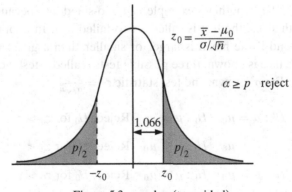

Figure 5.3. *p*-value (two-sided)

Table 5.1

	Null hypothesis (H_0)	Alternative hypothesis (H_1)	Reject H_0 if	Test statistic
1	$\mu = \mu_0$ (σ known)	$\mu \neq \mu_0$ $\mu < \mu_0$ $\mu > \mu_0$	$|z_0| > z_{\alpha/2}$ $z_0 < -z_\alpha$ $z_0 > z_\alpha$	$z_0 = \dfrac{\bar{x} - \mu_0}{\sigma/\sqrt{n}}$
2	$\mu = \mu_0$ (σ unknown)	$\mu \neq \mu_0$ $\mu < \mu_0$ $\mu > \mu_0$	$|t_0| > t_{\alpha/2, n-1}$ $t_0 < -t_{\alpha, n-1}$ $t_0 > t_{\alpha, n-1}$	$t_0 = \dfrac{\bar{x} - \mu_0}{s/\sqrt{n}}$
3	$\mu_1 = \mu_2$ (σ_1, σ_2 known)	$\mu_1 \neq \mu_2$ $\mu < \mu_0$ $\mu > \mu_0$	$|z_0| > z_{\alpha/2}$ $z_0 < -z_\alpha$ $z_0 > z_\alpha$	$z_0 = \dfrac{\bar{x}_1 - \bar{x}_2}{\sqrt{\sigma_1^2/n_1 + \sigma_2^2/n_2}}$
4	$\mu_1 = \mu_2$ (σ_1, σ_2 unknown)	$\mu_1 \neq \mu_2$ $\mu < \mu_0$ $\mu > \mu_0$	$|t_0| > t_{\alpha/2, \nu}$ $t_0 < -t_{\alpha, \nu}$ $t_0 > t_{\alpha, \nu}$	$t_0 = \dfrac{\bar{x}_1 - \bar{x}_2}{s_D}$ s_D, ν from Eq. (5.12) or (5.13)
5	$\sigma^2 = \sigma_0^2$	$\sigma^2 \neq \sigma_0^2$ $\sigma^2 < \sigma_0^2$ $\sigma^2 > \sigma_0^2$	$\chi_0^2 > \chi_{\alpha/2, n-1}^2$ or $\chi_0^2 < \chi_{1-\alpha/2, n-1}^2$ $\chi_0^2 < \chi_{1-\alpha, n-1}^2$ $\chi_0^2 > \chi_{\alpha, n-1}^2$	$\chi_0^2 = \dfrac{(n-1)s^2}{\sigma_0^2}$
6	$\sigma_1^2 = \sigma_2^2$	$\sigma_1^2 \neq \sigma_2^2$ $\sigma_1^2 < \sigma_2^2$ $\sigma_1^2 > \sigma_2^2$	$F_0 < F_{1-\alpha/2, n_1-1, n_2-1}$ or $F_0 > F_{\alpha/2, n_1-1, n_2-1}$ $F_0 < F_{1-\alpha, n_1-1, n_2-1}$ $F_0 > F_{\alpha, n_1-1, n_2-1}$	$F_0 = \dfrac{s_1^2}{s_2^2}$

The hypothesis example is a two-sided one because the rejection region is on both sides; the test is called a two-tailed test. In another test it may be of interest to find if the mean is larger or smaller than a given value. When the population variance is known, three possible tests (called z-tests) can be used.

For σ known, and test statistic $z_0 = \frac{\bar{x}-\mu_0}{\sigma/\sqrt{n}}$,

$H_0 : \mu = \mu_0 \quad H_1 : \mu \neq \mu_0 \quad$ Reject H_0 for $z_0 > z_{\alpha/2} \quad$ or $p = 2\Phi(-z_0)$

$H_0 : \mu = \mu_0 \quad H_1 : \mu < \mu_0 \quad$ Reject H_0 for $z_0 < -z_\alpha \quad$ or $p = \Phi(z_0) \quad$ (5.22)

$H_0 : \mu = \mu_0 \quad H_1 : \mu > \mu_0 \quad$ Reject H_0 for $z_0 > z_\alpha \quad$ or $p = 1 - \Phi(z_0)$.

Table 5.1 gives the test statistic and the criteria for the rejection of the null hypothesis for various cases discussed in the previous section on confidence intervals. In Table 5.1, examples 1 and 3 are referred to as z-tests, and examples 2 and 4 are referred to as t-tests.

Example 5.8. Automobiles with 2.5-liter 4-cylinder engines from two different manufacturers are tested for their fuel economy rating. Ten cars from one manufacturer gave an average rating of 12.5 kilometers per liter (KPL) and a standard deviation of 1.5 KPL and 15 cars from a different manufacturer gave an average rating of 14.5 KPL and a standard deviation of 2 KPL. Are the KPL ratings for the two brands equal at a level of significance of 0.05?

Solution. We make use of the hypotheses from row 4 of Table 5.1:

$$H_0: \mu_1 = \mu_2, \quad H_1: \mu_1 \neq \mu_2.$$

We have $n_1 = 10$, $\bar{x}_1 = 12.5$ KPL, and $s_1 = 1.5$ for the first manufacturer, and $n_2 = 15$, $\bar{x}_2 = 14.5$, and $s_2 = 2$ for the second manufacturer:

$$s_D = \sqrt{\frac{s_1^2}{n_1} + \frac{s_2^2}{n_2}} = \sqrt{\frac{1.5^2}{10} + \frac{2^2}{15}} = 0.7012$$

$$\nu = \frac{\left(\frac{1.5^2}{10} + \frac{2^2}{15}\right)^2}{(1.5^2/10)^2/9 + (2^2/15)^2/14} = 22.58 \approx 23$$

$$t_0 = \frac{\bar{x}_1 - \bar{x}_2}{s_D} = \frac{12.5 - 14.5}{0.7012} = -2.852.$$

From Table A.6, $t_{0.025,23} = 2.0687$. Since $|t_0| = 2.852 > 2.0687$, we reject H_0. The two means are not equal; the 95% confidence interval for $\mu_1 - \mu_2$ from the previous section gives $(-3.45, -0.55)$. Since both limits are negative it shows that the first mean is lower at a 5% level of significance. The relationships given in Table 5.1 are implemented in the program *HypthesisTesting.xls*.

In the hypothesis testing we are trying to determine if the null hypothesis is rejected or not. Even when the null hypothesis holds, there is a chance of rejecting it with a probability of α, the level of significance. What if the alternative hypothesis, H_1, holds and we do not reject H_0? There are two types of errors that may be committed, which are discussed in the next section.

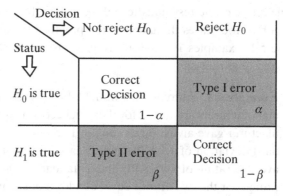

Figure 5.4. Type I and Type II errors

5.10 Type I and Type II Errors

If H_0 is true and we decide not to reject H_0, or H_1 is true and we decide to reject H_0, there is no problem. If H_0 is true and we make a decision to reject H_0 then we have committed an error. This is called a *Type I error*. When a Type I error is committed, good products that are produced are rejected based on the sample, which is also called the *producer's risk*. The error is committed when the distribution is favorably positioned about the value specified by the null hypothesis and the rejection is at the tails. The probability of committing a Type I error is equal to the level of significance, α. When the alternative hypothesis, H_1, is true and we conclude that H_0 is not rejected, we have committed the second type of error, or *Type II error*. When a Type II error is committed, products with a shift of the mean are accepted and there is a resulting chance of some defective parts reaching the consumer, also called the *consumer's risk*. The evaluation of the probability of committing a Type II error, β, is somewhat involved and will be discussed in some detail. Fig. 5.4 summarizes the decision table which defines the errors.

We already stated that the probability of committing a Type I error is the same as the level of significance, α. The probability distribution is in one unique position and the rejection areas of the distribution are well defined. A Type II error implies a deviation from this position. Small shifts are to be accepted with a greater probability and larger shifts with a low probability. The probability of committing a Type II error is β, and $1 - \beta$ is called the *power* of the (hypothesis) test. The power, $1 - \beta$, is the probability of making the correct decision when H_1 is true (the bottom right-hand block in Fig. 5.4). Let us evaluate β for the normal probability distribution when σ is known.

When H_0 holds, we evaluate Z_0 as $\frac{\overline{X} - \mu_0}{\sigma/\sqrt{n}}$. If H_1 holds, let $\overline{Y}$ be the variable with shift δ; then $\overline{Y}$ is set as $\overline{X} + \sigma$. The normal curve is now symmetrically placed with

5.10 Type I and Type II Errors

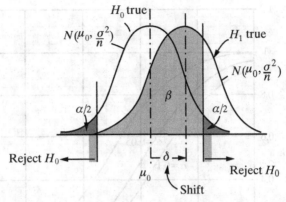

Figure 5.5. Type II error

respect to $\mu_0 + \delta$ as shown in Fig. 5.5. The standard normal variable is calculated using

$$\frac{\overline{X} + \delta - \mu_0}{\sigma/\sqrt{n}} = \frac{\overline{X} - \mu_0}{\sigma/\sqrt{n}} + \frac{\delta\sqrt{n}}{\sigma} = Z_0 + \frac{\delta\sqrt{n}}{\sigma}. \qquad (5.23)$$

Equation (5.23) shows that the δ-shift in μ_0 results in a shift of the standard normal curve by $\frac{\delta\sqrt{n}}{\sigma}$, which is shown in Fig. 5.6. The probability β of a Type II error is given by

$$\beta = \Phi\left(z_{\alpha/2} - \frac{\delta\sqrt{n}}{\sigma}\right) - \Phi\left(-z_{\alpha/2} - \frac{\delta\sqrt{n}}{\sigma}\right). \qquad (5.24)$$

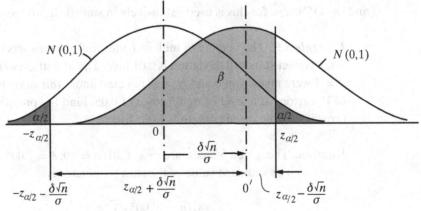

Figure 5.6. Shift of standard normal

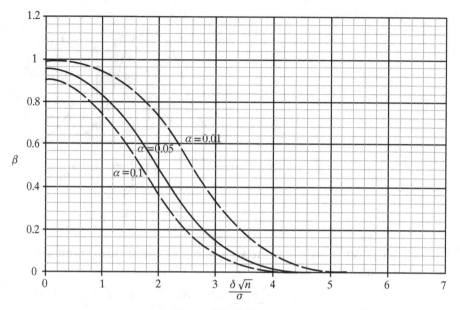

Figure 5.7. OC curve

The plot of β versus the shift δ is referred to as the *operating characteristic (OC) curve*. There is an OC curve for each value of the level of significance, α. The curve also changes for various values of n and σ. Fig. 5.7 shows OC curves for $\alpha = 0.01$, 0.05, and 0.1. The x-axis is taken as the value of the parameter $\frac{\delta\sqrt{n}}{\sigma}$. Given δ, n, and σ, the parameter can be evaluated and the value of β can be determined. Thus, this one curve serves all possibilities. The program *OCCurveCh5.xls* generates the OC curve and may be used to plot OC curves for other values of α. We observed before that if the level of significance is chosen for a two-sided test as $\alpha = 0.0027$, $z_{\alpha/2} = 3$ and the OC curve for this is used extensively in statistical process control.

Example 5.9. The volume of milk in 1-liter milk cartons was studied in a dairy. The product standard deviation is 0.01 liter. The hypotheses H_0: $\mu = 1$ and H_1: $\mu \neq 1$ were investigated and H_0 was rejected under this study based on a sample of 10 cartons at a level of significance of 0.05. Find the probability of a Type II error if the true mean content is 1.005 liters.

Solution. The given values are $\sigma = 0.01, n = 10, \delta = 1.005 - 1 = 0.005$, and $\alpha = 0.05$. The standard value of the shift parameter is

$$\frac{\delta\sqrt{n}}{\sigma} = \frac{0.005\sqrt{10}}{0.01} = 1.5811.$$

At $\alpha = 0.05$, $z_{1-\alpha/2} = 1.96$ and $z_{\alpha/2} = -1.96$ (from Table A.4). The probability of error is

$$\beta = \Phi\left(z_{\alpha/2} - \frac{\delta\sqrt{n}}{\sigma}\right) - \Phi\left(-z_{\alpha/2} - \frac{\delta\sqrt{n}}{\sigma}\right)$$
$$= \Phi(1.96 - 1.5811) - \Phi(-1.96 - 1.5811)$$
$$= \Phi(0.3789) - \Phi(3.5411)$$
$$= 0.6476 - 0.0002$$
$$= 0.6474.$$

The cumulative normal probabilities were obtained from Table A.3 at the last step. The probability of a Type II error is $\beta = 0.6474$.

Next the method of evaluation of a Type II error is considered when the population standard deviation is not known. For this case, a t-test is performed to investigate the hypotheses. Let δ be the shift in the mean value of the variable. If the shifted variable is $\overline{Y}$, then after setting $\overline{Y} = \overline{X} + \delta$ we get

$$\frac{\overline{Y} - \mu_0}{s/\sqrt{n}} = \frac{\overline{X} + \delta - \mu_0}{s/\sqrt{n}} = \frac{\sqrt{n}(\overline{X} - \mu_0 + \delta)}{s} = \frac{\sqrt{n}(\overline{X} - \mu_0) + \delta\sqrt{n}}{s}$$
$$= \frac{\frac{\sqrt{n}(\overline{X} - \mu_0)}{\sigma} + \frac{\delta\sqrt{n}}{\sigma}}{s/\sigma} = \frac{\frac{(\overline{X} - \mu_0)}{\sigma/\sqrt{n}} + \frac{\delta\sqrt{n}}{\sigma}}{\sqrt{(n-1)s^2/(n-1)\sigma^2}}.$$

In the final expression, the denominator is the square root of a chi-squared distribution divided by its degrees of freedom. We recognize that this represents a noncentral t-distribution with noncentrality parameter $\frac{\delta\sqrt{n}}{\sigma}$. Thus,

$$\frac{\overline{X} - \mu_0 + \delta}{s/\sqrt{n}} \sim t_{n-1}\left(\frac{\delta\sqrt{n}}{\sigma}\right). \tag{5.25}$$

The functions `tnc(x,df,delta)` and `tncinv(p,df,delta)` provided in *NoncentralDistributions.xls* give the cumulative probability and its inverse for the noncentral t-distribution.

We set the acceptance limits under the assumption that H_0 holds. The limits are $-t_{n-1,\alpha/2}$, and $-t_{n-1,\alpha/2}$. If the noncentrality parameter is $\Delta = \delta\sqrt{n}/\sigma$, then the probability β of a Type II error is given by

$$\beta = \text{tnc}(t_{n-1,\alpha/2}, n-1, \Delta) - \text{tnc}(-t_{n-1,\alpha/2}, n-1, \Delta), \tag{5.26}$$

which is illustrated in Fig. 5.8.

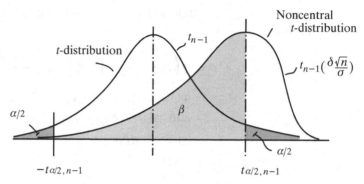

Figure 5.8. β for t-distribution

Example 5.10. An experiment is set up as $H_0: \mu = \mu_0$, $H_1: \mu \neq \mu_0$ and is tested at a level of significance of 0.1. The sample size is 15. Determine the probability of rejecting H_0 if the true μ has shifted 0.25 standard deviation to the right of μ_0.

Solution. We need to find the probability of rejecting H_0, which is $1 - \beta$. Here σ is not known. The sample size is $n = 15$, and δ/σ is given as 0.25. The shift is $\delta\sqrt{n}/\sigma = 0.25\sqrt{15} = 0.9682$. The level of significance is $\alpha = 0.1$. From Table A.6 for critical values of the t-distribution, $t_{n-1,\alpha/2} = t_{14,0.95} = 1.7613$. From *NoncentralDistributions.xls*, tnc(1.7613, 14, 0.9682) = 0.7652, and tnc(−1.7613, 14, 0.9682) = 0.00512. From Eq. (5.24), we have $\beta = 0.7652 - 0.00512 = 0.7601$.

The probability of a Type II error is 76.01% and the probability of rejecting H_0 is 23.99%.

OC curves can be constructed by using a noncentral t-distribution when the population standard deviation is unknown.

5.11 Sample Size Determination

We note from the central limit theorem that the standard deviation of the sample mean is $\sigma/\sqrt{n}$. When σ is known, the confidence interval for the mean with a level of significance, α, is $(\overline{X} - z_{\alpha/2}\frac{\sigma}{\sqrt{n}}, \overline{X} + z_{\alpha/2}\frac{\sigma}{\sqrt{n}})$. The change with respect to the mean value, ρ, is

$$\rho = z_{\alpha/2}\frac{\sigma}{\sqrt{n}}, \qquad (5.27)$$

where ρ depends on α, n, and σ. If the smallest recognized change in the mean, ρ, is known, we can choose the sample size using

$$n = \frac{z_{\alpha/2}^2 \sigma^2}{\rho^2}. \tag{5.28}$$

In a similar manner, the variation on the proportion p at a confidence level of $1 - \alpha$ is $\rho = z_{\alpha/2}\sqrt{p(1-p)/n}$. This leads to

$$n = \frac{z_{\alpha/2}^2 p(1-p)}{\rho^2}, \tag{5.29}$$

where $p(1-p)$ has a maximum value when $p = \frac{1}{2}$. A quick estimate when p is not known is

$$n = \frac{z_{\alpha/2}^2}{4\rho^2}. \tag{5.30}$$

For 95% confidence, $z_{\alpha/2} = z_{0.025} = 1.96$. For this case n is nearly $1/\rho^2$. If $\rho = 0.05$, the sample size is 400. This is the type of example used at the beginning of the confidence interval section.

5.12 Summary

In this chapter, the central limit theorem was introduced, which is the key step in performing sampling studies. Basic ideas involving confidence intervals and hypothesis testing have been introduced. A clear understanding of the various steps discussed in this chapter helps the quality analyst in formulating and solving a variety of problems.

The computer programs discussed in this chapter include the following:

ConfidenceInt.xls
HypothesisTesting.xls
OCCurveCh5.xls.

EXERCISE PROBLEMS

5.1. A random sample of 12 silicon wafers has an average thickness of 270 μm. The standard deviation is known to be 8 μm. Determine a two-sided 99% confidence interval.

5.2. A metal manufacturing company produces ingots of nickel at its plant. A random sample of 25 ingots yielded $\bar{x} = 2.56$ kg. Production records from the previous three years show that the variance of the mass of nickel ingots is $\sigma^2 = 0.0049$ kg^2. Construct a two-sided 80% confidence interval on the mean mass of nickel ingots from this company.

5.3. A sample of 5 randomly chosen tensile test bars gave a mean yield strength of 269 MPa. The process standard deviation is known to be 3.5 MPa. Determine a 90% confidence interval for the mean yield strength. Having set the interval, comment if a mean value of 273 MPa of another sample of 5 is acceptable.

5.4. A sample of 16 randomly chosen cylindrical shafts gave a mean diameter of $\bar{x} = 30.45$ mm. The standard deviation of the sample is $s = 0.96$ mm. Determine a 95% confidence interval for the mean diameter.

5.5. A random sample of 5 beverage cans is taken at a beverage manufacturing plant and the volume of fluid is measured in milliliters as 335, 338, 341, 342, 339 ml. Determine a two-sided 95% confidence interval for the mean beverage volume.

5.6. A confidence interval for the difference in the mean processing times between two processes for producing a part is to be established. A sample of 12 units from the first process gives an average processing time of 4.4 minutes with a standard deviation of 0.5 minutes. A sample of 8 units from the second process has an average processing time of 4.9 minutes with a standard deviation of 0.6 minutes. Determine a 90% confidence interval for the difference in the processing times assuming that the population variabilities may not be equal.

5.7. A random sample of 140 silicon wafers has 12 nonconforming units. Estimate the process fraction that are nonconforming and construct a two-sided 95% confidence interval for the true proportion that are nonconforming.

5.8. A random sample of 16 units produced a standard deviation of 2.5. Determine a 90% confidence interval for the population variance.

5.9. A random sample of 15 units from a process has a standard deviation of 3.5, and a random sample of 20 units from an alternative process has a standard deviation of 2.8. Determine a 95% confidence interval on the ratio of the population variances.

5.10. In Problem 5.1, test the hypothesis that the true mean is 275 μm at a level of significance of 0.05. What is the p-value?

5.11. In Problem 5.3, test the hypothesis that the true mean exceeds 265 MPa with a level of significance of 0.1. Also determine the p-value.

5.12. In Problem 5.4, test the hypothesis that the true mean is 30 mm with a level of significance of 0.05. Also determine the p-value.

5.13. In Problem 5.7, test the hypothesis that the true nonconforming proportion is 0.08 at a level of significance of 0.05.

5.14. In Problem 5.8, test the hypothesis that the true variance is 2.3 at a level of significance of 0.1. Determine the *p*-value.

5.15. The process standard deviation is known to be 1.5. If a two-sided 95% confidence interval for the mean is to be established as $\pm 0.5\sigma$, determine the sample size.

6

Data Presentation: Graphs and Charts

6.1 Introduction

Data gathered for statistical analysis may fill volumes. The collected data generally consist of numbers gathered over time, and the period covered might include ups and downs of production cycles. The data need to be sifted through and must be brought into a presentable form. Numbers may make sense to the statistical analyst but they must be communicated to others in the department, division, plant, or business organization. Graphs and charts are convenient means for this communication. The graphs and charts covered here are stem-and-leaf plots, histograms, cause-and-effect diagrams, Pareto charts, box plots, and run charts.

6.2 Stem-and-Leaf Plots

A *stem-and-leaf plot* provides a visual display of the variation in a parameter. If the parameter represents the diameter of a shaft produced by an automatic lathe, or the voltage of a power supply, the values have a small variation about a mean value. Let us consider the development of the stem-and-leaf plot through an example. The data in Table 6.1 are forty measurements of the diameter of a cylindrical pin, in millimeters.

Inspection of the data shows that the first digit is 5 for each of the values. If we arrange the data in ascending order using a spreadsheet and consider only the digits after the decimal point, we get the numbers 28, 36, 37, 37, 38, 39, 39, 39, 40, 40, 41, 41, 42, 43, 43, 44, 45, 46, 46, 46, 47, 47, 47, 47, 47, 48, 48, 49, 49, 50, 52, and so on. We may then make a visual display of the numbers following the first digit 2, 3, 4,..., 7. If there are several numbers associated with each digit, we may form the subgroups 0–4 and 5–9, or 0–1, 2–3, 4–5, 6–7, and 8–9. The

Table 6.1. *Diameter of a cylindrical pin, in millimeters*

5.47	5.36	5.59	5.47	5.46
5.37	5.52	5.42	5.52	5.46
5.28	5.43	5.57	5.55	5.44
5.50	5.61	5.39	5.46	5.41
5.58	5.37	5.39	5.43	5.62
5.52	5.48	5.47	5.56	5.68
5.49	5.52	5.39	5.41	5.48
5.38	5.59	5.47	5.54	5.71
5.59	5.47	5.53	5.40	5.49
5.78	5.53	5.40	5.58	5.45

stem-and-leaf display with two subgroups for each digit, 0–4 and 5–9, is given as follows:

2	8
3	6778999
4	00112334
4	5666777778899
5	02222334
5	56788999
6	12
6	8
7	1
7	8

This is the stem-and-leaf plot. It provides a visual display of the distribution.

6.3 Histograms

A *histogram* provides a visual display of the distribution of the population. If data are collected for a continuous variable, such as the length of a part or current flow in a circuit, the range for this variable (max − min) is divided into several bins of equal width. The suggested number of bins is about $\sqrt{n}$, where n is the number of data points. If there are 100 data points, the suggested number of bins is 10, which may be used as a rough guideline. An alternative is to use Sturgis' rule, which suggests using an integer N that approximately satisfies $2^{N-1} = n$. For discrete data such as the frequency of occurrence of the number of nonconformities, the intervals may be taken as integers or integer groups. Once the bins are chosen, the number of data points falling inside a bin has to be determined. For bin i, we may choose the interval $x_{i-1} < x \leq x_i$, where strict inequality is set at the left. We follow this in our construction of the histogram. The "frequency" function in Excel also uses this approach. Some references place the strict inequality at the right. In our formulation if we divide the min-to-max interval into n bins, we add a bin to the left to form $n+1$ bins in order to cover all the data points. The steps involved in the construction of the histogram are shown by means of an example.

Example 6.1. Prepare a histogram for the 50 data points for the cylindrical pin diameter given in Table 6.1.

Solution. The data are entered in an Excel spreadsheet (see *Histogram.xls* included in the CD). We used 10 intervals defining 11 bins:

| Max | 5.78 | #Bins | 11 |
| Min | 5.28 | Interval | 0.05 |

Bin#	<=(bin)	Frequency
1	5.28	1
2	5.33	0
3	5.38	4
4	5.43	10
5	5.48	12
6	5.53	9
7	5.58	6
8	5.63	5
9	5.68	1
10	5.73	1
11	5.78	1

The middle column shows the right-end values for each bin. The frequency is obtained by following these steps. First highlight the range where frequency values are to be obtained. Next key in the formula =FREQUENCY(freq cell range, bin cell range) and press Shift+Ctrl+Enter. After the frequency values are obtained we choose the column-type chart option in the chart wizard. The frequency histogram is shown in Fig. 6.1.

If the frequency for each bin is divided by the total number of data points we get the relative frequency. This relative frequency approaches the probability distribution function as the bin width approaches 0. The cumulative frequency of bin i is the sum of frequencies of bins 1 through i. The cumulative frequency may also be plotted in the histogram.

6.4 Cause-and-Effect Diagrams

Cause-and-effect diagrams are also called *Ishikawa diagrams* after Kaoru Ishikawa, who proposed them in 1943. They are also referred to as *fishbone diagrams* because

6.4 Cause-and-Effect Diagrams

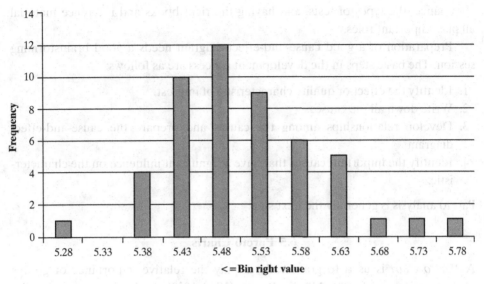

Figure 6.1. Histogram

of their unique structure. The cause-and-effect diagram shows the relationship between a quality characteristic and its factors. The quality characteristic may be dimensional variation, defects in the product, power fluctuation, and so forth. The main causes may be related to equipment, materials, people, methods, measurement, environment, and so on. As an example consider the problem of a student's low score on an exam, the main causes of which are illustrated in Fig. 6.2. The student deals with equipment such as calculators and computers and interacts with faculty and other students. The way the student collects notes, the regularity of

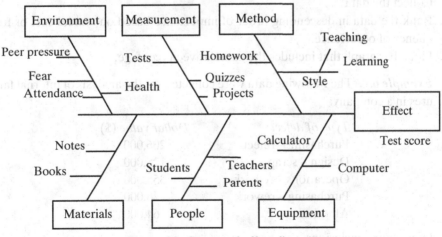

Figure 6.2. Cause-and-effect diagram

attendance, the types of tests, and having the right books and reference material all play important roles.

Preparation of a good cause-and-effect diagram needs a good brainstorming session. The basic steps in the development process are as follows:

1. Identify the effect or quality characteristic of interest.
2. Write down all the causes.
3. Develop relationships among the causes and prepare the cause-and-effect diagram.
4. Identify the important causes that have a significant influence on the characteristic.

Pareto analysis is generally the next step.

6.5 Pareto Charts

A *Pareto chart* is used to graphically display the relative importance of groups of data. It is named after Alfredo Pareto (1848–1923), an Italian economist who conducted research on the distribution of wealth in Europe. His observation that the majority of wealth (~80%) is held by a relatively small segment of the population (~20%) led to the 80–20 rule. Juran extended the application of this rule to quality: 80% of the problems (related to quality) are due to 20% of the causes. The numbers may not be exact for every case but the rule serves as a useful guideline.

We identify the important causes by first performing a cause-and-effect analysis. The basic idea of Pareto analysis is to identify the relative importance of these causes. The main steps in Pareto analysis are stated here:

1. Identify the problem and decide about the type and method of data collection.
2. Collect the data.
3. Rank the data in descending order of importance based on annual cost or frequency of occurrence.
4. Plot a bar graph that includes the cumulative percentage.

Example 6.2. The following data were collected for the analysis of internal failures in a company:

Type of defect	Dollar value ($)
Purchasing – rejects	206,000
Design – scrap	125,000
Operations – rework	355,000
Purchasing – rework	26,000
All other	64,000

Draw your conclusions.

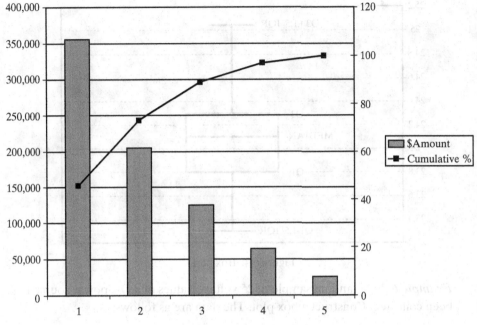

Figure 6.3. Pareto diagram

Solution. First we enter the data in Excel. The "data sort" command is used to arrange the data in descending order. A column is then added for cumulative percent. The chart utility is then used to provide a bar graph for the dollar amount and a line form is used for the cumulative percentage. The Pareto diagram is shown in Fig. 6.3. The program implementation is in *ParetoChart.xls*.

6.6 Box Plots

A *box plot* shows the central tendency, dispersion of data, and skewness, all in a simple graph. The plot also indicates the location of outliers, if any. In constructing the box plot, the first quartile (Q_1), the third quartile (Q_3), and the median (second quartile or Q_2) are evaluated, as well as the maximum (x_{max}) and the minimum (x_{min}) values of the data points. The interquartile range is IQR $= Q_3 - Q_1$. A box of convenient width is drawn with the upper line at Q_3 and the lower line at Q_1. A line is also drawn in the box at the median level. Two whiskers are drawn at min(Q_3 + 1.5 IQR, x_{max}) and max($Q_1 - 1.5$ IQR, x_{min}) with a central vertical line extending from the box to the cross lines, as shown in Fig. 6.4. The box plot calculations and plotting steps are included in the program *BoxPlot.xls*. The program shows the plot for the data given in the example problem.

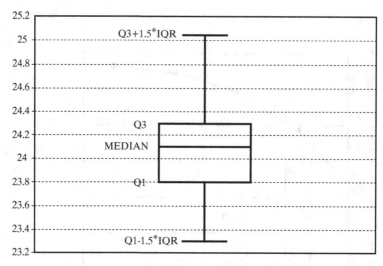

Figure 6.4. Box plot

Example 6.3. A random sample of 25 voltage values of a DC power source has been collected. Construct a box plot. The data are as follows:

24.0	24.6	24.1	24.7	23.3
24.1	24.1	24.4	24.1	23.8
23.8	24.3	25.5	23.5	24.1
23.4	24.0	24.0	23.6	24.3
23.5	24.3	24.2	24.3	24.3

Solution. The data are entered into the file *BoxPlot.xls*. The quartile values and the whisker levels are calculated:

MAX	25.5	1 ⇐ Number of points above
Upper whisker	25.05	
Q3	24.3	
Median	24.1	
Q1	23.8	
Lower whisker	23.3	
MIN	23.3	0 ⇐ Number of points below

The resulting box plot is shown in Fig. 6.4. The plot clearly shows that the median is not central with respect to the first and third quartiles, which indicates skewness. The data set has one outlier. Box plots are a quick and easy way of visualizing the distribution of data.

6.7 Normal Probability Plot

In statistical quality studies, the underlying assumption is that the data come from a normal distribution. Can we test the data to find out if it is indeed from a normal

6.7 Normal Probability Plot

distribution? A *normal probability plot* is commonly used to check this. The basic idea is simple: a random set of data points are arranged in ascending order. Let $x_1, x_2, \ldots, x_n$ be the n data points arranged in ascending order. The probability value for the ith data point is taken as

$$\Phi_i = \frac{i - 0.5}{n}, \tag{6.1}$$

where x_i on the horizontal axis and the corresponding Φ_i on the probability (vertical) axis form a point on the normal probability plot. The normal probability form (NormalProbPaper.pdf) is provided in the program directory of the CD. The normal probability paper has a built-in normal scale so that normal data will plot as a straight line.

Another method that is convenient is to find the z-value corresponding to Φ_i using the standard normal inverse ($z_i = \text{NORMSINV}(\Phi_i)$) and to plot x_i versus z_i. If the data are from a normal distribution, the plot of x_i versus z_i must form a straight line. We use this approach for the example. The mean from the best-fit line is obtained as the x-value at $z = 0$. The standard deviation is obtained as (x at $z = 1$) − (x at $z = 0$).

Example 6.4. The diameters in millimeters of a random sample of 16 shafts in a rough turning operation were 22.01, 21.80, 21.94, 22.13, 21.71, 22.26, 22.16, 21.76, 22.39, 21.97, 21.74, 21.96, 22.07, 21.48, 21.97, and 22.16. Construct a normal probability plot and check if the data follow a normal distribution.

Solution. The data are arranged in ascending order and the probability and z-values are obtained using Excel formulas:

i	Diameter	$(i - 0.5)/16$	z-value
1	21.48	0.03125	−1.8627
2	21.71	0.09375	−1.3180
3	21.74	0.15625	−1.0010
4	21.76	0.21875	−0.7764
5	21.80	0.28125	−0.5791
6	21.94	0.34375	−0.4022
7	21.96	0.40625	−0.2372
8	21.97	0.46875	−0.0784
9	21.97	0.53125	0.07841
10	22.01	0.59375	0.2372
11	22.07	0.65625	0.4022
12	22.13	0.71875	0.5791
13	22.16	0.78125	0.7764
14	22.16	0.84375	1.0010
15	22.26	0.90625	1.3180
16	22.39	0.96875	1.8627

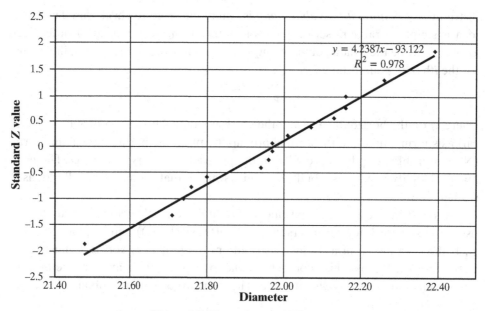

Figure 6.5. Normal probability plot

The scatter plot in the chart wizard is then used to plot the diameter versus the z-value (see *NormalProbPlot.xls*). The normal probability plot is shown in Fig. 6.5. We have the option of obtaining the equation for the straight-line fit and the coefficient of determination R^2: the R^2-value is obtained as 0.978, which indicates that the straight-line fit is a good one. The best-fit line is obtained as $z = 4.2387x - 93.122$. The mean and standard deviation from the best-fit line are obtained as 21.97 and 0.236, respectively.

This plot also adequately illustrates how a *scatter diagram* is constructed. A scatter diagram is used for plotting one variable on the x-axis and the other variable on the y-axis. Different curve formats may be used. The OC curves presented earlier were also prepared using a scatter diagram.

6.8 Run Charts

A *run chart* is used to plot a quality characteristic as a function of time. The consecutive measurements are plotted. The target value may be shown on the chart. A run chart illustrates something that can be practiced in day-to-day life. Suppose that we make it a practice to fill the fuel tank in a vehicle used for transportation to the full level at each visit to the station. At each visit, the amount of fuel in liters and the kilometers driven from the previous fill are recorded. It is then a simple matter to monitor the vehicle's performance (a quality characteristic) in kilometers per liter

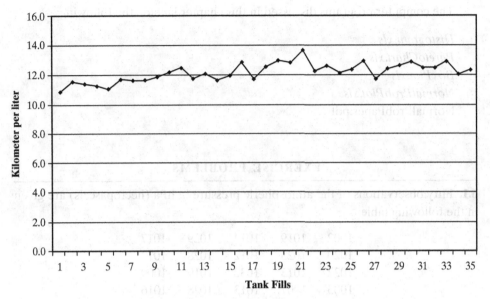

Figure 6.6. Run chart

(KPL) for each visit. Then each visit is a time-dependent variable and the visit number versus KPL value can be plotted. This becomes a run chart in practice. Fig. 6.6 shows the plot of the record kept by a driver during a nine-month period involving 35 tank-filling trips. The average KPL of this vehicle with a four-cylinder engine is 12.2 KPL with a standard deviation of 0.613 KPL. A closer look at the run chart shows that during the first fifteen visits the average KPL was below 12 and for the latter half it showed good improvement. A check of the record showed that, during the first few months, driving involved stop-start conditions in the city; during the latter half, longer trips were made on the highway. A quality monitoring system, as discussed in the example, may help in identifying the condition of the engine and the performance of the vehicle. Whenever there is a large variation, we may be able to look for the cause: is it a chance cause that is within a permissible variation in the process? Or is it a special cause that could be identified and corrective action taken? We will study these in the next chapter. Statistical process control uses run charts where means and ranges from randomly chosen samples are used in preparing control charts.

6.9 Summary

In this chapter we have taken a closer look at a number of data presentation techniques. The use of a computer has been stressed in preparing the charts. Charts are a good way to present data to upper management and to those who work on a shop floor. These techniques help quality improvement at all levels in an organization.

The computer programs discussed in this chapter include the following:

Histogram.xls
ParetoChart.xls
BoxPlot.xls
NormalProbPlot.xls
NormalProbPaper.pdf

EXERCISE PROBLEMS

6.1. Fifty observations of the atmospheric pressure in hPa (hectopascals) are given in the following table:

1002	1019	1014	1029	1017
1020	1023	1012	1028	1013
1027	1013	1032	1019	1009
1003	1007	1013	1028	1016
1018	1019	1022	1020	1009
1024	1025	1025	1028	1015
1020	1031	1028	1015	1027
1032	1016	1014	1021	1024
1010	1010	1026	1027	1022
1001	1015	1022	1018	1010

Construct a histogram and state your conclusions.

6.2. For the observations given in Problem 6.1, prepare a stem-and-leaf plot and draw your conclusions.

6.3. Choose a process of your choice and prepare a cause-and-effect diagram.

6.4. The following data were collected for the analysis of internal failures in a company:

Type of defect	Dollar value ($)
Purchasing – rejects	205,000
Design – scrap	120,000
Operations – rework	355,000
Purchasing – rework	25,000
All other	65,000

    a. Give the steps used in the construction of a Pareto diagram and construct the diagram.
    b. What are your conclusions?

6.5. A random sample of 30 voltage readings from a transducer are given in the following table:

12.4	11.3	12	12.2	12
11.5	12	12.2	11.7	11.9
12.1	12.1	12.1	12	11.5
12.3	12	12	12	12.2
12	12.2	11.8	11.8	11.8
12	11.8	11.9	12	11.7

Construct a box plot for the data and draw your conclusions.

6.6. Construct a normal probability plot for the data in Problem 6.1. Draw your conclusions from this plot.

7

Statistical Process Control

7.1 Introduction

Statistical process control (SPC) relies on the use of control charts. A *control chart* (also called a *Shewhart control chart*) provides a visible means of identifying the activity of an ongoing process. Walter A. Shewhart first proposed the use of a control chart in 1924 at Bell Telephone Laboratories. He developed it to distinguish and separate large variations due to assignable causes from those due to chance causes – more about these causes are discussed in a later section. A control chart is a run chart in which a characteristic of a variable or an attribute is monitored over time. The characteristic may be the average of a sample or a range of a sample or another parameter. It consists of a centerline and two control limits, one on either side of the centerline. The characteristic value plotted on the chart shows the condition of the process. We will discuss various important charts for variables and attributes and how they are used to monitor a process. The main idea of SPC is to bring the statistical theory to the shop-floor level, but first we need some preliminary results from order statistics to develop the SPC concepts.

7.2 Order Statistics and Other Preliminaries

To set control limits in a process, information is needed on the process standard deviation. In the absence of a known population standard deviation, we need to estimate it using the sample data. The sample standard deviation is an unbiased estimate of the population standard deviation. The range of a sample is an easier parameter to determine. We first establish the relationship between the population standard deviation and the mean value of the sample range, using results from order statistics.

7.2 Order Statistics and Other Preliminaries

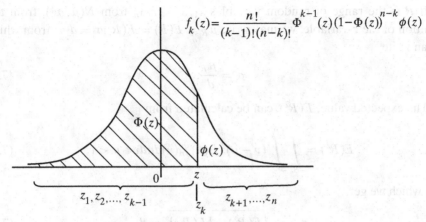

Figure 7.1. Distribution of z_k

Let $z_1, z_2, \ldots, z_k, \ldots, z_n$ be a random sample from a standard normal distribution [$N(0, 1)$], which has been ordered such that $z_1 \le z_2 \le \cdots \le z_k \le \cdots \le z_n$ and where z_k is the kth term in ascending order. We use the standard notation that $\phi(z)$ is the probability distribution function (pdf), and $\Phi(z)$ is the cumulative distribution function at z as shown in Fig. 7.1. The pdf $f_k(z)$, for which z is of order k, is formed by using the multinomial distribution such that each of z_1 through z_{k-1} is chosen with a probability of $\Phi(z)$, and each of the $n - k$ variables z_{k+1} through z_n is chosen with a probability of $1 - \Phi(z)$. The kth variable is chosen from the interval $(z, z + dz)$ with a probability $\phi(z)dz$. The pdf is given by

$$f_k(z) = \frac{n!}{(k-1)!(n-k)!} \Phi^{k-1}(z)[1 - \Phi(z)]^{n-k}\phi(z). \tag{7.1}$$

Let us make an attempt to find d_2, the expected value of the range of the sample. Since the expected values of the minimum (z_1) and the maximum (z_n) are placed symmetrically with respect to zero, the expected value of the range is given by

$$d_2 = 2E(z_n) = 2\int_{-\infty}^{\infty} zf_n(z)dz = 2n\int_{-\infty}^{\infty} z\Phi^{n-1}(z)\phi(z)dz, \tag{7.2}$$

where $f_n(z)$ is obtained by setting $k = n$ in Eq. (7.1). The calculation of d_2 has been implemented as the function `rds(n)` in the file *ControlChartFactors.xls* included in the APPENDIX directory on the CD provided with the book. The integral in Eq. (7.2) needs to be performed from -8 to $+8$ since the integrand has a negligible value outside these limits. We obtain rds(2) = 1.128, rds(3) = 1.693, and so on. The d_2-values in the control chart factors in Appendix Table A.8 are generated using this function.

If R is the range of random variables $x_1, x_2, \ldots, x_n$ from $N(\mu, \sigma^2)$, from the definition of the z-variable, we may write $\mu_R = E(R) = E(R_z)\sigma = d_2\sigma$, from which we can write

$$\sigma = \frac{\mu_R}{d_2}. \tag{7.3}$$

The expected value, $E(R_z^2)$, can be calculated from

$$E(R_z^2) = \int_{-\infty}^{\infty} \int_{-\infty}^{v} (v-u)^2 f_1(u) f_n(v) du dv \quad u < v, \tag{7.4}$$

from which we get

$$\sigma_{R_z} = \sqrt{E(R_z^2) - [E(R_z)]^2} = d_3. \tag{7.5}$$

This has been implemented as the function srds(n) in the file *ControlChart-Factors.xls* included on the CD. We have srds(2) = 0.853, srds(3) = 0.888, and so on. Various n-values are listed in the factors for control charts in Appendix Table A.8. Using this on the x-variables from $N(\mu, \sigma^2)$, we have the relation

$$\sigma_R = d_3\sigma, \tag{7.6}$$

since s^2 is an unbiased estimate of σ^2, but the expected value of s is not σ. The unbiased estimate μ_s of s is given by

$$\mu_s = c_4\sigma, \tag{7.7}$$

where $c_4 = \sqrt{\frac{2}{n-1}} \frac{\Gamma(n/2)}{\Gamma((n-1)/2)}$, and $\Gamma(x) = \int_0^\infty y^{x-1} e^{-y} dy$. The function GAMMALN(x) is available in Excel; we use $\Gamma(x)$ = exp(GAMMALN(x)).

The standard deviation of s for a population with a normal distribution is

$$\sigma_s = \sigma\sqrt{1 - c_4^2}. \tag{7.8}$$

We make use of these relationships using sample information to develop the control limits.

7.3 Causes of Variation

Variation is an inherent property of a production process, and variations occur no matter how much care is exercised. If we consider a product such as a shaft, its dimensional variations may be due to variations in the material of the shaft, variations in the tool used, variations in rigidity of support, and so on. These variations are inherent in the system. This natural variability is due to *chance causes* or common causes. If variations result only from chance causes in a process, the process is said to be *in control*. The system is considered a stable system and the variations are a product of noise. On the other hand, for the shaft example, it is likely that variations may occur due to tool breakage, operator errors, defective raw material, improper positioning, and so forth. These types of variations are not acceptable. These causes are called *assignable causes* or special causes. The variations due to such causes may generally be larger or carry their own signature. Variations due to assignable causes are avoidable. A process operating under the presence of assignable causes is said to be *out of control*. Assignable causes must be detected and eliminated. A process in control allows us to determine control limits for a variable that is measured. If the variable values fall outside these limits or if they show a certain trend we may recognize that the process is out of control.

7.4 Statistical Process Control Concepts

We stated in the previous section that a process is said to be out of control when it operates with variations due to assignable causes. Even in a process in control, the assignable causes eventually find their place. In the shaft production example, if we do not have a continuous tool-wear compensation system, the wear of the tool will result in an increase of the mean diameter – clearly a case of a process out of control. We may detect this through a trend. The following steps describe how we set up the limits to identify a process in control.

Step 1. Identify a quality characteristic to be monitored.
Step 2. Gather information by carefully collecting rational subgroups of samples. A *rational subgroup* is one in which the variation within the subgroup is due to chance causes only and variation between subgroups may be due to assignable causes. A rational subgroup sample is collected under similar conditions – it may be a predetermined number of consecutive parts produced every hour in a shift, or randomly chosen parts from a previous hour's production. The number in the sample is generally from 4 to 10 parts.
Step 3. Estimate the mean $\tilde{\mu}$, and standard deviation $\tilde{\sigma}$, for the control parameter, for the process based on the collected data. Note that care is taken to avoid the influence of assignable causes during this process.
Step 4. Set the centerline (CL), upper control limit (UCL), and the lower control limit (LCL):

$$\begin{aligned} \text{UCL} &= \tilde{\mu} + k\tilde{\sigma} \\ \text{CL} &= \tilde{\mu} \\ \text{LCL} &= \tilde{\mu} - k\tilde{\sigma}. \end{aligned} \quad (7.9)$$

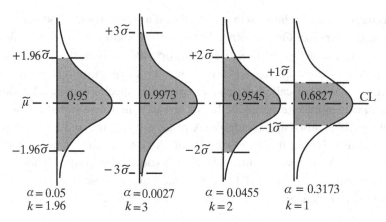

Figure 7.2. Control limit selection

The standard deviation is multiplied by the factor k, which is determined based on an interval for a confidence level of $1 - \alpha$. For the two-sided limit, $k = z_{1-\alpha/2}$. Fig. 7.2 shows various choices of k. The choice of $k = 3$, which corresponds to a level of significance of $\alpha = 0.0027$ or $1 - \alpha = 0.9973$, is widely accepted for setting the control limits for process control. This is referred to as $\pm 3\sigma$ limits. With this, the probability of committing a Type I error is 0.0027 or about 3 out of 1000 parts produced. Warning limits of $\pm 2\tilde{\sigma}$ and $\pm 1\tilde{\sigma}$ may be used to check quality characteristic trends.

Once the limits are set, a shift in the mean is in fact a case of an out-of-control process. Fig. 7.3 shows that the area β of the normal curve inside the control limits is the probability of acceptance when the process has shifted – β is the probability

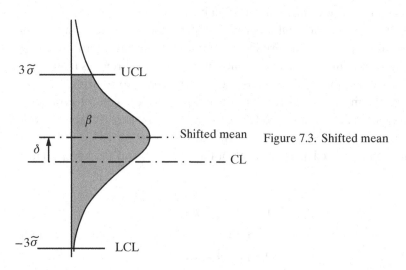

Figure 7.3. Shifted mean

of committing a Type II error. A plot of shift versus β is the operating characteristic (OC) curve for the process.

Example 7.1. A sample of five parts is used to monitor the mean diameter of a shaft in a production process. The mean and standard deviation for the diameter for the process are established as 55 and 0.02 mm, respectively. Set the control limits for the sample mean and determine (a) the probability of a Type I error and (b) probability of a Type II error if the mean has shifted to 55.01 mm.

Solution. For the problem the expected mean value of the sample is $\tilde{\mu} = \mu = 55$ mm, $n = 5$, and the standard deviation of the sample mean, $\tilde{\sigma}$, is $\tilde{\sigma} = \sigma/\sqrt{n} = 0.02/\sqrt{5} = 0.00894$. From Eqs. (7.9),

$$\text{UCL} = 55 + 3 \times 0.00894 = 55.027 \text{ mm}$$
$$\text{CL} = 55 \text{ mm}$$
$$\text{LCL} = 55 - 3 \times 0.00894 = 54.973 \text{ mm}.$$

a. By the $\pm 3\sigma$ limits used, the probability of a Type I error is $\alpha = 0.0027$.
b. For the shifted mean, the z-value at UCL is

$$z_U = \frac{\text{UCL} - \mu_{shifted}}{\tilde{\sigma}} = \frac{55.027 - 55.01}{0.00894} = 1.9016$$

and the z-value at LCL is

$$z_L = \frac{\text{LCL} - \mu_{shifted}}{\tilde{\sigma}} = \frac{54.973 - 55.01}{0.00894} = -4.1387.$$

The probability of a Type II error, β, is now calculated as

$$\beta = \Phi(z_u) - \Phi(z_L) = 0.9714 - 0 = 0.9714.$$

Having established the control limits in Eqs. (7.9), the basic rule is that the process is out of control when the sample mean falls outside these limits. However, small shifts result in other patterns that may indicate that the process is out of control. The rules for recognizing an out-of-control process are presented in the next section.

Out-of-Control Process Rules

The Western Electric Company formulated rules to recognize when a process is out of control. It included them in its handbook in 1956 and they are referred to as WECO rules. The key rule is that the process is out of control if there is a point outside the $\pm 3\sigma$ limits. Warning limits are also set at $\pm 2\sigma$ and $\pm 1\sigma$. The rules are formulated based on the patterns of the sample mean with respect to these limits and

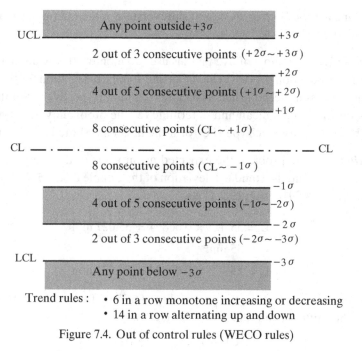

Figure 7.4. Out of control rules (WECO rules)

are summarized in Fig. 7.4. These rules must be checked to identify if the process is out of control. If the process is out of control, the following three steps must be followed.

Three Steps for an Out-of-Control Process

Step 1. The process must be stopped.
Step 2. The assignable cause or causes must be identified and removed.
Step 3. The out-of-control point or points must be removed from the data and calculations must be revised.

From the knowledge of committing a Type I error we find that 1 out of 370 ($= 1/0.0027$) points may fall outside the control limits. This may raise a false signal but it is worth investigating to ensure that the process is in control. If p is the probability that a point falls outside the control limits ($p = \alpha$ for a process in control, $p = 1 - \beta$ for a process with a shift), the *average run length* (ARL) is given by

$$\text{ARL} = \frac{1}{p}. \tag{7.10}$$

The ARL is the average run length before a point falls outside the control limits ($p = 0.0027$ for limits based on $\pm 3\sigma$). Thus, the ARL for a process in control is 370.

7.5 Control Charts for Variables

We now present the commonly used control charts for variables. *Variables* are quality characteristics that are measurable on a numerical scale. Examples of variables are diameter, length, thickness, temperature, pressure, voltage, current, tensile strength, and so forth. We need to control the mean value of the characteristic and also its variability. The most widely used charts for variables are the $\bar{x}$–R chart and the $\bar{x}$–s chart. Sometimes, a target value for a control may be known.

$\bar{x}$- and R-Charts

We collect samples, each of size n, to form m rational subgroups. The sample size may be generally 4 to 10 and the rational subgroups are generally more than 20. If $x_1, x_2, \ldots, x_n$ are the members of a sample subgroup, we evaluate the mean of the subgroup and the range of the subgroup as

$$\bar{x} = \frac{x_1 + x_2 + \cdots + x_n}{n} \tag{7.11}$$

$$R = \max(x_1, x_2, \ldots, x_n) - \min(x_1, x_2, \ldots, x_n). \tag{7.12}$$

We evaluate a mean for each subgroup and a range for each subgroup. Let $\bar{x}_1, \bar{x}_2, \ldots, \bar{x}_m$ represent the means of the subgroups and let $R_1, R_2, \ldots, R_m$ be the corresponding ranges of the subgroups. We now evaluate $\bar{\bar{x}}$, the average of the mean values (sometimes called the grand mean), and $\bar{R}$, the average value of the subgroup ranges:

$$\bar{\bar{x}} = \frac{\bar{x}_1 + \bar{x}_2 + \cdots + \bar{x}_m}{m} \tag{7.13}$$

$$\bar{R} = \frac{R_1 + R_2 + \cdots + R_m}{m}. \tag{7.14}$$

The value of $\bar{\bar{x}}$ may be taken as the best estimator of the process average μ, and $\bar{R}$ may be taken as the expected value μ_R of the range (for a random sample of n). Using this in Eq. (7.3), developed earlier, we get the estimator for process standard deviation:

$$\sigma = \frac{\bar{R}}{d_2}, \tag{7.15}$$

where d_2 is dependent on n and is listed in the control chart factors in Appendix Table A.8. The d_2-values have been generated using the function `rds(n)`, which is included in *ControlChartFactors.xls*.

The control limits for $\bar{x}$ are

$$\begin{aligned}
\text{UCL}_{\bar{x}} &= \bar{\bar{x}} + 3\frac{\sigma}{\sqrt{n}} = \bar{\bar{x}} + 3\frac{\overline{R}}{d_2\sqrt{n}} = \bar{\bar{x}} + A_2\overline{R} \\
\text{CL}_{\bar{x}} &= \bar{\bar{x}} \\
\text{LCL}_{\bar{x}} &= \bar{\bar{x}} - 3\frac{\sigma}{\sqrt{n}} = \bar{\bar{x}} - 3\frac{\overline{R}}{d_2\sqrt{n}} = \bar{\bar{x}} - A_2\overline{R}
\end{aligned} \quad (7.16)$$

for which A_2 values are listed in Appendix Table A.8.

After introducing Eq. (7.15) into (7.6) we get

$$\sigma_R = \frac{d_3 \overline{R}}{d_2}. \quad (7.17)$$

The control limits for $\overline{R}$ can be set as

$$\begin{aligned}
\text{UCL}_R &= \overline{R} + 3\sigma_R = \overline{R} + 3\frac{d_3 \overline{R}}{d_2} = \left(1 + \frac{3d_3}{d_2}\right)\overline{R} = D_4 \overline{R} \\
\text{CL}_R &= \overline{R} \\
\text{LCL}_R &= \overline{R} + 3\sigma_R = \left(1 - \frac{3d_3}{d_2}\right)\overline{R} = D_3 \overline{R}.
\end{aligned} \quad (7.18)$$

D_3-values are taken to be zero whenever $(1 - \frac{3d_3}{d_2})$ is negative since LCL_R for the range is not less than zero. The results, Eqs. (7.17) and (7.18), are now combined to define the $\bar{x}$–R chart.

$\bar{x}$-Chart:

$$\begin{aligned}
\text{UCL}_{\bar{x}} &= \bar{\bar{x}} + A_2 \overline{R} \\
\text{CL}_{\bar{x}} &= \bar{\bar{x}} \\
\text{LCL}_{\bar{x}} &= \bar{\bar{x}} - A_2 \overline{R}
\end{aligned} \quad (7.19a)$$

R-Chart:

$$\begin{aligned}
\text{UCL}_R &= D_4 \overline{R} \\
\text{CL}_R &= \overline{R} \\
\text{LCL}_R &= D_3 \overline{R}
\end{aligned} \quad (7.19b)$$

Values for A_2, D_3, and D_4 are listed in Appendix Table A.8 for various values of n.

7.5 Control Charts for Variables

Table 7.1.

SubGr #	Width of key (mm)					$\bar{x}$	R
	x_1	x_2	x_3	x_4	x_5		
1	19.03	19.14	18.97	19.23	18.89	19.05	0.34
2	19.26	18.98	19.25	19.26	19.09	19.17	0.28
3	19.12	18.98	19.16	19.03	19.18	19.09	0.20
4	19.10	18.91	18.93	18.77	18.85	18.91	0.33
5	18.91	18.99	18.87	19.07	19.05	18.98	0.20
6	19.33	18.55	18.94	19.15	19.01	19.00	0.78
7	19.15	18.92	19.17	19.05	18.91	19.04	0.26
8	19.08	18.92	19.01	19.13	19.19	19.07	0.27
9	19.00	18.88	19.11	18.95	18.92	18.97	0.23
10	18.86	19.01	18.82	18.97	19.06	18.94	0.24
11	19.31	18.91	18.72	19.00	18.96	18.98	0.59
12	19.22	18.98	18.89	18.81	18.94	18.97	0.41
13	18.75	19.06	19.04	18.93	19.29	19.01	0.54
14	19.18	19.11	19.07	19.22	19.02	19.12	0.20
15	19.26	18.64	18.90	18.95	19.14	18.98	0.62
16	19.01	19.02	19.08	19.11	18.91	19.03	0.20
17	18.93	19.15	18.84	18.81	19.04	18.95	0.34
18	18.84	19.00	18.99	18.99	19.22	19.01	0.38
19	19.17	18.93	19.14	19.41	18.93	19.12	0.48
20	19.15	18.58	18.82	19.09	19.17	18.96	0.59
21	19.03	19.15	19.08	18.89	19.09	19.05	0.26
22	18.70	18.87	18.75	19.32	19.12	18.95	0.62
23	18.89	19.06	19.10	18.97	19.19	19.04	0.30
24	19.00	19.21	18.72	19.26	18.82	19.00	0.54
25	19.04	19.05	19.10	18.86	19.05	19.02	0.24
						$\bar{\bar{x}} = 19.02$	$\bar{R} = 0.38$

Example 7.2. SPC is planned to control the width of a key used in a gear-shaft assembly. Five random samples were taken from production during the previous hour. Readings from 25 successive subgroups are listed in Table 7.1. Set up the control limits for $\bar{x}$, R and plot the $\bar{x}$–R chart.

Solution. The values are entered into a spreadsheet (see *XbarRChart.xls*) and $\bar{x}$, R, $\bar{\bar{x}}$, and $\bar{R}$ are evaluated: $\bar{\bar{x}} = 19.02$ and $\bar{R} = 0.38$. From Appendix Table A.8, the values for the control chart constants for a sample size of 5 are read as $A_2 = 0.577$, $D_3 = 0$, and $D_4 = 2.114$. The control limits are now calculated:

$$\text{UCL}_{\bar{x}} = \bar{\bar{x}} + A_2 \bar{R} = 19.02 + 0.577 \times 0.38 = 19.239$$
$$\text{CL}_{\bar{x}} = \bar{\bar{x}} = 19.02$$
$$\text{LCL}_{\bar{x}} = \bar{\bar{x}} - A_2 \bar{R} = 19.02 - 0.577 \times 0.38 = 18.801$$

$$\text{UCL}_R = D_4 \overline{R} = 2.114 \times 0.38 = 0.8033$$
$$\text{CL}_R = \overline{R} = 0.38$$
$$\text{LCL}_R = D_3 \overline{R} = 0.$$

The $\bar{x}$-chart and the R-chart for the data are shown in Figs. 7.5a and 7.5b. A review of the charts shows that all the points in both charts are within the control limits. We also see that there is no trend with respect to the warning limits. We conclude that the process is in control and we continue to use these limits for monitoring the process.

If the process is out of control, the three steps for an out-of-control process must be followed. The revised limits are to be followed for future monitoring.

$\bar{x}$- and R-Charts for Given Target or Standard Values

Sometimes the target or standard values, μ_0 and σ_0, are known. These may be set by the management as target values to be achieved. In this case the control limits are set by using these values. The control limits for $\bar{x}$ are easily set as follows:

$$\text{UCL}_{\bar{x}} = \mu_0 + 3\frac{\sigma_0}{\sqrt{n}} = \mu_0 + A\sigma_0$$
$$\text{CL}_{\bar{x}} = \mu_0 \tag{7.20}$$
$$\text{LCL}_{\bar{x}} = \mu_0 - 3\frac{\sigma_0}{\sqrt{n}} = \mu_0 - A\sigma_0.$$

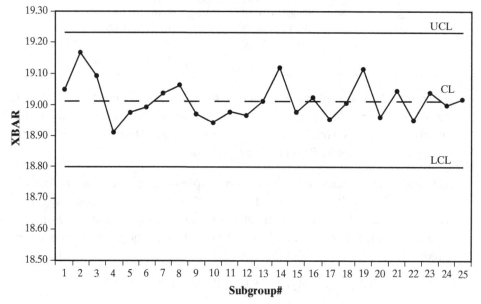

Figure 7.5a. $\bar{x}$-chart

7.5 Control Charts for Variables

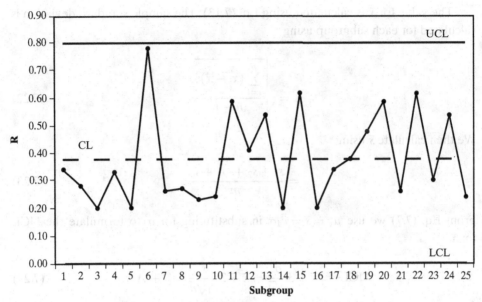

Figure 7.5b. R-chart

In setting the control limits for R we use $\overline{R} = d_2\sigma$ from Eq. (7.15) and $\sigma_R = \frac{d_3\overline{R}}{d_2} = d_3\sigma$ from Eq. (7.17) to get

$$\begin{aligned} \text{UCL}_R &= \overline{R} + 3\sigma_R = (d_2 + 3d_3)\sigma_0 = D_2\sigma_0 \\ \text{CL}_R &= d_2\sigma_0 \\ \text{LCL}_R &= \overline{R} - 3\sigma_R = (d_2 - 3d_3)\sigma_0 = D_1\sigma_0, \end{aligned} \quad (7.21)$$

where A, D_1, and D_2 are provided in Appendix Table A.8. Care must be exercised when target values are used. If the process values differ from the target values, points may fall outside the control limits without any assignable causes.

$\overline{x}$- and s-Charts

When the sample sizes are 10 or more, $\overline{x}$- and s-charts are generally preferred. The $\overline{x}$- and R-values are more popular since the range calculation for a sample is much easier than calculating the standard deviation. We have seen that estimating the population standard deviation from the expected value of the sample range uses results from order statistics. Whereas the sample variance is an unbiased estimator of the population variance, the sample standard deviation is not an unbiased estimator of σ. We make use of relationships (7.7) and (7.8) here.

The value for $\bar{\bar{x}}$ is calculated using Eq. (7.13). The sample standard deviation is calculated for each subgroup using

$$s = \sqrt{\frac{\sum_{i=1}^{n}(x_i - \bar{x})^2}{n-1}}. \tag{7.22}$$

We then calculate $\bar{s}$ using

$$\bar{s} = \frac{s_1 + s_2 + \cdots + s_m}{m}. \tag{7.23}$$

From Eq. (7.7) we use $\mu_s = \bar{s} = c_4 \sigma$ in substituting for σ to formulate the UCL for $\bar{x}$

$$\text{UCL}_{\bar{x}} = \bar{\bar{x}} + 3\frac{\sigma}{\sqrt{n}} = \bar{\bar{x}} + 3\frac{\bar{s}}{c_4\sqrt{n}} = \bar{\bar{x}} + A_3\bar{s}. \tag{7.24}$$

Now we use Eq. (7.8) in the development of the control limits for s:

$$\begin{aligned}\text{UCL}_s &= \bar{s} + 3\sigma_s = \bar{s} + 3\sigma\sqrt{1-c_4^2} = \left(1 + \frac{3\sqrt{1-c_4^2}}{c_4}\right)\bar{s} = B_4\bar{s} \\ \text{LCL}_s &= \bar{s} - 3\sigma_s = \left(1 - \frac{3\sqrt{1-c_4^2}}{c_4}\right)\bar{s} = B_3\bar{s} \quad (B_3 \geq 0).\end{aligned} \tag{7.25}$$

Combining Eqs. (7.24) and (7.25) gives the follwing.

$\bar{x}$-Chart:

$$\begin{aligned}\text{UCL}_{\bar{x}} &= \bar{\bar{x}} + A_3\bar{s} \\ \text{CL}_{\bar{x}} &= \bar{\bar{x}} \\ \text{LCL}_{\bar{x}} &= \bar{\bar{x}} - A_3\bar{s}\end{aligned} \tag{7.26a}$$

s-Chart:

$$\begin{aligned}\text{UCL}_s &= B_4\bar{s} \\ \text{CL}_s &= \bar{s} \\ \text{LCL}_s &= B_3\bar{s}.\end{aligned} \tag{7.26b}$$

The constants A_3, B_3, and B_4 are listed in Appendix Table A.8.

$\bar{x}$- and s-Charts for Given Target or Standard Values

When the target or standard values, μ_0 and σ_0, are given, the $\bar{x}$-chart is as defined in Eqs. (7.20). The s-chart is obtained from

$$\text{UCL}_s = B_6 \sigma$$
$$\text{CL}_s = c_4 \sigma \quad (7.27)$$
$$\text{LCL}_s = B_5 \sigma,$$

where B_5 and B_6 are listed in Appendix Table A.8.

Example 7.3. Plot the $\bar{x}$- and s-charts for the data in Example 7.2.

Solution. The data are entered into Excel (see *XbarSChart.xls* on the CD). Values for $\bar{x}$, s, $\bar{\bar{x}}$, and $\bar{s}$ are evaluated: $\bar{\bar{x}} = 19.02$ and $\bar{s} = 0.15$. From Appendix Table A.8, the values for the control chart constants for a sample size of 5 are read as $A_3 = 1.427$, $B_3 = 0$, and $B_4 = 2.089$. The control limits are now calculated:

$$\text{UCL}_{\bar{x}} = \bar{\bar{x}} + A_3 \bar{s} = 19.02 + 1.427 \times 0.15 = 19.234$$
$$\text{CL}_{\bar{x}} = \bar{\bar{x}} = 19.02$$
$$\text{LCL}_{\bar{x}} = \bar{\bar{x}} - A_3 \bar{s} = 19.02 - 1.427 \times 0.15 = 18.806$$

$$\text{UCL}_s = B_4 \bar{s} = 2.089 \times 0.15 = 0.3133$$
$$\text{CL}_s = \bar{s} = 0.15$$
$$\text{LCL}_s = B_3 \bar{s} = 0.$$

The $\bar{x}$-chart and the s-chart for the data are shown in Figs. 7.6a and 7.6b. A review of the charts shows that all the points in both charts are within the control limits. We also see that there is no trend with respect to the warning limits. We conclude that the process is in control and continue to use these limits for monitoring the process.

The three steps stated earlier are followed when the process is out of control. Process monitoring continues after that. A comparison of the $\bar{x}$–s and the $\bar{x}$–R charts shows that the points are very similar in both plots. The upper and lower limits for the $\bar{x}$-chart based on s are slightly tighter.

x- and MR-Charts for Individual Measurements

In monitoring a process when production is slow, it is general practice to measure the controlled characteristic on every part produced. In these situations the sample size is $n = 1$. The individual characteristic is monitored by setting the upper and lower control limits and its variability is monitored by the moving range (MR).

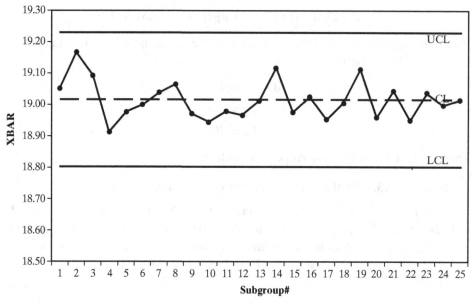

Figure 7.6a. $\bar{x}$-chart

Let $x_1, x_2, \ldots, x_m$ be m observations. The MR is the range of two successive observations. The moving range of observation j, MR_j is abs $(x_j - x_{j-1})$ and starts with the second observation. MR parameters are based on twenty to thirty observations. The average of the moving range, $\overline{MR}$, is evaluated based on all the MR values for

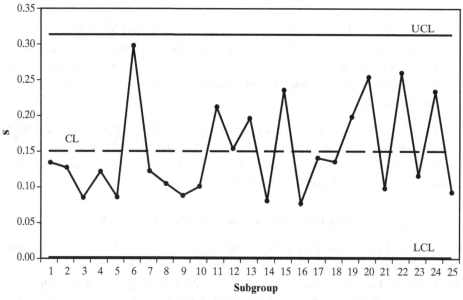

Figure 7.6b. s-chart

these observations; $\bar{x}$ is the average of these measurements. The process standard deviation is evaluated using the d_2-value for range established by samples of two. The control limits for the x- and MR-charts are as follows.

x-Chart:

$$\text{UCL}_x = \bar{x} + 3\,\frac{\overline{\text{MR}}}{d_2} \quad (d_2 = 1.128)$$

$$\text{CL}_x = \bar{x} \quad (7.28)$$

$$\text{LCL}_x = \bar{x} - 3\,\frac{\overline{\text{MR}}}{d_2}$$

MR-Chart:

$$\text{UCL}_{MR} = D_4\,\overline{\text{MR}} \quad (D_4 = 3.267)$$

$$\text{CL}_{MR} = \overline{\text{MR}}$$

$$\text{LCL}_{MR} = D_3\,\overline{\text{MR}} \quad (D_3 = 0).$$

Care must be exercised in using x–MR control charts. Since all the data are used in the order collected, there is a good chance that a correlation exists among the data.

Example 7.4. The first 20 values of kilometers per liter (KPL) data used in the run chart given at the end of Chapter 6 are 10.9, 11.6, 11.4, 11.3, 11.1, 11.7, 11.6, 11.7, 11.9, 12.2, 12.5, 11.8, 12.1, 11.6, 12.0, 12.9, 11.7, 12.5, 13.0, and 12.8. Determine the control limits for the x-chart and MR-chart based on the MR approach.

Solution. The first range is $11.6 - 10.9 = 0.7$, the second is $11.4 - 11.6 = -0.2$, and so on. The 19 moving ranges are 0.7, 0.2, 0.1, 0.2, 0.6, 0.1, 0.0, 0.2, 0.3, 0.3, 0.7, 0.3, 0.5, 0.3, 0.9, 1.2, 0.8, 0.4, 0.2. The average of the 20 x-values is $\bar{x} = 11.9$ and the average of the 19 MRs is $\overline{\text{MR}} = 0.428$.

x-Chart:

$$\text{UCL}_x = \bar{x} + 3\,\frac{\overline{\text{MR}}}{1.128} = 13.04$$

$$\text{CL}_x = \bar{x} = 11.9$$

$$\text{LCL}_x = \bar{x} - 3\,\frac{\overline{\text{MR}}}{d_2} = 10.76$$

MR-Chart:

$$\text{UCL}_{MR} = 3.267\,\overline{\text{MR}} = 1.3983$$

$$\text{CL}_{MR} = \overline{\text{MR}} = 0.428$$

$$\text{LCL}_{MR} = 0.$$

The maximum and minimum of x are 13.0 and 10.9, respectively. The maximum of the MR is 1.2. The KPL data show a process in control. Determining the graphic display for this problem is given as an exercise.

The $\bar{x}$- and r-charts and $\bar{x}$- and s-charts, which are referred to as Shewhart control charts, are good for detecting large shifts in the mean. They are less sensitive to shifts of 1.5σ or less. Smaller shifts can only be detected by looking at the form of the plot and its changes relative to the warning limits. There are other charts, such as exponentially weighted moving average (EWMA) charts and cumulative sum (CUSUM) charts, which use information from all or more recent observations. These charts are capable of detecting smaller shifts. We present details of these charts in the following sections.

Exponentially Weighted Moving Average (EWMA) Control Chart

The EWMA is also called the *geometric moving average* (GMA). The EWMA is used for individual measurements $x_j (j = 1, 2, \ldots)$ or for the average of samples of size n, $\bar{x}_j, (j = 1, 2, \ldots)$. For all calculations involving the process standard deviation, σ is used for the first case and $\sigma/\sqrt{n}$ is used for the latter. We present all equations for the individual measurements and we start with the following known parameters:

Target value of the mean $= \mu$
Process standard deviation $= \sigma$
Weighting parameter $= \lambda$ $\quad (0 < \lambda \leq 1)$.

The EWMA is calculated at each data point. We denote E_i as the EWMA at data point i. We start with

$$E_0 = \mu$$

and evaluate E_i using

$$E_i = \lambda x_i + (1 - \lambda) E_{i-1}. \tag{7.29}$$

E_i is a weighted average of all previous measurements. On expanding E_{i-1} successively, we have

$$E_i = \lambda x_i + \lambda(1-\lambda)x_{i-1} + \lambda(1-\lambda)^2 x_{i-2} + \lambda(1-\lambda)^3 x_{i-3} + \cdots + (1-\lambda)^i E_0. \tag{7.30}$$

Using the summation formula for a geometric series, it is possible to show that the sum of all the coefficients is 1. The weight of the current sample is λ and the weight of the previous sample is obtained as $(1 - \lambda)$ times the current sample and so on, resulting in a geometric series from i through 1. The EWMA is also called the geometric moving average for this reason; $\lambda = 1$ corresponds to the Shewhart control chart.

The variance of E_i given by Eq. (7.30) is obtained as

$$\sigma_{E_i} = \sigma\sqrt{\left(\frac{\lambda}{2-\lambda}\right)[1-(1-\lambda)^{2i}]}, \qquad (7.31)$$

where $(1-\lambda)^{2i}$ may be taken as zero when i is large and the limiting value may be used as a single σ_E for setting the control limits.

The EWMA control limits are set as follows:

$$\begin{aligned} \text{UCL}_i &= \mu + K\sigma_{E_i} \\ \text{CL} &= \mu \\ \text{LCL}_i &= \mu - K\sigma_{E_i}. \end{aligned} \qquad (7.32)$$

We note that $K = 3$ is used for the Shewhart control charts where $\lambda = 1$. For an ARL of 370, which is the expected ARL for the zero-shift Shewhart chart, the recommended values of λ and K for EWMA are 0.1 and 2.7, respectively. Other combinations for the same ARL are $(0.05, 2.5)$, $(0.2, 2.9)$, and $(0.3, 3)$. These values provide the same ARL when there is no shift. The EWMA control chart has been found to work well for non-normal distributions.

The control limits given in Eqs. (7.31) and (7.32) are implemented in the program *EWMA.xls*. The data for the following example are shown in sheet1 which gives the EWMA chart and the $\bar{x}$-chart. Some editing may be needed if the number of data points is larger or smaller than that in the example.

Example 7.5. The following table shows mean values of 30 samples.

i	$\bar{x}_i$	i	$\bar{x}_i$
1	5.36	16	5.53
2	5.45	17	5.74
3	5.55	18	5.60
4	5.43	19	5.48
5	5.36	20	5.65
6	5.51	21	5.59
7	5.50	22	5.66
8	5.44	23	5.61
9	5.51	24	5.60
10	5.47	25	5.58
11	5.74	26	5.66
12	5.44	27	5.67
13	5.62	28	5.52
14	5.52	29	5.65
15	5.50	30	5.51

Each sample is an average value of 5 parts. The target value is 0.55 and the process standard deviation is 0.22. Prepare an EWMA control chart for the data shown in the table and compare it with the $\bar{x}$-chart. Use $\lambda = 0.1$ for the exponential parameter and $K = 2.7$ for setting the limits.

Solution. The calculated values are given in the following table:

Sample i	X_i	$EWMA_i$	UCL	LCL
1	5.36	5.486	5.527	5.473
2	5.45	5.482	5.536	5.464
3	5.55	5.489	5.542	5.458
4	5.43	5.483	5.547	5.453
5	5.36	5.471	5.550	5.450
6	5.51	5.475	5.552	5.448
7	5.50	5.477	5.554	5.446
8	5.44	5.474	5.556	5.444
9	5.51	5.477	5.557	5.443
10	5.47	5.477	5.558	5.442
11	5.74	5.503	5.559	5.441
12	5.44	5.497	5.559	5.441
13	5.62	5.509	5.560	5.440
14	5.52	5.510	5.560	5.440
15	5.50	5.509	5.561	5.439
16	5.53	5.511	5.561	5.439
17	5.74	5.534	5.561	5.439
18	5.60	5.541	5.561	5.439
19	5.48	5.535	5.561	5.439
20	5.65	5.546	5.561	5.439
21	5.59	5.550	5.562	5.438
22	5.66	5.561	5.562	5.438
23	5.61	5.566	5.562	5.438
24	5.60	5.570	5.562	5.438
25	5.58	5.571	5.562	5.438
26	5.66	5.580	5.562	5.438
27	5.67	5.589	5.562	5.438
28	5.52	5.582	5.562	5.438
29	5.65	5.589	5.562	5.438
30	5.51	5.581	5.562	5.438

The EWMA chart is shown in Fig. 7.7a. The corresponding $\bar{x}$-chart is shown in Fig. 7.7b. The EWMA chart shows that the process is out of control at data

7.5 Control Charts for Variables

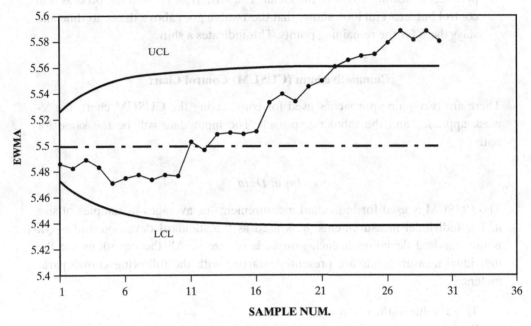

Figure 7.7a. EWMA control chart

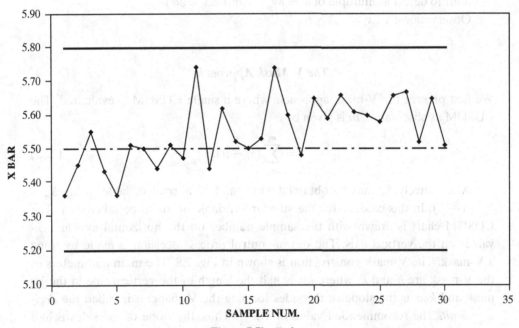

Figure 7.7b. $\bar{x}$-chart

point 22, indicating a shift of the mean. The $\bar{x}$-chart shows that the process is in control but a careful look shows that the x-value goes above the centerline and stays there for the remaining points. This indicates a shift.

Cumulative Sum (CUSUM) Control Chart

There are two main approaches used for constructing the CUSUM chart: the V-mask approach and the tabular approach. The input data will be the same for both.

Input Data

The CUSUM is used for individual measurements or averages of samples of size n. For individual measurements, σ is used as the standard deviation and $\sigma/\sqrt{n}$ is the standard deviation if each sample is of size n. All the equations for the individual measurements are presented, starting with the following known parameters:

Target value of the mean $= \mu$
Process standard deviation $= \sigma$
Shift to detect as multiple of $\sigma = \delta$ (shift $\Delta x = \delta\sigma$)
Observations: $x_1, x_2, \ldots$.

The V-Mask Approach

We first present the V-mask approach, where a single CUSUM is evaluated. The CUSUM at observation m is given by

$$S_m = \sum_{i=1}^{m}(x_i - \mu). \tag{7.33}$$

Alternatively S_m may be obtained with standard normal variables using $S'_m = \sum_{i=1}^{m}(\frac{x_i-\mu}{\sigma})$. In this case, σ' for the standard variable needs to be taken as 1. The CUSUM chart is drawn with the sample number on the horizontal axis and S_i-values on the vertical axis. The out-of-control process decision is made by using a V-mask. The V-mask construction is shown in Fig. 7.8. The main parameters of the V-mask are h and k, where $h\sigma$ is half the length of the vertical side in the V-mask and $k\delta\sigma$ is the slope of the sides forming the V-shape; h is called the *decision limit*. The recommended value of k is $\frac{1}{2}$. Thus, the slope of the sides is half of the shift to be detected. The value of h is chosen around 4 or 5. The V-mask is placed at every point as shown in Fig. 7.8. In the V-mask design, d may be calculated

7.5 Control Charts for Variables

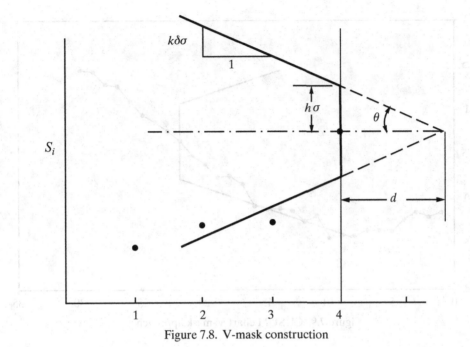

Figure 7.8. V-mask construction

directly using α, the probability of Type I error, as

$$d = -\frac{2\sigma \ln \alpha}{\delta^2}. \tag{7.34}$$

When d from Eq. (7.34) is used, h is calculated using $h = dk\delta$. The angle θ may be calculated using the geometry given in the figure.

In Fig. 7.8, the process is seen to be out of control at sample 4. If the process is out of control, corrective action has to be taken as discussed for other control charts. The V-mask is now ready for use on subsequent samples.

For a mean shift of 0.5σ, and $h = 4$, the CUSUM chart approach gives an ARL of about 27 compared to 155 for the Shewhart control chart.

The CUSUM control chart using the V-mask approach is implemented in sheet1 of *CUSUM.xls*. The V-mask can be moved from point to point using the spin button application. The $\bar{x}$-chart is also plotted on the sheet. The use of this program will be clear through the following example.

Example 7.6. For the problem stated in Example 7.5, prepare a CUSUM chart using the V-mask approach.

Solution. The data are entered in sheet1 of *CUSUM.xls*. The CUSUM chart from the program is shown in Fig. 7.9. The process is out of control at point 22 as shown in the figure. This out-of-control condition continues to be so for later

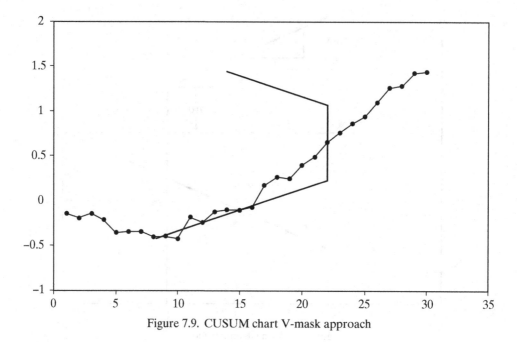

Figure 7.9. CUSUM chart V-mask approach

points. The V-mask can be moved using the spin button at the top to see if the process is in control. The $\bar{x}$-chart is the same as that in Fig. 7.7b.

We now proceed to the tabular approach.

Tabular Approach

In the tabular approach, two cumulative sums – *upper cusum* and *lower cusum* – are defined. If $\delta\sigma$ is the shift to be detected, we define a factor k to get a *reference value* Δ:

$$\Delta = k\delta\sigma. \tag{7.35}$$

The upper cumulative sum U_j and lower cumulative sum L_j, for samples $j = 1$ to m, are defined using starting values $U_0 = 0$ and $L_0 = 0$:

$$\begin{aligned} U_0 &= 0 \quad L_0 = 0 \\ U_j &= \max(0, x_j - \mu - \Delta + U_{j-1}) \\ L_j &= \min(0, x_j - \mu + \Delta + L_{j-1}) \\ j &= 1 \text{ to } m. \end{aligned} \tag{7.36}$$

7.5 Control Charts for Variables

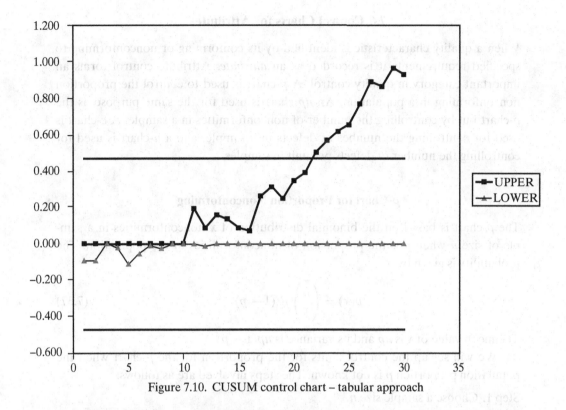

Figure 7.10. CUSUM control chart – tabular approach

Breaking the cumulative sums into upper and lower values will enable us to make a decision of whether the process is out of control using one decision parameter, h. The decision is made when U_j is below $h\sigma$ and L_j is above $-h\sigma$. Values of h of 4 or 5 are generally used.

When the CUSUM value crosses the line, we backtrack to previous successive nonzero values to determine the point where the shift occurred. We may also look at the $\bar{x}$-chart carefully to check the trend. The CUSUM control chart using the tabular approach is implemented in sheet2 of *CUSUM.xls*. The $\bar{x}$-chart is also plotted on the sheet.

Example 7.7. For the problem stated in Example 7.5, prepare the CUSUM chart using the tabular approach.

Solution. The data are entered in sheet2 of *CUSUM.xls*. The CUSUM chart from the program is shown in Fig. 7.10. The process is out of control at point 22, as shown in the figure. The rise of the upper CUSUM starts at point 10. A careful look at the $\bar{x}$-chart (Fig. 7.7b) shows that the shift may be more distinct at point 19.

7.6 Control Charts for Attributes

When a quality characteristic is identified by its conforming or nonconforming to specified requirements it is recorded as an *attribute*. Attribute control forms an important category in quality control. A *p*-chart is used to control the proportion nonconforming in a population. An *np*-chart is used for the same purpose as the *p*-chart but by controlling the number of nonconformities in a sample. A *c*-chart is used for controlling the number of defects in a sample, and a *u*-chart is used for controlling the number of defects per unit in samples.

p-Chart for Proportion Nonconforming

The *p*-chart is based on the binomial distribution of x nonconformities in a sample of size n when p is the proportion of nonconformities in the population. The probability is given by

$$p(x) = \binom{n}{x} p^x (1-p)^{n-x}. \tag{7.37}$$

The mean value of x is np and its variance is $np(1-p)$.

We will set up the control limits for the proportion for the *p*-chart when the population proportion p is not known. The steps involved are as follows:

Step 1. Choose a sample size n.
Step 2. Let $x_1, x_2, \ldots, x_m$ be the number of defective units in samples $1, 2, \ldots, m$.
Step 3. Evaluate the proportion of the samples that are defective using

$$p_i = \frac{x_i}{n} \quad (i = 1, 2, \ldots, m). \tag{7.38}$$

Step 4. Evaluate the mean $\bar{p}$ of the m proportions:

$$\bar{p} = \frac{p_1 + p_2 + \cdots + p_m}{m}. \tag{7.39}$$

The expected value of the standard deviation of the proportion is calculated using

$$\sigma_p = \sqrt{\frac{\bar{p}(1-\bar{p})}{n}}. \tag{7.40}$$

Step 5. Set the control limits for the *p*-chart:

$$\begin{aligned} \text{UCL}_p &= \bar{p} + 3\sqrt{\frac{\bar{p}(1-\bar{p})}{n}} \\ \text{CL}_p &= \bar{p} \\ \text{LCL}_p &= \max\left(0, \bar{p} - 3\sqrt{\frac{\bar{p}(1-\bar{p})}{n}}\right). \end{aligned} \tag{7.41}$$

7.6 Control Charts for Attributes

These control limits are first used as trial control limits. After reviewing if the process is found to be out of control, the cause must be identified and rectified. The out-of-control points are then removed from the control limit calculations. The points may be retained in the p-chart with the cause identified at the out-of-control points.

If the target value p_0 is specified, setting $\bar{p} = p_0$ in Eq. (7.41) will give the control limits. In this case a trial run and adjustment step are not necessary. If the target value is low and the actual proportion of nonconformities is higher, the process may be out of control.

Example 7.8. A sample size of 50 was used in establishing a control chart for styrofoam cups produced in a company. Twenty-five samples were taken and the number of nonconformities in each sample were counted. The proportion data are listed in the following table.

Sample #	Number nonconforming x_i	Sample fraction, p_i
1	3	0.06
2	2	0.04
3	1	0.02
4	6	0.12
5	4	0.08
6	5	0.10
7	3	0.06
8	2	0.04
9	4	0.08
10	3	0.06
11	2	0.04
12	2	0.04
13	3	0.06
14	4	0.08
15	5	0.10
16	6	0.12
17	0	0.00
18	2	0.04
19	3	0.06
20	4	0.08
21	5	0.10
22	6	0.12
23	4	0.08
24	3	0.06
25	4	0.08

Calculate the control limits and plot the p-chart.

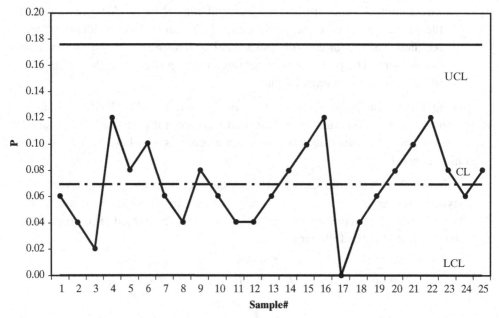

Figure 7.11. *p*-chart

Solution. The third column in the table lists calculated data. The sample size and the second column are entered into Excel (see *PNPChart.xls*) and we obtain

$$\overline{p} = 0.069$$

$$\sqrt{\frac{\overline{p}(1-\overline{p})}{n}} = \sqrt{\frac{0.069(1-0.069)}{50}} = 0.036.$$

The control limits are $CL_p = 0.069$, $UCL_p = 0.069 + 3(0.036) = 0.176$, and $UCL_p = 0.069 - 3(0.036) = -0.039 => 0$. The *p*-chart is shown in Fig. 7.11. The chart shows that the process is in control.

If the sample size is changed for each subgroup, the curve of the *p*-chart is similar except that the limits UCL_p and LCL_p in Eq. (7.33) are calculated for each subgroup *i* using the sample size n_i. The control limits are then discontinuous stepped lines.

np-Chart

We can also control the number of nonconforming parts (for a given sample size) instead of the proportion that are nonconforming. The control limits for the *np*-chart are given by

$$\text{UCL}_{np} = n\overline{p} + 3\sqrt{n\overline{p}(1-\overline{p})}$$
$$\text{CL}_{np} = n\overline{p} \qquad (7.42)$$
$$\text{LCL}_{np} = \max(0, n\overline{p} - 3\sqrt{n\overline{p}(1-\overline{p})}).$$

If a target value p_0 is given, $\overline{p}$ can be replaced by p_0. The *np*-chart for the data in Example 7.4 is included in the program *PNPChart.xls*, which is in the Programs directory of the CD included with the book.

c-Chart

The number of nonconformities may be monitored in some situations. The measurement may be the number of nonconformities on one unit or the number of nonconformities for a set number of units. The number of nonconformities, x, may be modeled with a Poisson distribution, where the parameter λ represents the average number of nonconformities. The probability of occurrence of x nonconformities is given by

$$p(x) = \frac{e^{-\lambda}\lambda^x}{x!}. \qquad (7.43)$$

The mean and the variance of the Poisson distribution are given by λ.

The number of nonconformities is measured in each of the m units as $c_1, c_2, \ldots, c_m$. The average number of nonconformities is calculated using

$$\overline{c} = \frac{c_1 + c_2 + \cdots + c_m}{m}. \qquad (7.44)$$

If no standard or target value is given, we use $\overline{c}$ as the approximation for λ, which is the mean and variance of the distribution. The control limits for the *c*-chart are now given using the three sigma limits:

$$\text{UCL}_c = \overline{c} + 3\sqrt{\overline{c}}$$
$$\text{CL}_c = \overline{c} \qquad (7.45)$$
$$\text{LCL}_c = \max(0, \overline{c} - 3\sqrt{\overline{c}}).$$

These are used as the trial control units. After the assignable causes have been identified and rectified, the control limits are finalized for a stable process. If the target value of c_0 is specified for the number of nonconformities, c_0 may be substituted for $\overline{c}$ in Eqs. (7.45). A trial run is not necessary when a target value is used.

u-Chart

If the number of nonconformities, x, is measured on n units in the sample and the number of nonconformities per unit is calculated, this value need not be an integer. As an example the number of nonconformities may be 5 in 2 square meters of a cloth, or 2.5 nonconformities per square meter. In inspecting unit defects, some units may be without defects. We may find x defects per n units. In such cases the parameter of interest is the number of nonconformities per unit, u:

$$u = \frac{x}{n} \qquad (7.46)$$

where u may be larger than 1 or it may be a fraction. We now determine the mean $\bar{u}$ and its standard deviation, given by $\sqrt{\bar{u}/n}$. The control limits for the u-chart are given by

$$\begin{aligned} \text{UCL}_u &= \bar{u} + 3\sqrt{\frac{\bar{u}}{n}} \\ \text{CL}_u &= \bar{u} \\ \text{LCL}_u &= \max\left(0, \bar{u} - 3\sqrt{\frac{\bar{u}}{n}}\right). \end{aligned} \qquad (7.47)$$

Example 7.9. A sample unit size of 50 is used in establishing the control limits for a c-chart. Twenty-five measurements for the number of nonconformities per 50 units are as follow: 5, 4, 3, 7, 4, 3, 7, 3, 5, 3, 5, 1, 5, 7, 5, 4, 3, 2, 5, 6, 2, 1, 3, 4, 2. Set up the control limits and plot the c-chart.

Solution. The data have been entered into the Excel file *C&UCharts.xls* included in the Programs directory of the CD. The mean value for the 25 samples is $\bar{c} = 3.96$. Substituting in Eqs. (7.45), we obtain

$$\begin{aligned} \text{UCL}_c &= 3.96 + 3\sqrt{3.96} = 9.93 \\ \text{CL}_c &= 3.96 \\ \text{LCL}_c &= \max(0, \bar{c} - 3\sqrt{\bar{c}}) = 0. \end{aligned}$$

The c-chart is shown in Fig. 7.12. The Excel file also includes the u-chart.

7.7 Operating Characteristic (OC) Curves

OC Curve for Variable Control Charts

We see from Fig. 7.3 in Section 7.4 how a shift of the mean increases the probability β of committing a Type II error. Let the control limits be set at $\pm k$ on the standard

7.7 Operating Characteristic (OC) Curves

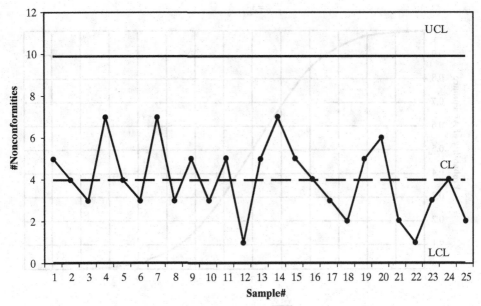

Figure 7.12. c-chart

distribution. A δ-shift in the mean value corresponds to a shift of Δ on the standard normal curve given by (see Chapter 5, Eq. (5.24))

$$\Delta = \frac{\delta \sqrt{n}}{\sigma}. \tag{7.48}$$

We note that $\sigma = \overline{R}/d_2$ for the $\overline{x}$–R chart and $\sigma = \overline{s}/c_4$ for the $\overline{x}$–s chart. The β-value is given by

$$\beta = \Phi(k - \Delta) - \Phi(-k - \Delta), \tag{7.49}$$

where $k = 3$ corresponds to the standard control limits. The OC curve, β versus Δ, for $k = 3$ is shown in Fig. 7.13. We may refer to this as the OC curve for variables. The file *OCCurve.xls* shows the general plot. Changes in the curve for other values of k can also be observed by changing k in sheet1 of the program.

We note from the variables in the parameter Δ that, for a given δ, a larger value of n indicates a smaller value for the Type II error, β. Sample size is an important factor in detecting the shift.

Example 7.10. A statistical process control setup uses the $\overline{x}$–R chart to control shaft sizes in a shop. The sample size is 5 and $\pm 3\sigma$ control limits are used. The value of $\overline{R}$ for the operation is 0.38 mm. If there is a shift of 0.1 mm in the mean due to an error in the tool setup, calculate the probability of a Type II error.

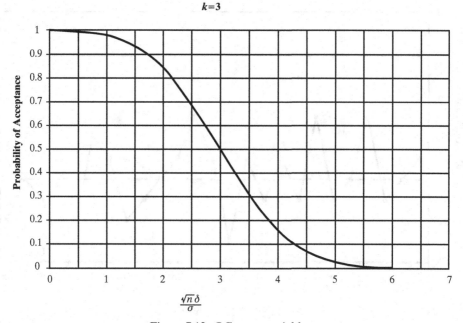

Figure 7.13. OC curve variables

Solution. The shift is $\delta = 0.1$ mm, $\overline{R} = 0.38$ mm, and $d_2 = 2.326$ from Appendix Table A.8. The estimated σ for the process is

$$\sigma = \frac{\overline{R}}{d_2} = \frac{0.38}{2.326} = 0.1634.$$

The standard shift is given by

$$\Delta = \frac{\delta \sqrt{n}}{\sigma} = \frac{0.1\sqrt{5}}{0.1634} = 1.3685.$$

Now β is calculated using Eqs. (7.41):

$$\beta = \Phi(k - \Delta) - \Phi(-k - \Delta) = \Phi(3 - 1.3685) - \Phi(-3 - 1.3685)$$
$$= \Phi(1.6315) - \Phi(-4.3685)$$
$$= 0.949 - 0 = 0.949.$$

The probability of a Type II error is 0.949.

OC Curve for Attribute Control Charts

For the control limits set for the p-chart, we can find the probability of committing a Type II error, β, by choosing various values of p for the shifted probability

7.7 Operating Characteristic (OC) Curves

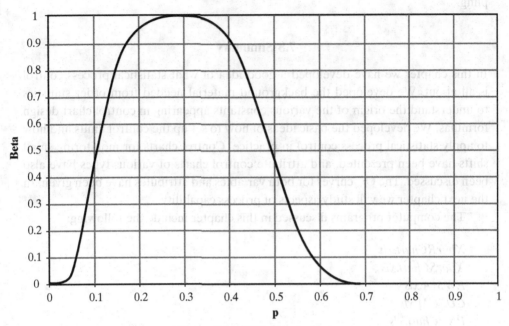

Figure 7.14. OC curve p-chart

of nonconformities in the population. Let $\bar{p}$ be the control centerline and let p be the shifted value of the probability. If n is the sample size, we evaluate $n_u = \text{int}(np.\text{UCL}_p)$ and $n_l = \text{int}(np.\text{LCL}_p)$. The probability of Type II error β is given by

$$\beta = \sum_{i=1}^{n_u} \binom{n}{i} p^i (1-p)^{n-i} - \sum_{i=1}^{n_l} \binom{n}{i} p^i (1-p)^{n-i}. \tag{7.50}$$

The two terms in (7.50) are cumulative binomial probabilities.

Alternatively β may be evaluated using normal approximation with $\sigma = \sqrt{\frac{p(1-p)}{n}}$ and $\beta = \Phi(\frac{UCL_p - p}{\sigma}) - \Phi(\frac{LCL_p - p}{\sigma})$. This calculation (7.50) has been implemented in the program *OC_PChart.xls*, which is included in the Programs directory of the CD. The OC curve for $p = 0.3$ is shown in Fig. 7.14 and is a skewed curve. The plots for various p-values can be obtained using the included program.

7.8 Summary

In this chapter we have developed a good idea of what statistical process control is all about. We developed the background material needed from order statistics to understand the origin of the various constants appearing in control chart design formulas. We developed the basic ideas of how to set up the control limits and how to apply statistical process control in practice. Control charts for monitoring small shifts have been presented, and attribute control charts of various types have also been discussed. The OC curves for both variables and attributes have been given. In the next chapter we will study aspects of process capability.

The computer programs discussed in this chapter include the following:

XbarRChart.xls
XbarSChart.xls
EWMA.xls
CUSUM.xls
PNPChart.xls
C&UCharts.xls
OCCurve.xls
OC_PChart.xls

EXERCISE PROBLEMS

7.1. The diameter of a bronze bushing used in a journal bearing is a characteristic of interest. Random samples of size 5 are taken. The average in millimeters and range in μm for each sample are calculated and the tabulation is shown here:

Sample	1	2	3	4	5	6	7	8
$\bar{x}$	49.52	49.53	49.52	49.50	49.49	49.37	49.45	49.48
R	75	60	90	95	70	45	80	90

a. Find the control limits for the $\bar{x}$–R charts and comment if the process is in control.
b. Estimate the process mean and the process standard deviation.

7.2. Piston-ring thickness is the parameter that is controlled in a grinding operation. Data are given here for 20 random samples of 4 rings each. Set up the $\bar{x}$- and R-charts. Draw your conclusions if the process is in control.

Sample	Thickness (mm)			
1	3.01	3.01	3.16	3.07
2	2.98	2.98	2.98	3.00
3	3.05	2.93	2.94	2.88
4	3.01	2.98	2.95	2.92
5	2.98	2.94	2.91	3.14
6	2.96	3.06	2.95	3.11
7	3.06	3.05	2.88	2.95
8	2.90	3.04	3.02	2.89
9	2.95	3.01	2.95	3.04
10	3.00	2.95	3.01	3.00
11	3.13	2.90	3.02	2.90
12	2.99	3.18	3.01	3.11
13	2.88	2.96	2.94	3.06
14	2.99	2.94	3.04	3.00
15	3.07	2.94	2.95	3.04
16	3.05	2.99	2.97	2.98
17	3.14	3.00	3.12	2.87
18	3.00	3.12	2.92	2.93
19	3.03	3.16	3.07	3.08
20	2.85	3.03	3.08	3.02

7.3. For the piston-ring thickness data in Problem 7.2, set up the $\bar{x}$- and s-charts. Draw your conclusions.

7.4. The thickness, in millimeters, of sheet metal used for making automobile bodies is a characteristic of interest. Random samples of size 5 are taken. The average and standard deviation for each sample are calculated and the tabulation is shown here:

Sample	1	2	3	4	5	6	7	8	9	10
$\bar{x}$	10.19	9.80	10.12	10.54	9.86	9.45	10.06	10.13	9.82	10.17
s	0.15	0.12	0.18	0.19	0.14	0.09	0.16	0.18	0.14	0.13

    a. Find the control limits for the $\bar{x}$–s charts. Comment if the process is in control.
    b. Estimate the process mean and the process standard deviation.

7.5. Each of the following subgroups is an average of 4 sample voltages in an electronic device:

22.5	22.6
22.6	22.8
22.5	22.4
22.4	23.1
22.6	22.7
22.6	22.4
22.5	22.8
22.8	23.0
22.4	22.3
22.4	22.7

The target value of the voltage is 22.5. The process standard deviation for the voltage is 0.32. Prepare the EWMA chart for $\lambda = 0.05$ and $\lambda = 0.1$. Draw your conclusions about the process.

7.6. For the data presented in Problem 7.5, draw the CUSUM chart using the V-mask approach. What is your choice for the parameters?

7.7. For the data presented in Problem 7.5, draw a CUSUM chart using the tabular approach. What is your choice for the parameters?

7.8. A sample size of 100 is used in preparing a control chart for hexagonal nuts produced by a manufacturer. Twenty samples were taken and the number of nonconforming items was measured in each sample:

Sample #	Number nonconforming x_i	Sample #	Number nonconforming x_i
1	10	11	6
2	2	12	8
3	8	13	10
4	11	14	9
5	5	15	4
6	9	16	7
7	3	17	6
8	7	18	9
9	6	19	8
10	5	20	7

Prepare a *p*-chart showing the control limits. Draw your conclusions.

7.9. Prepare an *np*-chart for the data in Problem 7.5.

7.10. A sample size of 5 is used in preparing a *c*-chart for the number of nonconformities. Twenty measurements for the number of nonconformities per 5 units are given:

$$7, 3, 9, 3, 5, 7, 4, 8, 6, 0, 6, 1, 4, 6, 7, 6, 8, 2, 3, 6.$$

Plot the *c*-chart showing the control limits. Draw your conclusions.

7.11. Prepare a *u*-chart for the data in Problem 7.7 and draw your conclusions.

7.12. An automatic screw machine turns out round-headed bolts with a specified shank diameter of 9.00 ± 0.04 mm. The process has been operating in control at an estimated μ of 9.00 mm and an $\overline{R}$ of 0.0208 mm. The subgroup size is 4.

    a. Calculate the $\bar{x}$- and *R*-chart control limits

    b. If the mean of the process shifts to 9.01 mm, compute the probability of a Type II error that the shift will not be detected in the first subgroup after it occurs.

    c. What proportion of the defective product is being produced at the shifted mean?

7.13. Cylindrical pins are produced on an automatic machine. The targeted mean value for the process is 12 mm. The standard deviation of the process is 0.08 mm. Set the control limits for $\bar{x}$ using a sample size of 5. If there is a shift in the process mean and it has a value of 12.04 mm, determine the probability of a Type II error that the shift will not be detected in a sample.

8

Process Capability Analysis

8.1 Introduction

In Chapter 2 we studied the international tolerance system and how tolerances are specified. Design engineers select a tolerance for a quality characteristic based on functional requirements, management guidelines, or customer demands. However, knowledge of the process used for the production and its capability are also necessary. The design specifications must match the process capabilities. To this end, the trend is toward a closer integration between design and production in today's manufacturing industries. A design engineer should aim at giving as large a tolerance as possible without compromising the performance of the product. The manufacturing engineer, on the other hand, must aim for production with as low a variation as is economically possible. The topics in this chapter, namely process capability, measurement capability, and rational determination of tolerance intervals, aim at unifying these diverse objectives. In Chapter 7 we discussed the aspects of process control. A process in control is one in which all variations are due to chance causes. All special or assignable causes are identified as soon as they appear and are rectified. However, there is always a chance that small shifts may not be detected; this leads to the possibility of Type II errors. Process capability analysis identifies the conditions that provide a cushion for this possibility.

8.2 Process Capability

Specification limits are needed to define process capability. *Specification limits* are boundaries, set by the management, engineers, or customers, within which the system must operate. The *upper specification limit* (USL) and the *lower specification limit* (LSL) are the specified limit values. These are generally specified by the designer based on the functional requirements and management guidelines. These

8.2 Process Capability

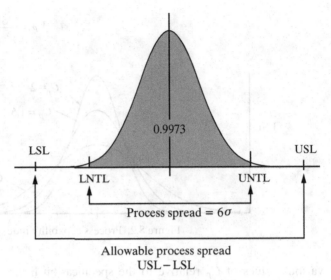

Figure 8.1. Process spread

are also referred to as *tolerance limits*. The upper and lower control limits on the sample mean of a quality characteristic were defined in statistical process control. When these control limits are adjusted for the sample size, they define the natural tolerance limits. The *natural tolerances* are closely tied to the process and are expressed in terms of the process mean μ and the standard deviation σ. The *upper natural tolerance limit* (UNTL) is at $\mu + 3\sigma$, and the *lower natural tolerance limit* (LNTL) is at $\mu - 3\sigma$: 99.73% of production falls in this interval. The difference, UNTL − LNTL = 6σ, is referred to as the *process spread*. The relationship among these is shown in Fig. 8.1.

A *capable process* is one in which all the measured characteristic values fall inside the specification limits. There are several indices (C_p, C_{pk}, C_{pm}) that are used to measure the capability of a process.

Process Capability Indices: C_p, C_{pu}, C_{pl}, C_{pk}

The *process capability index*, C_p, is the ratio of USL − LSL and 6σ:

$$C_p = \frac{\text{USL} - \text{LSL}}{6\sigma}. \tag{8.1}$$

The process capability index does not take into account any shift. The conforming proportion is a maximum when the process spread is placed symmetrically with respect to the central value 0.5 (USL + LSL). Fig. 8.2 shows the normal curves with

$$C_p = \frac{\text{USL} - \text{LSL}}{6\sigma}$$

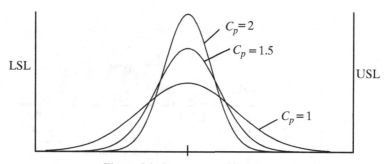

Figure 8.2. Process capability index

various values of C_p relative to the specification limits. The conforming proportion drops as the shift from the center increases. The effect of the shift is shown in Fig. 8.3 for various values of C_p. All shifts and the specification limit values shown are multiples of σ. The curve for $C_p = 1$ corresponds to the operating characteristic curve with the standard shift as seen earlier. In manufacturing industries, $C_p > 1$ is

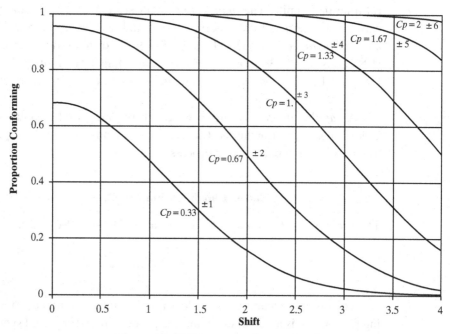

Figure 8.3. Shift vs proportion conforming

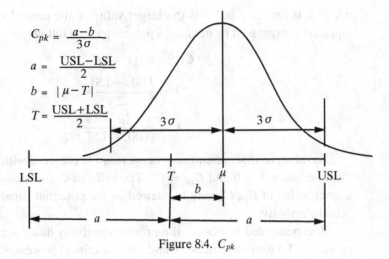

Figure 8.4. C_{pk}

desired; $C_p = 2$ corresponds to USL $-$ LSL $= 12\sigma\,(6\sigma)$. For this case even a shift of 1.5σ will only result in 3.4 parts per million nonconformities. This is the underlying principle of the *Six Sigma* quality philosophy. The *process capability ratio* (C_r) is defined as $1/C_p$.

C_{pu} is defined for upper specification only:

$$C_{pu} = \frac{\text{USL} - \mu}{3\sigma}, \tag{8.2}$$

and C_{pl} is defined for lower specification only:

$$C_{pl} = \frac{\mu - \text{LSL}}{3\sigma}. \tag{8.3}$$

For a two-sided specification, the index C_p does not take into account if there is a shift in the mean. For an off-center process, we define C_{pk} as

$$C_{pk} = \min\left(\frac{\text{USL} - \mu}{3\sigma}, \frac{\mu - \text{LSL}}{3\sigma}\right) = \min\left(C_{pu}, C_{pl}\right). \tag{8.4}$$

From Fig. 8.4, C_{pk} may also be expressed as

$$C_{pk} = \frac{a - b}{3\sigma}$$

$$a = \frac{\text{USL} - \text{LSL}}{2}$$

$$b = |\mu - T|$$

$$T = \frac{\text{USL} + \text{LSL}}{2},$$

where b is the shift and T is the target value or the central value. The preceding equation is expressed by defining a parameter k as follows:

$$C_{pk} = C_p(1 - k)$$
$$T = \frac{\text{USL} + \text{LSL}}{2} \tag{8.5}$$
$$k = \frac{|\mu - T|}{(\text{USL} - \text{LSL})/2}.$$

We observe that, when the process mean is centered with respect to the specification limits, $k = 0$ and $C_{pk} = C_p$. The value of C_{pk} decreases as k increases for a given value of C_p; C_p may be viewed as the potential capability and C_{pk} as the actual capability.

Recommended values for the process capability index are 1.33 for established processes, 1.5 for new processes, and 1.67 for critical processes.

Example 8.1. Holes of 20-mm diameter are drilled and reamed in a production process. The process standard deviation σ is known to be 15 μm. The lower and upper specification limits for the hole are 20.01 and 20.11 mm, respectively. The mean diameter of 100 holes measured is found to be 20.07 mm. Determine the process capability indices, C_p and C_{pk}.

Solution. We have $\sigma = 15$ μm $= 0.015$ mm, USL $= 20.11$ mm, LSL $= 20.01$ mm, and $\mu = 20.07$ mm:

$$C_p = \frac{\text{USL} - \text{LSL}}{6\sigma} = \frac{20.11 - 20.01}{6 \times 0.015} = 1.11$$

$$C_{pu} = \frac{\text{USL} - \mu}{3\sigma} = \frac{20.11 - 20.07}{3 \times 0.015} = 0.89$$

$$C_{pl} = \frac{\mu - \text{LSL}}{3\sigma} = \frac{20.07 - 20.01}{3 \times 0.015} = 1.33$$

$$C_{pk} = \min(C_{pu}, C_{pl}) = 0.89.$$

C_{pk} can reach a value of $C_p = 1.11$ by setting the process mean at the mean of the specification limits.

All process capability indices (C_p, C_{pu}, C_{pl}, and C_{pk}) are defined using σ. In practice, however, σ is not known; in that case we use an estimate for $\hat{\sigma}$.

The candidates for $\hat{\sigma}$ estimates are

1. s evaluated from a large sample in a histogram experiment;
2. the characteristic value at $z = 1$ minus the characteristic value at $z = 0$, from a normal probability plot, or the 84th percentile minus 50th percentile if a percentile plot is used;

8.2 Process Capability

3. $\frac{\bar{R}}{d_2}$ for a process in control from an $\bar{x}$–R chart; and
4. $\frac{\bar{s}}{c_4}$ for a process in control from an $\bar{x}$–s chart.

The characteristic values evaluated with $\hat{\sigma}$ are denoted $\hat{C}_p$, $\hat{C}_{pu}$, $\hat{C}_{pl}$, and $\hat{C}_{pk}$.

When an experiment is conducted to evaluate process capability, all the machine, operator, and process variable settings need to be recorded for future reference. The data order and the collection process have to be carefully monitored.

Example 8.2. Estimate the process capability parameters for the following data from an $\bar{x}$–R chart for a process in control: $\bar{\bar{x}} = 125.4$ mm, $\bar{R} = 3.2$ mm, and sample size $n = 5$. The specification limits are 125 ± 6 mm.

Solution. We have $\mu = \bar{\bar{x}} = 125.4$ mm, USL $= 125 + 6 = 131$ mm, and LSL $= 125 - 6 = 119$ mm, and from Appendix Table A.8 we have $d_2 = 2.326$:

$$\hat{\sigma} = \frac{\bar{R}}{d_2} = \frac{3.2}{2.324} = 1.377$$

$$\hat{C}_p = \frac{\text{USL} - \text{LSL}}{6\hat{\sigma}} = \frac{131 - 119}{6 \times 1.377} = 1.453.$$

We use Eqs. (8.5) to calculate $\hat{C}_{pk}$:

$$T = \frac{\text{USL} + \text{LSL}}{2} = 125 \text{ mm}$$

$$|\mu - T| = |125 - 125.4| = 0.4$$

$$k = \frac{|\mu - T|}{(\text{USL} - \text{LSL})/2} = \frac{0.4}{6} = 0.0667$$

$$\hat{C}_{pk} = \hat{C}_p (1 - k) = 1.356.$$

Process Capability Index: C_{pm}

The process capability index C_{pm} is also referred to as the Taguchi capability index. The index is calculated using

$$C_{pm} = \frac{\text{USL} - \text{LSL}}{6\tau}, \tag{8.6}$$

where τ^2 is the expected value of $(X - T)^2$, with the target value $T = (\text{USL} + \text{LSL})/2$. Since $\tau^2 = E[(X - T)^2] = E[(X - \mu + \mu - T)^2] = E[(X - \mu)^2] + (\mu - T)^2$, we have

$$\tau = \sqrt{\sigma^2 + (\mu - T)^2}.$$

By denoting

$$\xi = \frac{\mu - T}{\sigma}, \tag{8.7}$$

we get $\tau = \sigma\sqrt{1 + \xi^2}$, and Eq. (8.6) becomes

$$C_{pm} = \frac{C_p}{\sqrt{1 + \xi^2}}. \tag{8.8}$$

To estimate $\hat{C}_{pm}$ in an experiment, we use $\hat{C}_p$ in Eq. (8.8) and $\hat{\xi}$ is evaluated using $\hat{\xi} = \frac{\bar{x} - T}{s}$. We may also evaluate the estimates using $\bar{x}$–R or $\bar{x}$–s chart data.

Example 8.3. Evaluate the process capability index C_{pm} for the data in Example 8.1.

Solution. We have $\sigma = 0.015$ mm, USL $= 20.11$ mm, LSL $= 20.01$ mm, $\mu = 20.07$ mm, and $T = (\text{USL} + \text{LSL})/2 = 20.06$. We evaluated $C_p = 1.11$. Using Eq. (8.7),

$$\xi = \frac{\mu - T}{\sigma} = \frac{20.07 - 20.06}{0.015} = 0.667.$$

Introducing the values in Eq. (8.8),

$$C_{pm} = \frac{C_p}{\sqrt{1 + \xi^2}} = \frac{1.11}{\sqrt{1 + 0.667^2}} = 0.9234.$$

This value is different from $C_{pk} = 0.89$, evaluated earlier.

Since ξ depends on the standard deviation, C_{pm} is a better indicator of the off-center variation relative to the process standard deviation.

As seen in the examples, we have estimated the value of the process capability index. The question now is – what is the confidence interval for process capability? Some relationships for the confidence intervals are provided in the next section.

Confidence Intervals

Once we evaluate the process capability in an experiment, we have only obtained an estimated value. A process capability study must include a confidence interval for the true process capability index. There is a concentrated ongoing effort to arrive at reliable results in this area. Two-sided confidence interval relations that are widely used in process capability studies are given here. These relations are recommended for sample sizes (n) of at least 25.

When $\hat{C}_p$ is evaluated using s as the estimator for σ as $\hat{C}_p = \frac{\text{USL} - \text{LSL}}{6s}$, note that $\frac{(n-1)C_p^2}{\hat{C}_p^2} = \frac{(n-1)s^2}{\sigma^2}$ has a chi-squared distribution with $n - 1$ degrees of freedom. Thus,

8.2 Process Capability

the two-sided $1-\alpha$ confidence interval for C_p is given by

$$\hat{C}_p\sqrt{\frac{\chi^2_{1-\alpha/2,n-1}}{n-1}} \leq C_p \leq \hat{C}_p\sqrt{\frac{\chi^2_{\alpha/2,n-1}}{n-1}}. \tag{8.9}$$

The chi-squared values are obtained from Appendix Table A.5. The approximate confidence interval formula for C_{pk} is

$$\hat{C}_{pk} - z_{\alpha/2}\sqrt{\frac{1}{9n} + \frac{\hat{C}_{pk}^2}{2(n-1)}} \leq C_{pk} \leq \hat{C}_{pk} + z_{\alpha/2}\sqrt{\frac{1}{9n} + \frac{\hat{C}_{pk}^2}{2(n-1)}}. \tag{8.10}$$

The z-values are obtained from the inverse normal distribution values in Appendix Table A.4.

Example 8.4. For a statistical process in control with specification limits 112 ± 4, a sample size of 40 was used. The mean was 111 and the sample standard deviation, s, was 0.9. Find the 90% confidence intervals for C_p and C_{pk}.

Solution. Given USL $= 116$, LSL $= 108$, $s = 0.9$, $\bar{x} = 111$, $n = 40$, and $T = 0.5\,(\text{USL} + \text{LSL}) = 112$,

$$\hat{C}_p = \frac{\text{USL} - \text{LSL}}{6s} = \frac{116 - 108}{6 \times 0.9} = 1.4815.$$

We use Eqs. (8.5) to calculate $\hat{C}_{pk}$:

$$|\bar{x} - T| = |111 - 112| = 1$$

$$k = \frac{|\bar{x} - T|}{(\text{USL} - \text{LSL})/2} = \frac{1}{4} = 0.25$$

$$\hat{C}_{pk} = \hat{C}_p(1-k) = 1.11.$$

Next we find the chi-squared distribution values for 39 degrees of freedom for 0.05, and 0.95 probabilities for the 90% confidence interval:

$$\chi^2_{0.95,39} = 25.7, \quad \chi^2_{0.05,39} = 54.572.$$

From Eq. (8.9),

$$\hat{C}_p\sqrt{\frac{\chi^2_{1-\alpha/2,n-1}}{n-1}} \leq C_p \leq \hat{C}_p\sqrt{\frac{\chi^2_{\alpha/2,n-1}}{n-1}}$$

$$1.4815\sqrt{\frac{25.7}{39}} \leq C_p \leq 1.4815\sqrt{\frac{54.572}{39}}$$

$$1.203 \leq C_p \leq 1.753.$$

For the confidence interval on C_{pk}, the z-value is obtained at $\alpha/2 = 0.05$ from Appendix Table A.4 as $z_{0.05} = 1.645$. From Eq. (8.10),

$$\hat{C}_{pk} - z_{\alpha/2}\sqrt{\frac{1}{9n} + \frac{\hat{C}_{pk}^2}{2(n-1)}} \leq C_{pk} \leq \hat{C}_{pk} + z_{\alpha/2}\sqrt{\frac{1}{9n} + \frac{\hat{C}_{pk}^2}{2(n-1)}}$$

$$1.11 - 1.645\sqrt{\frac{1}{9 \times 40} + \frac{1.11^2}{2 \times 39}} \leq C_{pk} \leq 1.11 + 1.645\sqrt{\frac{1}{9 \times 40} + \frac{1.11^2}{2 \times 39}}$$

$$0.886 \leq C_{pk} \leq 1.334.$$

The lower limit of the 90% confidence interval for C_p is larger than 1. We may conclude that the process is good when it is centered. The second case shows that there is a chance that C_{pk} is lower than 1, which suggests that ways should be found to center the process.

8.3 Measurement System Analysis – Gage Repeatability and Reproducibility Study

Process control and process capability studies involve analysis of the process data. These data come from the measurement of quality characteristics. Previous analyses dealt with measured values. A measured value is made up of its true value and an error component, and these errors may come from the instrument or gage used. Error may also be contributed from operators, setup, and the environment. Calibration is another key area; all gages must be properly calibrated. The true value and the measurement error may be treated as independent random variables. The variance of the measured value is the sum of the variance of the true value and the variance of the error:

$$\sigma^2_{\text{measured value}} = \sigma^2_{\text{true value}} + \sigma^2_{\text{measurement error}}. \tag{8.11}$$

The aim here is to determine the true variability of the process. This will be accomplished by finding the variability due to measurement.

When we speak of measuring instruments, we speak of precision. Precision implies that the measurement variability should be a small part of the variability of the measured value. This brings us to review the terms accuracy, trueness, bias, precision, repeatability, and reproducibility.

Accuracy is the closeness of agreement between a test result (measurement) and the accepted reference value. The International Standards Organization notes that a systematic or bias component may be included in results that are termed accurate. *Trueness* is the closeness of agreement between the average value obtained from a large series of test results (measurements) and an accepted reference value.

8.3 Measurement System Analysis – Gage Repeatability and Reproducibility Study

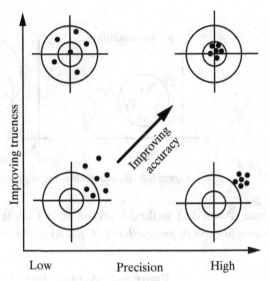

Figure 8.5. Precision vs trueness

Bias is the difference between the expectation of the test results and an accepted reference value. *Precision* is the closeness of agreement between independent test results (measurements) obtained under stipulated conditions. It is measured as the standard deviation of test results. Trueness may correctly contrast with precision, not accuracy. The precision–trueness contrast is illustrated in Fig. 8.5.

Repeatability is precision estimated under conditions where independent test results (measurements) are obtained with the same method on identical test items in the same laboratory by the same operator using the same equipment within a short interval of time. Gage repeatability is the variation obtained from one gage and one operator when measuring the same part several times.

Reproducibility is precision under conditions where independent test results are obtained with the same method on identical test items in different laboratories with different operators using different equipment over extended periods of time. Gage reproducibility is the variability of measurements by different operators using the same gage (or different conditions). Repeatability and reproducibility are illustrated in Fig. 8.6. The variance of the gage measurement error is the sum of variances attributable to repeatability and reproducibility.

The measurement variance in Eq. (8.11) can be expressed as

$$\sigma^2_{\text{measurement error}} = \sigma^2_{\text{GR\&R}} = \sigma^2_{\text{repeatability}} + \sigma^2_{\text{reproducibility}}. \qquad (8.12)$$

We now present the range and average method for computing the standard deviation of the gage repeatability and reproducibility (GR&R). An analysis of

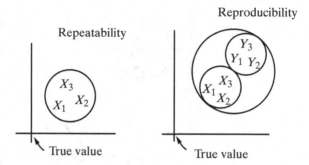

Figure 8.6. Repeatability and reproducibility (R&R)

variance (ANOVA) method is presented in Chapter 10, Section 10.3, under "*Gage Repeatability and Reproducibility (GR&R) for Measurement Systems.*"

Range and Average Method for GR&R

Before undertaking the evaluation of GR&R, we must ensure that the gage has been calibrated. The recommended method is to use ten parts, two or three appraisers, and two or three trials, which amounts to sixty measurements. For each trial the parts are run in a randomized order and measured.

All the measurements taken are tabulated as shown in Example 8.5. All the calculations discussed here are implemented for the example problem in *GageR&R.xls* included in the Programs directory of the CD. Data in the program may be modified for solving other problems. The data are tabulated in the following format:

	Operators A,B,C					
	A		B		C	
Part	Trial1	Trial2	Trial1	Trial2	Trial1	Trial2
x	x	x	x	x	x	x

The ranges between trials are evaluated for each part and operator. The steps in the evaluation process are developed as follows.

Steps for GR&R Evaluation

1. Calculate the mean of all trial ranges, $\overline{R}$.
2. D_4, repeated here, from Appendix Table A.8, is used to calculate UCL_R for the data:

$$\text{UCL}_R = D_4 \overline{R}. \tag{8.13}$$

8.3 Measurement System Analysis – Gage Repeatability and Reproducibility Study

	Number of Trials			
	2	3	4	5
D_4	3.267	2.575	2.282	2.115

Check if all the trial ranges are below UCL_R. The lower control limit is 0 for up to 5 trials. If a range falls outside the limit, the assignable cause must be found and resolved before proceeding further.

3. Calculate R_a, which is the range of the averages for the appraisers or operators. If $\overline{X}_A$, $\overline{X}_B$, and $\overline{X}_C$, are the average measurements for the operators, the range is the maximum minus the minimum for the three values.
4. Calculate R_p, which is the range of average measurements of the parts.
5. Calculate the standard repeatability error:

$$\sigma_{\text{repeatability}} = \frac{5.15\overline{R}}{d_2} = k_1 \overline{R}, \tag{8.14}$$

where 5.15 is $2z_{0.005}$ (spread of 99% area under the normal curve), and d_2 is a constant from Appendix Table A.8. The factor k_1 is given here:

	Number of Trials			
	2	3	4	5
k_1	4.57	3.04	2.5	2.21

6. Find the reproducibility variance:

$$\sigma_{\text{reproducibility}} = \sqrt{(k_2 R_a)^2 - \frac{\sigma^2_{\text{repeatability}}}{nr}}, \tag{8.15}$$

where n = number of parts, r = number of trials, and k_2 is given by following:

Number of Appraisers	2	3	4	5
k_2	3.65	2.7	2.3	2.08

7. The GR&R is given by

$$\sigma_{\text{GR\&R}} = \sqrt{\sigma^2_{\text{repeatability}} + \sigma^2_{\text{reproducibility}}}. \tag{8.16}$$

8. The part standard error is given by

$$\sigma_{\text{part}} = k_3 R_p, \tag{8.17}$$

where k_3 is given by

Number of Parts	2	3	4	5	6	7	8	9	10
k_3	3.65	2.7	2.3	2.08	1.93	1.82	1.74	1.67	1.62

9. The total variation is given by

$$\sigma_{\text{total}} = \sqrt{\sigma_{\text{GR\&R}}^2 + \sigma_{\text{part}}^2}. \tag{8.18}$$

10. The percentage of the total variation is calculated using

$$\%\text{R\&R} = 100 \left(\frac{\sigma_{\text{GR\&R}}}{\sigma_{\text{total}}} \right), \tag{8.19}$$

where %R&R < 10% is considered satisfactory; 10–30% may be acceptable. If this condition prevails, more experienced operators must be used. A value higher than 30% is not satisfactory.

If reproducibility is large in comparison to repeatability, it indicates that the operator needs training or the gage needs calibration.

Comparing Eq. (8.18) to Eq. (8.11), we note that $\sigma_{\text{measured value}} = \sigma_{\text{total}}$, $\sigma_{\text{true value}} = \sigma_{\text{part}}$, and $\sigma_{\text{measurement error}} = \sigma_{\text{GR\&R}}$. Thus,

$$\sigma_{\text{true value}} = \sigma_{\text{measured value}} \sqrt{1 - \left(\frac{\%\text{R\&R}}{100} \right)^2}. \tag{8.20}$$

ANOVA methods are now used for GR&R evaluation. The two-factor factorial approach and its application to GR&R are presented in Chapter 10. We now consider an example where the aforementioned procedure is implemented. (Also see the Excel program *GageR&R.xls*, where all the preceding steps are implemented through an example.)

Example 8.5. Ten parts were measured by three operators each of whom took two measurements per part. Perform the GR&R study.

	Operators					
	A		B		C	
Part	Trial1	Trial2	Trial1	Trial2	Trial1	Trial2
1	5.62	5.61	5.60	5.61	5.65	5.64
2	5.51	5.51	5.52	5.50	5.52	5.52
3	5.50	5.51	5.49	5.50	5.53	5.52
4	5.47	5.48	5.52	5.51	5.44	5.44
5	5.50	5.50	5.54	5.52	5.48	5.46
6	5.62	5.63	5.62	5.61	5.61	5.61
7	5.54	5.56	5.56	5.57	5.54	5.53
8	5.57	5.56	5.55	5.55	5.59	5.60
9	5.41	5.41	5.41	5.40	5.44	5.44
10	5.55	5.54	5.58	5.57	5.54	5.55

8.3 Measurement System Analysis – Gage Repeatability and Reproducibility Study

Solution. We have the number of parts, $n = 10$; the number of trials is 2; and the number of operators is 3. First the ranges are noted for each pair of measurements for operators A, B, and C and $\bar{R}$ is calculated:

	Ranges	
A	B	C
0.01	0.01	0.01
0	0.02	0
0.01	0.01	0.01
0.01	0.01	0
0	0.02	0.02
0.01	0.01	0
0.02	0.01	0.01
0.01	0	0.01
0	0.01	0
0.01	0.01	0.01

For $\text{UCL}_R = D_4 \bar{R} = 0.028$, using $D_4 = 3.267$ for a number of trials of 2 and $\text{UCL}_R = 0$, we find $\bar{R} = 0.009$. All individual ranges are well inside the control limit:

$$\bar{X}_A = 5.530, \quad \bar{X}_B = 5.537, \quad \bar{X}_C = 5.533.$$

The appraiser range, R_a, is $5.537 - 5.530 = 0.006$.

The part averages (for each row) are calculated as 5.622, 5.513, 5.508, 5.477, 5.500, 5.617, 5.550, 5.570, 5.418, and 5.555; the part range is evaluated as $R_p = 0.204$. The total variation is found as follows:

$$\sigma_{\text{repeatability}} = k_1 \bar{R} = 4.57 \times 0.009 = 0.040$$

$$\sigma_{\text{reproducibility}} = \sqrt{(k_2 R_a)^2 - \frac{\sigma_{\text{repeatability}}^2}{nr}} = \sqrt{(2.7 \times 0.006)^2 - \frac{0.040^2}{(10)(2)}} = 0.017$$

$$\sigma_{\text{GR\&R}} = \sqrt{\sigma_{\text{repeatability}}^2 + \sigma_{\text{reproducibility}}^2} = \sqrt{0.040^2 + 0.017^2} = 0.043$$

$$\sigma_{\text{part}} = k_3 R_p = 1.62 \times 0.204 = 0.330$$

$$\sigma_{\text{total}} = \sqrt{\sigma_{\text{GR\&R}}^2 + \sigma_{\text{part}}^2} = \sqrt{0.043^2 + 0.330^2} = 0.333$$

and

$$\%\text{R\&R} = 100 \left(\frac{\sigma_{\text{GR\&R}}}{\sigma_{\text{total}}} \right) = 100 \left(\frac{0.043}{0.333} \right) = 12.90\%.$$

This is higher than 10%; therefore the gage is marginally acceptable. Complete calculations are included in the program *GageR&R.xls*.

Gage Capability and Process Capability

The *precision of a gage* is defined as $5.15\sigma_{GR\&R}$. We have already seen that 5.15 came from the spread of 99% ($2z_{0.005}$) of the area under a standard curve.

The gage capability is defined in terms of the *precision-to-tolerance ratio* (P/T).

$$P/T = \frac{5.15\sigma_{GR\&R}}{\text{USL--LSL}} = \frac{5.15\sigma_{GR\&R}}{6\sigma C_p}$$

$$\% P/T = \frac{0.86\,(\%R\&R)}{C_p} \qquad (8.21)$$

8.4 Propagation of Errors

Let Y be a function of random variables X_i ($i = 1, 2, \ldots, n$):

$$Y = f(X_1, X_2, \ldots, X_n). \qquad (8.22)$$

For small changes about the mean, the Taylor's expansion of the function at $\mu_1, \mu_2, \ldots, \mu_n$ and for the mean values of $X_1, X_2, \ldots, X_n$ (neglecting the higher order terms) is given by

$$Y \cong f(\mu_1, \mu_2, \ldots, \mu_n) + \sum_{i=1}^{n}(X_i - \mu_i)g_i$$
$$g_i = \frac{\partial f}{\partial x_i} \text{ (evaluated at } \mu_1, \mu_2, \ldots, \mu_n). \qquad (8.23)$$

The mean and variance of Y are approximately

$$\mu_Y \cong f(\mu_1, \mu_2, \ldots, \mu_n) \qquad (8.24)$$

$$\sigma_Y^2 \cong \sum_{i=1}^{n} g_i^2 \sigma_i^2, \qquad (8.25)$$

where σ_i^2 is the variance of X_i. Eq. (8.25) is referred to as the *propagation of error formula*.

For a linear function, each g_i is a constant coefficient. A common dimensional problem discussed in Chapter 2 is $Y = \pm X_1 \pm X_2 \pm \cdots \pm X_n$. Then $g_i^2 = 1$ in Eq. (8.25). If the process for each component dimension has a similar process capability C_p, we may set the tolerance $t_i = k\sigma_i$. We get the following tolerance relation, Eq. (2.5) that was used in Chapter 2:

$$t_Y^2 = t_1^2 + t_2^2 + \cdots + t_n^2.$$

This rule states that the square of the resulting tolerance is the sum of component tolerances.

Example 8.6. The angular twist in radians of a solid round bar is given by $\theta = \frac{32Tl}{\pi d^4 G}$, where T is the torque in Newton-meters, l is the length of the bar in meters, d is the diameter of the bar in meters, and $G = 77$ GPa is the modulus of rigidity of the material. (Note: 1 GPa $= 10^9$ N/m^2.) The mean values of the variables are $T = 320$ N-m, $l = 0.12$ m, $d = 0.04$ m, and the approximate standard deviations are $\sigma_T = 3$ N-m, $\sigma_l = 0.01$ m, and $\sigma_d = 0.0004$ m. Determine the mean value of θ and the standard deviation σ_θ.

Solution. The mean value is evaluated using Eq. (8.24):

$$\theta = \frac{32Tl}{\pi d^4 G} = \frac{32 \times 320 \times 0.12}{3.1416 \times 0.04^4 \times 77 \times 10^9} = 0.00198 \text{ radians}.$$

To evaluate the variance of θ, we need to evaluate the coefficients g_i in Eqs. (8.23):

$$g_T = \frac{\partial \theta}{\partial T} = \frac{32l}{\pi d^4 G} = \frac{32 \times 0.12}{3.1416 \times 0.04^4 \times 77 \times 10^9} = 6.2 \times 10^{-6}$$

$$g_l = \frac{\partial \theta}{\partial l} = \frac{32T}{\pi d^4 G} = \frac{32 \times 320}{3.1416 \times 0.04^4 \times 77 \times 10^9} = 0.0165$$

$$g_d = \frac{\partial \theta}{\partial d} = \frac{(-4)32Tl}{\pi d^5 G} = \frac{-4 \times 32 \times 320}{3.1416 \times 0.04^5 \times 77 \times 10^9} = -0.198.$$

From Eq. (8.25), we have

$$\sigma_\theta^2 = g_T^2 \sigma_T^2 + g_l^2 \sigma_l^2 + g_d^2 \sigma_d^2$$
$$= (6.2 \times 10^{-6})^2 3^2 + 0.0165^2 \times 0.01^2 + (-0.198)^2 0.0004^2$$
$$= 3.4 \times 10^{-8}$$
$$\sigma_\theta = 0.000184.$$

Using this information, confidence intervals can be evaluated for a desired $1 - \alpha$.

8.5 Prediction and Tolerance Intervals for Normal Distribution

We developed confidence interval formulas for the mean of a population using the data from a single sample. The two-sided $1 - \alpha$ confidence interval for the mean when σ is known is given by

$$\bar{x} \pm z_{\alpha/2} \frac{\sigma}{\sqrt{n}}. \tag{8.26}$$

When σ is not known, the confidence interval is given by

$$\bar{x} \pm t_{n-1,\alpha/2} \frac{s}{\sqrt{n}}. \tag{8.27}$$

Prediction Intervals

We now ask a question – if we make one future measurement, what will be the $1 - \alpha$ confidence interval for the measurement? This interval is termed the *prediction interval*. We have two cases: (1) when σ is known and (2) when σ is not known.

Known σ

Let us first develop the upper one-sided prediction interval when σ is known. If x is the future measurement, we would like to find k such that

$$P(x \leq \bar{x} + k\sigma) = 1 - \alpha \Rightarrow P\left[\left(\frac{x - \bar{x}}{\sigma\sqrt{1 + \frac{1}{n}}}\right)\sqrt{1 + \frac{1}{n}} \leq k\right] = 1 - \alpha$$

$$\Rightarrow P\left(z\sqrt{1 + \frac{1}{n}} \leq k\right) = 1 - \alpha$$

$$\Rightarrow z_\alpha \sqrt{1 + \frac{1}{n}} = k,$$

where we use the property that $x - \bar{x} \sim N[0, \sigma^2(1 + \frac{1}{n})]$ to convert to the standard variable z. From the last step, since the value for k has been found, the prediction interval for the future measurement x is given by

$$x \leq \bar{x} + z_\alpha \sigma \sqrt{1 + \frac{1}{n}}. \tag{8.28}$$

The lower prediction interval is

$$\bar{x} - z_\alpha \sigma \sqrt{1 + \frac{1}{n}} \leq x. \tag{8.29}$$

The two-sided $1 - \alpha$ prediction interval is given by

$$\bar{x} - z_{\alpha/2} \sigma \sqrt{1 + \frac{1}{n}} \leq x \leq \bar{x} + z_{\alpha/2} \sigma \sqrt{1 + \frac{1}{n}}. \tag{8.30}$$

Unknown σ

Using similar steps as before, we get the prediction intervals. The upper prediction interval is

$$x \leq \bar{x} + t_{\alpha, n-1} s \sqrt{1 + \frac{1}{n}}. \tag{8.31}$$

8.5 Prediction and Tolerance Intervals for Normal Distribution

The lower prediction interval is

$$\bar{x} - t_{\alpha,n-1} s \sqrt{1 + \frac{1}{n}} \leq x. \tag{8.32}$$

The two-sided $1 - \alpha$ prediction interval is given by

$$\bar{x} - t_{\alpha/2,n-1} s \sqrt{1 + \frac{1}{n}} \leq x \leq \bar{x} + t_{\alpha/2,n-1} s \sqrt{1 + \frac{1}{n}}. \tag{8.33}$$

Tolerance Intervals

The objective of a tolerance interval is different from that of a prediction interval or a confidence interval. In confidence intervals, the bounds reach $\bar{x}$ as n tends to infinity. In the two-sided case the interval tends to zero. A *tolerance interval* is an interval for which it can be stated with a given level of confidence, $1 - \gamma$, that it contains at least a proportion, $1 - p$, of the population. The tolerance interval is sometimes referred to as a *statistical coverage interval*. In the prediction interval, the confidence of assurance, $1 - \gamma$, is lacking. A tolerance interval is a good way of checking if specification limits are meaningful. If the difference USL − LSL is smaller than the corresponding $(1 - p, 1 - \gamma)$ tolerance interval based on a sample of n parts, we cannot say with confidence that the production meets the requirement; maybe the specification limits are to be revisited or maybe a new process must be adopted. We once again have the cases where σ is either known or unknown.

Known σ

We would like to find constants κ_1 and κ_2 such that, for given $1 - \gamma$, proportion $1 - p$ of the population is

$$\begin{aligned} &\text{below (upper limit) } \bar{x} + \kappa_1 \sigma \\ &\text{above (lower limit) } \bar{x} - \kappa_1 \sigma \\ &\text{in the interval (two-sided limit) } \bar{x} \pm \kappa_2 \sigma \end{aligned} \tag{8.34}$$

Let us first consider the one-sided (lower) limit. If μ is the population mean, the proportion $1 - p$ is contained in the interval above $\mu - z_p \sigma$. We would like this part of the distribution to be larger than $\bar{x} - \kappa_1 \sigma$ with a probability $1 - \gamma$:

$$P(\bar{x} - \kappa_1 \sigma \leq \mu - z_p \sigma) = 1 - \gamma \Rightarrow P\left[\frac{\bar{x} - \mu}{\sigma/\sqrt{n}} \leq \sqrt{n}(\kappa_1 - z_p)\right] = 1 - \gamma$$
$$\Rightarrow z_\gamma = \sqrt{n}(\kappa_1 - z_p).$$

176   Process Capability Analysis

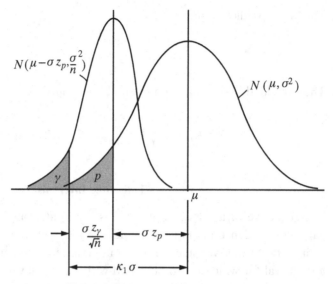

Figure 8.7. Tolerance interval, σ-known

The second part is a rearrangement to bring it into a standard normal form. From the last step, we have

$$\kappa_1 = z_p + \frac{z_\gamma}{\sqrt{n}}. \tag{8.35}$$

The same constant works for the upper one-sided limit. This case is illustrated in Fig. 8.7. Using similar steps, we get κ_2 for the two-sided interval:

$$\kappa_2 = z_{p/2} + \frac{z_{\gamma/2}}{\sqrt{n}}. \tag{8.36}$$

Unknown σ

In most practical problems involving tolerance interval decisions, the population σ is not known. We use the standard deviation, s. We need to find constants k_1 and k_2, such that, for given $1 - \gamma$, proportion $1 - p$ of the population is

$$\begin{aligned} &\text{below (upper limit) } \bar{x} + k_1 s \\ &\text{above (lower limit) } \bar{x} - k_1 s \\ &\text{in the interval (two-sided limit) } \bar{x} \pm k_2 s. \end{aligned} \tag{8.37}$$

Let us first consider the one-sided (lower) limit. If μ is the population mean, the proportion $1 - p$ is contained in the interval $\mu - z_p \sigma$. We would like this part of the

8.5 Prediction and Tolerance Intervals for Normal Distribution

distribution to be larger than $\bar{x} - k_1 s$ with a probability $1 - \gamma$:

$$P(\bar{x} - k_1 s \leq \mu - z_p \sigma) = 1 - \gamma \Rightarrow P\left[\frac{\sqrt{n}(\bar{x} - \mu + z_p \sigma)}{s} \leq k_1 \sqrt{n}\right] = 1 - \gamma$$

$$\Rightarrow P\left[\frac{\frac{\bar{x} - \mu}{\sigma/\sqrt{n}} + z_p \sqrt{n}}{s/\sigma} \leq k_1 \sqrt{n}\right] = 1 - \gamma$$

$$\Rightarrow t_{1-\gamma, n-1}(z_p \sqrt{n}) = k_1 \sqrt{n}.$$

In the first term on the second line in the previous equation, recognize that the denominator $\frac{s}{\sigma} = \sqrt{\frac{(n-1)s^2}{(n-1)\sigma^2}}$ is the square root of the chi-squared distribution with $n - 1$ degrees of freedom divided by $n - 1$ and the numerator is the standard normal with a shift parameter $z_p \sqrt{n}$. Thus, it is a noncentral t-distribution with $n - 1$ degrees of freedom and noncentrality parameter $\delta = z_p \sqrt{n}$. We use the inverse of the noncentral t-distribution to get the final step. We then have

$$k_1 = \frac{t_{1-\gamma, n-1}(z_p \sqrt{n})}{\sqrt{n}}. \tag{8.38}$$

The function tol1s(g,p,n) has been developed for Excel and is included in the program *TolerLimits.xls*. This function uses tncinv(p, nu, delta), which is included in the noncentral distributions. The functions use p and g as left-tail areas. Values for k_1 are provided in Appendix Table A.9 for $1 - p = 0.9, 0.95$, and 0.99 and for $1 - \gamma = 0.95$ and 0.99. The table is also available on sheet1 of the program. The program can be used for calculating intervals for other probabilities not included in the table.

The value of k_2 for the two-sided interval is determined using the following formula, which can be found at http://www.itl.nist.gov/div898/handbook/:

$$k_2 = r\sqrt{\frac{n-1}{\chi^2_{1-\gamma, n-1}}}, \tag{8.39}$$

with r such that

$$1 - p = \frac{1}{\sqrt{2\pi}}\left(\Phi\left(\frac{1}{\sqrt{n}} + r\right) - \Phi\left(\frac{1}{\sqrt{n}} - r\right)\right).$$

The function tol2s(g,p,n) has been developed for Excel and is included in the program *TolerLimits.xls*. The function uses g and p as left-tail areas. Values for k_2 are provided in Appendix Table A.9 for $1 - p = 0.9, 0.95$, and 0.99 and for $1 - \gamma = 0.95$ and 0.99. The table is also available on sheet2 of the program. The program can be used for calculating intervals for other probabilities not included in the table.

Example 8.7. A sample of 15 test specimens were used in a tensile test for strength. The mean tensile strength obtained was 290 MPa. The standard deviation of the sample was 12 MPa. Find a two-sided tolerance limit so that the interval contains 99% of the population with a confidence of 95%. A normal distribution may be assumed for the tensile strength.

Solution. We have $\bar{x} = 290$ MPa, $s = 12$ MPa, $n = 15$, $p = 0.99$, and $\gamma = 0.95$. From Appendix Table A.9 for the two-sided tolerance limits, $k_2 = 3.878$. The tolerance limits are $k_2 s = 3.878 \times 12 = 46.54$. The interval is $(290 - 46.54, 290 + 46.54) = (243.36, 336.54)$ MPa. We can say with 95% confidence that 99% of the population is contained in the interval.

Tolerance Intervals Based on the Extreme Values of Observations

The tolerance intervals discussed earlier were estimated by assuming a normal population distribution. The tolerance interval estimation presented here is based on the largest and smallest observations in the sample. This estimation is good for populations that are not necessarily normal. The sample size n for a proportion p of the population to be contained in the interval between the largest and smallest values with a confidence of γ is given by

$$n = \frac{1}{4}\left(\frac{1+p}{1-p}\right)\chi^2_{1-\gamma,4} + \frac{1}{2}. \tag{8.40}$$

For one-sided tolerance limits where the proportion p of the population is greater than the smallest sample value for the lower tolerance interval (or less than the largest sample value for the upper tolerance interval) with a confidence of γ, n is given by

$$n = \frac{\log(1-\gamma)}{\log p}. \tag{8.41}$$

Example 8.8. A sample size of 45 was used in a test to estimate a two-sided tolerance interval. Estimate the proportion of the population with a confidence of 95%.

Solution. We have $\gamma = 0.95$, $1 - \gamma = 0.05$. Using $\chi^2_{1-\gamma,4} = 9.488$ from Appendix Table A.5,

$$n = \frac{1}{4}\left(\frac{1+p}{1-p}\right)\chi^2_{1-\gamma,4} + \frac{1}{2} \Rightarrow \frac{1+p}{1-p} = \frac{4n-2}{\chi^2_{1-\gamma,4}} = \frac{4 \times 45 - 2}{9.488} = 18.761$$

$$\Rightarrow 1 + p = 18.761 - 18.761 p$$

$$p = \frac{17.761}{19.761} = 0.899.$$

Example 8.9. Calculate the sample size for a one-sided tolerance limit based on the largest and smallest values in a sample so that 90% of the population is below the largest value with a confidence of 95%.

Solution. We have $p = 0.9$ and $\gamma = 0.95$. Using Eq. (8.41),

$$n = \frac{\log(1-\gamma)}{\log p} = \frac{\log(0.05)}{\log(0.9)} = 28.5 \cong 29.$$

8.6 Summary

In this chapter we developed process capability concepts. Various capability indices were introduced, and confidence intervals for the estimated values of capability indices were discussed. Measurement system errors were presented, and gage R&R calculations included a complete example. The error propagation relation showed how probability aspects can be applied to a design problem. A discussion of prediction and tolerance intervals showed how each can be used for making meaningful decisions about tolerance limits.

The computer programs discussed in this chapter include the following:

GageR&R.xls
A9_ToleranceLimits.xls (also in the APPENDIX directory on the CD)

EXERCISE PROBLEMS

8.1. A shaft diameter is specified as 50 ± 0.03 mm. The standard deviation of the process is 0.012. A random sample of 9 shafts gives an average diameter of 50.01. Determine
   a. C_p.
   b. C_{pu}, C_{pl}, and C_{pk}.

8.2. For a statistical process control study, $\bar{x}$- and R-charts have been prepared using a sample size of 4. We have $\bar{\bar{x}} = 25.8$ and $\bar{R} = 0.6$. If the specification limits are given by 26 ± 0.8, estimate C_p, C_{pk}, and C_{pm}, C_{km}.

8.3. A runout tolerance of 40 μm full indicator movement is specified. The average runout for the process is 15 μm and the standard deviation is 2 μm. Determine C_{pk}. (Note: For runout specification, C_{pu} is C_{pk}.)

8.4. A painting contract specifies a paint drying time of 20 ± 6 minutes. Two brands of paint are evaluated. Large sample tests conducted on brands A and B gave $\bar{x}_A = 23$, $s_A = 2.5$ and $\bar{x}_B = 21$, $s_B = 3.5$, respectively. Calculate C_p, C_{pk}, and C_{pm} for the two processes. Which brand of paint would you choose?

8.5. The specification limits for a process are given as 50 ± 5. A random sample of 25 parts yields $\bar{x} = 48$ and $s = 0.9$. Determine

a. an estimate of C_{pk} and
b. a 90% confidence interval for C_{pk}.

8.6. In a GR&R study, two operators use the same gage to take 2 measurements on 8 parts. The data are shown in the following table:

	Operator A		Operator B	
Part	Trial1	Trial2	Trial1	Trial2
1	60	61	63	64
2	49	51	51	52
3	41	42	59	60
4	46	44	52	54
5	55	53	58	59
6	57	56	54	53
7	50	49	46	45
8	62	63	41	43

Perform the GR&R study. If C_p is 1.2, evaluate the precision-to-tolerance ratio (P/T).

8.7. The maximum stress S in a beam of rectangular cross section, with width b and depth h, subjected to a bending moment M is given by $S = \frac{32M}{bh^2}$. The mean values of the variables are $M = 50$ N-m, $b = 30$ mm, and $h = 50$ mm, and the approximate standard deviations are $\sigma_M = 0.5$ N-m, $\sigma_b = 1.0$ mm, and $\sigma_h = 1.5$ mm. Determine the mean value of the stress S and the standard deviation σ_S. (Note: Evaluate the stress S in MPa ($=$N/mm^2).)

8.8. For a random sample of 10 items, the mean is $\bar{x} = 32.2$ and the standard deviation is $s = 3.5$. Determine a two-sided 90% prediction interval for one future observation.

8.9. Shafts produced in a machine shop have a normal distribution. A sample of 25 shafts has a mean of $\bar{x} = 45.5$ and a standard deviation of $s = 0.75$. Determine a two-sided tolerance limit so that the interval contains 95% of the population with a confidence of 95%.

8.10. A tensile strength test was conducted on a random sample of 16 specimens. The mean value was 240 MPa and the standard deviation was 5 MPa. Determine an upper tolerance interval that contains 99% of the population with a confidence of 95%. Assume a normal distribution for tensile strength.

8.11. Voltage values measured on a power source have a normal distribution. A sample of 15 measurements gave a mean of $\bar{x} = 24.5$ V and a standard deviation of

$s = 1.5$. Determine a two-sided tolerance limit so that the interval contains 85% of the population with a confidence of 80%.

8.12. A sample size of 25 is used for the estimation of a two-sided tolerance interval. Estimate the proportion of the population in the interval with a confidence of 90%.

8.13. Determine the sample size for estimating a two-sided tolerance interval that contains 90% of the population with a confidence of 95%.

8.14. A sample size of 40 is used for the estimation of a one-sided tolerance interval. With what confidence can we say that the interval contains 90% of the population?

9

Acceptance Sampling

9.1 Introduction

Acceptance sampling is used on each lot produced to decide if the lot is acceptable or not. Attribute sampling plans are used to make these decisions based on the number of nonconforming parts in a sample. Sampling plans for variables involve the use of a sample average and standard deviation data in the decision process. Acceptance sampling is widely used in business and industry. Some quality experts have expressed the opinion that acceptance sampling may not be needed if statistical process control is properly applied; however, although process control improves quality, it does not prevent nonconforming parts from passing through the system. The probability of a Type II error is inherent in a system. Acceptance sampling is performed at the end of all operations. If a lot is rejected, we may subject that lot to total inspection. As a result of this inspection, we pass only good parts and also this study may suggest process improvements. Acceptance sampling is necessary when inspection involves destructive testing or inspection of every part is expensive and time consuming.

Acceptance sampling makes use of the ANSI/ASQC Z1.4 standard for attributes and the ANSI/ASQC Z1.9 standard for variables. These standards adapted the early work that led to the military standards MIL-STD 105E and MIL-STD 414. The ISO standards ISO 2859 and ISO 3951 (for attributes and variables, respectively), are also based on the military standards. We will discuss the basis of acceptance sampling and then visit the standards.

9.2 Acceptance Sampling for Attributes

Acceptance sampling plans are best suited for attributes. *Attributes* are quality characteristics that are either conforming or nonconforming. We may look at these as good parts or defective parts. The basic idea of a sample plan is simple. The plan is defined by the sample size n and an acceptance number c. We select a random sample of n units from the lot; then we check the number of nonconforming units in the sample. If this number of nonconforming units is c or less, we accept the lot; if the number exceeds c, we reject the lot. This is called a single sampling plan; there are others, which include double, multiple, and sequential sampling plans. We will discuss the single and double sampling plans in some detail in this chapter. A *lot* is a collection of units of a product of the same size, type, and style. In an inspection lot, the units in the lot must be homogeneous – produced by the same machines, by the same operators, using the same raw materials, and in the same time period. Lots may be small (tens to hundreds) or large (thousands to hundreds of thousands).

Single Sampling Plan

A single sampling plan is identified by two numbers; the sample size n, and the acceptance number c, as stated earlier. It is referred to as the lot acceptance sampling plan, denoted LASP(n,c). The steps in the single sampling are as follows:

1. Set sample size n and acceptance number c.
2. Select a random sample of n units.
3. Determine the number of nonconforming parts, k.
4. If $k \leq c \Rightarrow$ accept the lot; if $k > c$, reject the lot.

Let us now see how n and c are decided. These decisions are made using the operating characteristic (OC) curve – a plot with the lot fraction nonconforming, p, on the horizontal scale and the probability of acceptance, P_a, on the vertical scale. We also need the lot size, N. For a finite lot size, P_a is obtained using the hypergeometric distribution. The number of defective units in the population is taken as the integer part of pN:

$$P_a(p) = \sum_{x=0}^{c} \frac{\binom{pN}{x}\binom{N-pN}{n-x}}{\binom{N}{n}}. \tag{9.1}$$

When the lot size is large relative to the sample size, the binomial distribution is used:

$$P_a(p) = \sum_{x=0}^{c} \binom{n}{x} p^x (1-p)^{n-x}. \tag{9.2}$$

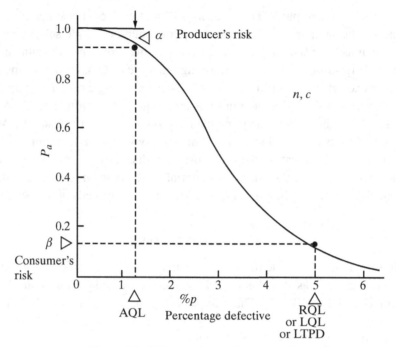

Figure 9.1. OC curve for attribute sampling

Note that P_a is 1 at $p = 0$. The OC curve starts from this point, as shown in Fig. 9.1. A percent scale is generally used for p. The OC curve based on (9.1) is called *Type A* OC curve and one based on (9.2) is called *Type B* OC curve. A Type A curve is in fact stepped because pN, the number nonconforming in the population, is taken as an integer in Eq. (9.1).

We now introduce the terminology used in the lot acceptance sampling field. *Acceptable quality level* (AQL) is the proportion or percentage of process average nonconforming items acceptable as a good lot. The *producer's risk*, α, is the probability of rejecting a good lot at the AQL. A generally used value for α is 0.05, or 5% at the AQL. The *rejectable quality level* (RQL), also called the *lot tolerance percent defective* (LTPD), is the largest percentage of defective items that will make the lot definitely unacceptable. The *consumer's risk*, β, is the probability of rejecting a bad lot at LTPD. A commonly used value for β is 0.1. If the OC curve has been chosen, the percentage defective at level β is called the *limiting quality level* (LQL). The abbreviations RQL, LTPD, and LQL are used to refer to the same point on the percentage defective line on the OC curve.

There are two distinct points marked on the OC curve shown in Fig. 9.1 – a point (AQL, $1 - \alpha$), where the producer accepts a good lot, and (RQL, β), where

9.2 Acceptance Sampling for Attributes

the lot is rejected to reduce the consumer's risk. The acceptance region (the region below the OC curve) must include (AQL, $1 - \alpha$) and exclude (RQL, β). The curve may pass through the two points in the limit. This requirement may be stated as follows:

$$P_a(p_1) \geq 1 - \alpha$$
$$P_a(p_2) \leq \beta, \quad (9.3)$$

where p_1 is the proportion at AQL and p_2 is the proportion at RQL. The inequalities in Eqs. (9.3) are easy to solve on a computer. For fixed values of n and p, the probability of acceptance, P_a, increases as c increases. Also, for fixed c and p, P_a decreases as n increases. This monotonicity with respect to n and c can be used in effectively solving Eqs. (9.3) using a computer. The vast data for the tables available in the standards were calculated as solutions to Eqs. (9.3).

We implemented a simple algorithm for the solution in *SP1_ATR.xls*. The input parameters for the program are p_1, α, p_2, and β; sheet1 gives a visible solution[1] based on the binomial distribution. The OC curve is plotted for some values of n and c seen on the screen. The curve can be manipulated interactively using the spin buttons that are linked to the cells where n and c can be changed. The curve is moved until it is above ($p_1, 1 - \alpha$) and below (p_2, β). This visible approach is both intuitive and interesting. Sheet2 contains the visible solution using the hypergeometric distribution. The curve is discontinuous for small batch sizes. Sheet2 has an additional spin button for changing the batch size. Sheet3 gives various possible solutions of Eqs. (9.3) based on the binomial distribution. Sheet4 provides solutions based on the hypergeometric distribution, which is recommended for batch sizes smaller than 800. The output gives various values of n and c that satisfy Eqs. (9.3).

Example 9.1. A manufacturer plans to introduce a single sampling plan for incoming material. The expected AQL is 2%, and the required LTPD is 8%. The producer's risk of 0.05 and the consumer's risk of 0.1 are to be used. The lot size is 2000. Recommend a solution.

Solution. We have $p_1 = 0.02$, $\alpha = 0.05$, $p_2 = 0.08$, and $\beta = 0.1$ for the problem. We use the binomial distribution approach. We input the data into sheet3 of *SP1_ATR.xls* and run it. The program gives solutions from $n = 98$ to 196, and some samples have the possibility of a choice of multiple c-values.

---

[1] S.H.K. Ng, Studying the Effect of the Parameters of an Attribute Acceptance Sampling Plan on Its Operating Characteristic Curve: A Visual Approach, *Journal of Statistics Education*, v11, n3 (2003).

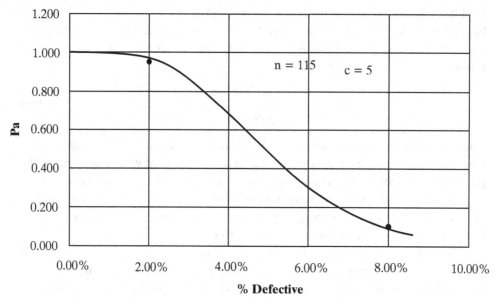

Figure 9.2. OC curve

The first four solutions are $n = 98$, $c = 4$; $n = 99$, $c = 4$; $n = 114$, $c = 5$; and $n = 115$, $c = 5$. The first solution with multiple values for c is $n = 130$, where $c = 5$ or $c = 6$. The OC curve for (98, 4) is shown in Fig. 9.2. The visible approach from sheet1 shows that at (97, 4), the condition at (p_2, β) is not satisfied and at (100, 4) the condition at (p_1, α) is not satisfied. The first set that satisfies the requirements at both points is (98, 4). For the present our choice is (98, 4). We will look at other parameters, such as the average outgoing quality limit, for various choices before making a final decision.

Average Outgoing Quality

In calculating the average outgoing quality, it is assumed that a rectifying inspection is applied to all rejected lots. The scheme for the rectifying inspection is showed in Fig. 9.3. In a *rectifying inspection*, 100% inspection (*screening*) is performed on a rejected lot and all defective parts are replaced and returned to the pipeline. Thus, the fraction of nonconforming parts moving through the system is the average fraction defective, called the *average outgoing quality* (AOQ). As shown in Fig. 9.3, we have

$$\text{AOQ} = \frac{P_a p (N - n)}{N}. \tag{9.4}$$

We know that P_a is a function of p. If AOQ is calculated over the range of p-values, the curve shows a peak and has a maximum. This maximum value is called the

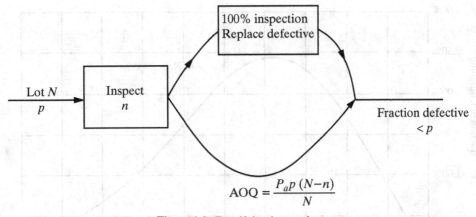

Figure 9.3. Rectifying inspection

average outgoing quality limit (AOQL). The *average total inspection* (ATI) is the average number of units per lot subjected to inspection. In each lot we inspect n units, and $N - n$ units are inspected with a probability of $1 - P_a$, as shown here and in Fig. 9.3:

$$\text{ATI} = n + (1 - P_a)(N - n). \quad (9.5)$$

Example 9.2. For the single sampling plan solution in Example 9.1, plot the AOQ curve and find the AOQL. Also calculate the ATI value at the process average (AQL).

Solution. For the problem with AQL $= 0.02$, $\alpha = 0.05$, RQL $= 0.08$, and $\beta = 0.1$, we found the first solution as $n = 98$, $c = 4$. The AOQ curve plotted in sheet1 of *SP1_ATR.xls* is given in Fig. 9.4. For this solution, AOQL $= 2.343\%$.

$$\text{ATI at } p = 0.02 \text{ is ATI} = n + (1 - P_a)(N - n)$$
$$= 98 + (1 - 0.953)(2000 - 98) = 187.4.$$

For the two other solutions, the AOQL values have been calculated: for $n = 99$ and $c = 4$, AOQL $= 2.315$; for $n = 114$ and $c = 5$, AOQL $= 2.470$. Therefore, (98, 4) is a good choice.

Dodge–Romig Plans for Single Sampling

H.F. Dodge and H.G. Romig of Bell Laboratories introduced sampling plan tables in 1959; here we discuss single sampling plan tables. The basic idea of the tables follows the single sampling plan theory discussed earlier. The tables are provided for

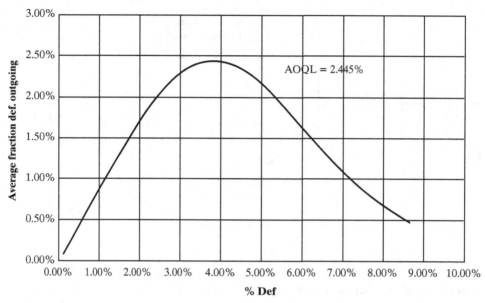

Figure 9.4. AOQ curve

batch sizes in 19 steps (1–30, 31–50, 51–100, 101–200, 201–300,..., 50001–100000) and AQL in six ranges (0–0.05%, 0.06–0.50%, 0.51–1.00%, 1.01–1.50%, 1.51–2.00%, 2.01–2.50%). At mid-value of the AQL, Dodge and Romig use $\alpha = 0.05$. In the first table they give n- and c-values at RQL of 5% and $\beta = 0.1$. The table also gives the values of AOQL for the suggested plan. The program *DodgeRomig.xls* provides the spin button approach to finding values of n and c using the Dodge–Romig approach.

Example 9.3. The Dodge–Romig table gives $n = 130, c = 3$ for a batch size of 1001–2000 and AQL of 0.51–1.00%, with an AOQL of 1.4. Check these data using batch sizes of 1,001, 1,500, and 2,000.

Solution. The results are checked using the program *DodgeRomig.xls*.

For a batch size of 1,001, $\beta = 10.58\%$ at LQL 5%, AOQL = 1.31%, and $\alpha = 1.45\%$ at AQL = 0.71%.

For a batch size of 1,500, $\beta = 10.58\%$ at LQL 5%, AOQL = 1.37%, and $\alpha = 1.45\%$ at AQL = 0.71%.

For a batch size of 2,000, $\beta = 10.58\%$ at LQL 5%, AOQL = 1.4%, and $\alpha = 1.45\%$ at AQL = 0.71%.

Only the AOQL value seems to change slightly. The producer's risk α is far below the 5% level.

9.2 Acceptance Sampling for Attributes

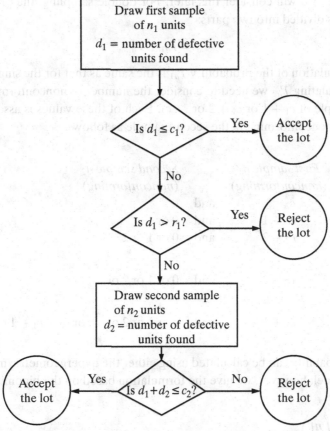

Figure 9.5. Double sampling plan

Dodge and Romig also provide another table where the (n, c)-values are provided for fixed AOQL of 3%. In this case the LQL is given in the table. These values range from 6 to 19%.

Double Sampling Plan

In the double sampling plan, first a sample of size n_1 is drawn. If the number of defective units in the sample is c_1 or fewer, the lot is accepted; if the number is greater than the rejection number r_1, the lot is rejected. Otherwise, a second sample of size n_2 is drawn. The number of defective units in the second sample is added to the previous number and if it is c_2 or fewer, the lot is accepted; it is rejected otherwise. The double sampling plan is shown in Fig. 9.5. We note that $c_1 < r_1 \leq c_2$. It is a usual practice to use $r_1 = c_2$. Double sampling may reduce the total number inspected, but we will consider this later. For double sampling, the probability of acceptance is divided into two parts:

$$P_a = P_{a1} + P_{a2} \tag{9.6}$$

The calculation of the probability P_{a1} is the same as that for the single sampling plan. In calculating P_{a2} we need to consider the number of nonconforming parts in the first sample of $c_1 + 1$ or $c_1 + 2$ or $\ldots c_2$. Each of these values is associated with the number nonconforming in the second sample as follows.

First sample n_1 (nonconforming)		Second sample n_2 (nonconforming)
c_2	and	0
$c_2 - 1$	and	0 or 1
$c_2 - 2$	and	0 or 1 or 2
$\vdots$		$\vdots$
$c_2 - i$	and	0 or 1 or 2 or $\ldots$ or i
$\vdots$		$\vdots$
$c_1 + 1$	and	0 or 1 or 2 or $\ldots$ or $c_2 - c_1 - 1$

The probability can be calculated using either the hypergeometric model or the binomial model; however, we give the formulation based on the binomial probability model:

$$\begin{aligned} P_{a1} &= \binom{n_1}{c_1} p^{c_1} (1-p)^{n_1 - c_1} \\ P_{a2} &= \sum_{i=0}^{c_2-c_1-1} \binom{n_1}{c_2 - i} p^{c_2 - i} (1-p)^{n_1 - c_2 + i} \sum_{j=0}^{i} \binom{n_2}{j} p^j (1-p)^{n_2 - j}. \end{aligned} \tag{9.7}$$

Denoting P_{r1} as the probability at which the lot will not be rejected at the first sample, we have

$$P_{r1} = \binom{n_1}{c_2} p^{c_2} (1-p)^{n_1 - c_2}. \tag{9.8}$$

The final requirement is that P_a should satisfy Eqs. (9.3). However, at the first sample stage, we need to satisfy the condition at the consumer's risk – the second part of Eq. (9.3) – and the probability P_{r1} should be higher than $1 - \alpha$ at AQL. The OC curves for the double sample plan are shown in Fig. 9.6. Equations (9.6)–(9.8) have been implemented in the Excel program *SP2_ATR.xls*. The screen shows the

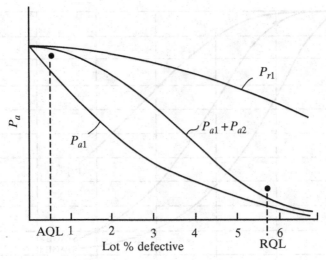

Figure 9.6. OC curve – double sampling plan

curves for P_{a1}, $P_{a1} + P_{a2}$, the two points at AQL and RQL, and the probability P_{r1}. The curves can be interactively manipulated using the spin button feature.

To compare with a single acceptance plan, we evaluate the average sample number for the various probabilities. Sample n_1 is always taken. Sample n_2 is taken only when no decision is made for the first sample. The probability of taking the second sample is $P_{r1} - P_{a1}$. The *average sample number* (ASN) at a specified percent defective lot is given by

$$\text{ASN} = n_1 + n_2(P_{r1} - P_{a1}). \tag{9.9}$$

The program *SP2_ATR.xls* also shows the plot of the ASN curve.

The ASN value for a single sampling plan is n_1. Since n_1 will be smaller than the sample size of the single sampling plan, ASN will be lower for some probability ranges. The ASN in Eq. (9.9) is calculated under the assumption that inspection of all n_2 components is carried out to find d_2 first (see Fig. 9.5), and then the decision is made. *Curtailment* may be applied at the second sample. Curtailment is the sample evaluation procedure where the inspection is stopped as soon as the number of nonconformities exceeds c_2. When curtailment is applied, d_2 does not exceed $c_2 + 1$. The ASN for the double sample plan is significantly lower when curtailment is used.

For the rectifying inspection, the AOQ is given by

$$\text{AOQ} = \frac{pP_{a1}(N - n_1) + pP_{a2}(N - n_1 - n_2)}{N}. \tag{9.10}$$

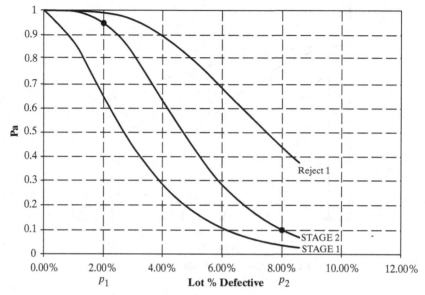

Figure 9.7a. OC curve

The ATI is given by

$$\text{ATI} = n_1 P_{a1} + (n_1 + n_2) P_{a2} + (1 - P_{a1} - P_{a2}) N. \tag{9.11}$$

Example 9.4. Select a double sampling plan for $p_1 = 0.02$, $\alpha = 0.05$, $p_2 = 0.08$, and $\beta = 0.1$ (the conditions in Examples 9.1 and 9.2). Plot the OC curves and the ASN curve.

Solution. By entering the data into the program *SP2_ATR.xls*, and by using the spin buttons, we obtain the double sample plan $n_1 = 62$, $c_1 = 1$, $n_2 = 39$, $c_2 = 4$. This sample plan satisfies the producer's and consumer's risk points. The OC curves and the ASN curve are shown in Figs. 9.7a and 9.7b, respectively. The ASN curve has a maximum of 87 at $p = 4.7\%$. When compared to the single sample plan solution of 98, 4 (ASN = 98) in Example 9.1, the average sample number is lower in this case.

Multiple and Sequential Sampling Plans

In a multiple sampling plan, more than two samples are taken. For each sample there is an acceptance number and a rejection number. The acceptance and rejection numbers of each successive sample are larger than the corresponding previous ones

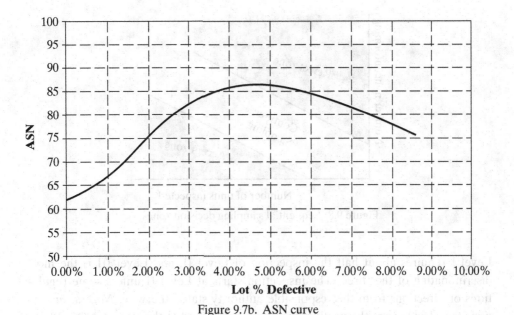

Figure 9.7b. ASN curve

by at least 1. The operation is similar to the double sample plan: in any sample the number of defective units obtained is added to the total previous defective units. If this number is less than or equal to the acceptance number for that sample, the lot is accepted. If the number is greater than the rejection number for that sample, it is rejected. Otherwise, the next sample is taken.

A sequential sampling plan combines ideas from double and multiple sampling plans. A sequence of samples is taken from the lot. The cumulative number of nonconforming units is plotted against the number of units inspected as shown in Fig. 9.8. The inspection operation is carried out until a decision is made for the lot.

Standard Sampling Plans for Attributes

Standard sampling plans were introduced in the 1940s. MIL-STD 105E, which was issued in 1989, is widely used. Both ANSI/ASQC Z1.4 and ISO 2859 retain the framework of MIL-STD 105E. Our current discussion is based on MIL-STD 105E.

AQL is the primary focus of the military standard. AQL values are 0.1% or less for critical, 1% for major, and 2.5% for minor defects. The sample size is related to the lot size, which is given in 15 steps (2–8, 9–15, 16–25,..., 500001 and over). There are three general levels of inspection – Level II is designated as normal,

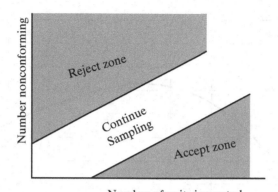

Figure 9.8. Sequential sampling decision zones

Level I requires about half the inspection of Level II, and Level III is the most discriminating of the three. The inspection starts at Level II unless some regulations or directions from the responsible authority state otherwise. *Normal inspection* is used as long as the product quality is maintained at the chosen AQL. When the recent quality history shows deterioration, *tightened inspection* is used. *Reduced inspection* is used when the quality from the producer has been outstanding over a sustained period.

MIL-STD 105E is now available online at http://www.sqconline.com. The main selection consists of the batch size, AQL, the inspection level, and the type of inspection, as shown in Fig. 9.9. Both single sampling and double sampling procedures are recommended by the standard. The table values for the input information from Fig. 9.9 are shown in Fig. 9.10. OC curves and ASN curves are shown on the screen.

Enter your process parameters:		
Batch size (N):	2 to 8	The number of items in the batch. more info...
AOL:	1.0%	The Acceptable Quality Level. more info...
Inspection Level:	II	Determines the discrimination power of the plan. more info...
Type of inspection:	Normal	Depends on the quality history. more info...
	Submit	

Figure 9.9. Input screen

For a lot of **2 to 8** items, and **AQL = 1%, normal** inspection plans are:

The *Single* sampling procedure is:
Sample **13*** items. If the number of non-conforming items is **0** --> accept the lot. **1 or more** --> reject the lot.

In this case, consider the **Single** sampling plan, or use the following plan:

The *Double* sampling procedure is:	
Step 1:	Sample **32*** items. If the number of non-conforming items is **0** --> accept the dot. **2 or more** --> reject the lot. Otherwise, continue to step 2.
Step 2:	Sample **32*** additional items. If the total number of non-conforming items is **1 or less** --> accept the lot. **2 or more** --> reject the lot.

Figure 9.10. Output screen

Output from the standard tables can be attempted in the programs *SP1_ATR.xls* and *SP2_ATR.xls* to make minor modifications to meet other requirements.

9.3 Sampling Plans for Variables

Variable sampling plans require that the quality characteristic has a normal distribution. Variable sampling plans are used for units that are submitted in lots, for example in bags, boxes, drums, or bins. There is a lot more information known about variables than attributes since measurement data may be available. Let us discuss a sampling plan for variables by considering the single specification limit.

Single Specification Limit – Known σ

Consider the case of the lower single specification limit with σ known to fix our ideas. We have the lower specification limit (LSL), denoted as L. We accept a proportion p_1 of out-of-limit units with a probability of $1 - \alpha$ (producer's risk of α) and accept a proportion p_2 ($p_2 > p_1$) of out-of-limit units with a probability of β (consumers' risk). As discussed in the attribute sampling section, p_1 may be taken

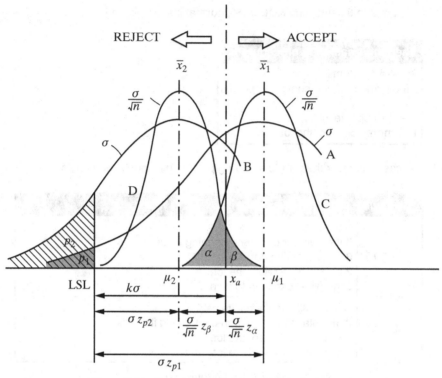

Figure 9.11. One-sided variable sampling (σ known)

as the AQL and p_2 as the RQL. These conditions are illustrated in Fig. 9.11. The normal curve A, $N(\mu_1, \sigma^2)$, shows a proportion p_1 nonconforming with the mean located at μ_1. The normal curve B, $N(\mu_2, \sigma^2)$, has a proportion p_2 nonconforming with the mean located at μ_2. We have

$$\mu_1 = L + z_{p_1}\sigma$$
$$\mu_2 = L + z_{p_2}\sigma. \tag{9.12}$$

If $\bar{x}_1 \, (= \mu_1)$ and $\bar{x}_2 \, (= \mu_2)$ are the means of two samples of n parts each, then we need to set the level of significance, α, as the probability of rejecting the batch with $\bar{x}_1$ and set β as the probability of accepting the batch with $\bar{x}_2$. The curves with $N(\mu_1, \sigma^2/n)$ and $N(\mu_2, \sigma^2/n)$ are also shown in Fig. 9.11. We find the position of such a point x_a located at $k\sigma$ from L.

From Fig. 9.11 we can write

$$z_{p_1}\sigma - z_{p_2}\sigma = \frac{z_\alpha}{\sqrt{n}}\sigma + \frac{z_\beta}{\sqrt{n}}\sigma$$
$$k\sigma = z_{p_1}\sigma - \frac{z_\alpha}{\sqrt{n}}\sigma.$$

9.3 Sampling Plans for Variables

On canceling σ and rearranging the terms, we get

$$n = \left(\frac{z_\alpha + z_\beta}{z_{p_1} - z_{p_2}}\right)^2$$
$$k = z_{p_1} - \frac{z_\alpha}{\sqrt{n}}. \tag{9.13}$$

We use n and k to define a sample plan. We take a sample of size n and find the mean value of the variable. We then convert it to the standard variable, z. If the standard variable is less than k, the lot is rejected; the lot is accepted otherwise.

Variable Sample Plan for Single Specification, σ Known (Form 1 or k Method)

Step 1. Calculate n using the first of Eqs. (9.13) and round it to the nearest integer.

Step 2. Determine k using the second of Eqs. (9.13).

Step 3. Take a sample of n parts and find $\bar{x}$ and $z = \frac{\bar{x}-L}{\sigma}$ for the lower specification and $z = \frac{U-\bar{x}}{\sigma}$ for the upper specification.

Step 4. Accept the lot if $z \geq k$; reject it otherwise.

Form 2 or M-Method

In the form 2 or M-method, we evaluate $Q = z\sqrt{\frac{n}{n-1}}$, which is an unbiased estimate, and $M = \Phi(-k\sqrt{\frac{n}{n-1}})$, $\hat{p} = \Phi(-Q)$. The decision rule is to reject if $\hat{p} > M$ and to accept otherwise.

Example 9.5. The upper specification limit for sulfur content in an alloy is 0.8%. The manufacturer wants to accept 2% nonconformities with a probability of 5%, and 12% nonconformities with a probability of 8%. The lots received are known to have a standard deviation of 0.15%. Develop a sampling plan. If a sample taken according to the plan has a mean value of 0.7% sulfur content, is the lot acceptable?

Solution. We have $p_1 = 0.02$, $\alpha = 0.05$, $p_2 = 0.12$, and $\beta = 0.08$. From Appendix Table A.4, we get $z_{p_1} = 2.054$, $z_\alpha = 1.645$, $z_{p_2} = 1.175$, $z_\beta = 1.405$, $U = 0.8\%$, and $\delta = 0.15\%$.

From Eqs. (9.13),

$$n = \left(\frac{z_\alpha + z_\beta}{z_{p_1} - z_{p_2}}\right)^2 = \left(\frac{1.645 + 1.405}{2.054 - 1.175}\right)^2 = 12.05 \approx 12$$

$$k = z_{p_1} - \frac{z_\alpha}{\sqrt{n}} = 2.054 - \frac{1.645}{\sqrt{12}} = 1.579.$$

If a sample of 12 units has an average of 0.7%,

$$z = \frac{U - \bar{x}}{\sigma} = \frac{0.8 - 0.7}{0.15} = 0.667,$$

by the form 1 approach; since $z < k$ the lot is rejected.

For the form 2 approach,

$$Q = z\sqrt{\frac{n}{n-1}} = 0.667\sqrt{\frac{12}{12-1}} = 0.697,$$

$$k\sqrt{\frac{n}{n-1}} = 1.579\sqrt{\frac{12}{12-1}} = 1.65,$$

$M = F(-1.65) = 0.0496$, and $\hat{p} = \Phi(-Q) = \Phi(-0.697) = 0.243$. Since $\hat{p} > M$, the decision is to reject the lot.

The plan for a known value of σ is implemented in sheet1 of the program *SP_VAR.xls*. The data are entered in the top section and the single specification limit section, and the sample size is manipulated by using the spin button until the curve satisfies the LQL point. The results confirm the previous calculations. If the sample size is chosen using a different basis, the value of n may be entered directly. In this case the LQL point is not monitored.

Double Specification Limit – Known σ

For the double specification limit the first step is to check if

$$\frac{U - L}{\sigma} < 2z_{p_1/2}. \tag{9.14}$$

If this condition is satisfied, the incoming quality is worse than AQL and the lot is rejected without even drawing a sample. But if it passes this test, the following procedure is used:

$$Q_L = \frac{\bar{x} - L}{\sigma}\sqrt{\frac{n}{n-1}}$$

$$Q_U = \frac{U - \bar{x}}{\sigma}\sqrt{\frac{n}{n-1}}$$

and $\hat{p} = \Phi(-Q_L) + \Phi(-Q_U)$ are evaluated. The entire p_1 is permitted at one extreme and $k = z_{p_1} - \frac{z_\alpha}{\sqrt{n}}$. We then define the two values $z_1 = \sqrt{\frac{n}{n-1}}k$ and

9.3 Sampling Plans for Variables

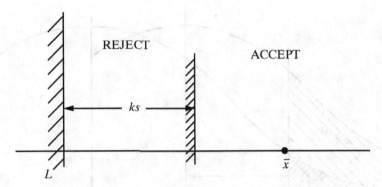

Figure 9.12. Acceptance criterion, (σ unknown)

$z_2 = \sqrt{\frac{n}{n-1}} \left(\frac{U-L}{\sigma} - k\right)$. The M-value is obtained as $M = \Phi(-z_1) + \Phi(-z_2)$. If $\hat{p} \leq M$, the lot is accepted.

Variable Sampling Plan for Single Specification – Unknown σ

The variable sampling plan for unknown σ, presented here, is based on the work of Lieberman and Resnikoff.[2] When σ is unknown, the sample standard deviation s is used, and the criterion for acceptance of the lower specification limit L is $\bar{x} \geq L + ks$ or $\frac{\bar{x}-L}{s} \geq k$, as shown in Fig. 9.12. The OC curve plots the probability of acceptance, P_p, for each level of the proportion of nonconforming parts, p. If μ is the mean for the batch of proportion p nonconforming parts, the probability of acceptance is given by

$$P_p = \Pr\left[\frac{\bar{x}-L}{s} \geq k\right]$$

$$= \Pr\left[\frac{\frac{\sqrt{n}(\bar{x}-\mu)}{\sigma} + \frac{\sqrt{n}(\mu-L)}{\sigma}}{s/\sigma} \geq \sqrt{n}k\right] \quad (9.15)$$

$$= \Pr[t_{n-1}(\delta) \geq \sqrt{n}k],$$

where $t_{n-1}(\delta)$ is the noncentral t-distribution with noncentrality parameter δ. We use the mean μ for the good-quality parts p_1; thus, the noncentrality parameter is given by

$$\delta = \frac{\sqrt{n}(\mu - L)}{\sigma} = \sqrt{n}z_p. \quad (9.16)$$

---

[2] G.J. Lieberman and G.J. Resnikoff, Sampling Plans for Inspection by Variables, *Journal of the American Statistical Association*, v50, n250 (1955).

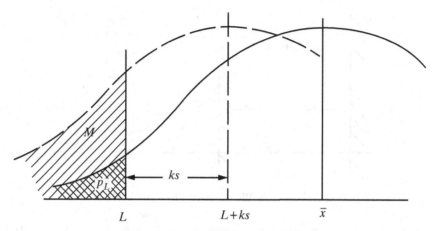

Figure 9.13. p_L and M for σ unknown

Since the probability of acceptance for AQL of p_1 is $1 - \alpha$, k may be obtained using the inverse of the noncentral distribution as

$$k = \text{tncinv}(1 - \alpha, n-1, \delta). \tag{9.17}$$

The points for the OC curve are constructed using this k-value in Eq. (9.17). These relations are implemented in sheet2 of *SP_VAR.xls*. The curve passes through the point at AQL and it can be adjusted by varying n to pass through the LQL point with the probability of acceptance set at β.

Lieberman and Resnikoff derive the following expression for estimating the proportion of nonconforming parts, p_L, for the sample mean $\bar{x}$:

$$p_L = I_v\left(\frac{n}{2} - 1, \frac{n}{2} - 1\right)$$
$$v = \max\left[0, \frac{1}{2} - \frac{\sqrt{n}}{n-1}\frac{\bar{x} - L}{s}\right], \tag{9.18}$$

where $I_x(a, b)$ is the beta distribution. The beta distribution is evaluated in Excel using the function BETADIST(x, a, b). The critical value M is calculated using Eqs. (9.18) when $\frac{\bar{x}-L}{s} = k$. Values for p_L and M are shown in Fig. 9.13. The decision is to accept when $p_L \leq M$.

For upper specification U, $\frac{\bar{x}-L}{s}$ is replaced by $\frac{U-\bar{x}}{s}$ in Eqs. (9.15) and (9.17). Thus, p_U is evaluated using Eqs. (9.18).

For the double specification case, both p_L and p_U are evaluated using the previous expressions: M_L is evaluated by setting $\frac{\bar{x}-L}{s} = k$ in Eqs. (9.18), and M_U

is evaluated by replacing $\frac{\bar{x}-L}{s}$ by $\frac{U-L}{s} - k$ in Eqs. (9.18). M is evaluated using $M = M_L + M_U$. The batch is accepted when $p_L + p_U \leq M$. The case when σ is unknown is implemented in sheet2 of *SP_VAR.xls*. The implementation of the previous equations ensures that the OC curve passes through $(p_1, 1-\alpha)$. The curve may be manipulated by changing n to make it pass through (p_2, β). Wallis[3] has given the following expressions for calculating k and n, which give remarkably good values:

$$k = \frac{z_\alpha z_{p_2} + z_\beta z_{p_1}}{z_\alpha + z_\beta}$$

$$n = \left(1 + \frac{k^2}{2}\right) \left(\frac{z_\alpha + z_\beta}{z_{p_1} - z_{p_2}}\right)^2.$$

(9.19)

The preceding formulas are based on the assumption that $\bar{x} \pm ks$ is approximately normally distributed as $N(\mu \pm k\sigma, \frac{\sigma^2}{n}(1+\frac{k^2}{2}))$.

Example 9.6. The electrical resistance specification of certain components is $630 \pm 25\,\Omega$. If 2.5% of the components are outside the limits, the manufacturer wishes to accept with a probability of 0.95; if 12% are outside the limits, the probability of acceptance is set at 0.1. Determine the sample size and the k-value for unknown σ.

Solution. The data $p_1 = 0.025$, $\alpha = 0.05$, $p_2 = 0.08$, and $\beta = 0.1$ are entered into sheet2 of *SP_VAR.xls* and the spin button is used to manipulate n. The curve passes through the two points for $n = 32$, and the corresponding value for k is 1.536.

Using the Wallis formula with $z_\alpha = 1.6449$, $z_{p_1} = 1.96$, $z_\beta = 1.282$, and $z_{p_2} = 1.175$ in Eqs. (9.19), we get $n = 30$ and $k = 1.519$. The calculations using the noncentral t-distribution give more accurate results.

Example 9.7. In Example 9.6, if a sample of 32 has a mean $\bar{x} = 640\,\Omega$, and the sample standard deviation is $s = 12.45\,\Omega$, draw a conclusion of whether the batch is acceptable.

---

[3] W.A. Wallis, Use of Variable in Acceptance Inspection for Percent Defective, in *Selected Techniques of Statistical Analysis and Industrial Research and Production and Management Engineering*, McGraw-Hill, New York (1947) pp. 3–93.

Solution. The output of *SP_VAR.xls* is as follows:

```
DOUBLE SPECIFICATION LIMIT
s              32       <====INPUT
L              605      <====INPUT
U              655      <====INPUT
Xbar           640      <====INPUT

(Xbar-L)/s     1.0938     u      0.4002
pL             0.1367
(U-Xbar)/s     0.4688     v      0.4572

pU             0.3208
pL+pU          0.4575

k1             1.5363     u      0.3598     Mu     0.0601
k2             0.0262     v      0.4976     Mv     0.4896

M              0.5497

Decision       ACCEPT
```

Since $p_L + p_U < M$, the decision is to accept the lot.

Standard sampling plans have been developed for variables. The standards are based on the AQL value. The tables provide sampling plans for cases where σ is either known or unknown. Plans for one-sided specification limits include both form 1 and form 2 approaches. For two-sided specification limits, only the form 2 approach is used and the plan suggested by MIL-STD 414 in the following section is recommended.

Standard Sampling Plans for Variables

MIL-STD 414 covers both one-sided and two-sided specifications. The standard also takes σ-known and σ-unknown cases into account. ANSI/ASQC Z1.9 and ISO 3951 have similar approaches.

AQL choices are from 0.04 to 15% in about 14 discrete choices. The lot sizes range over 17 size categories (from 3 to 8,..., 550001 and over). There are five inspection levels, I through V; level IV is designated as normal.

MIL-STD 414 is now available online at http://www.sqconline.com. The results for the M method are obtained by using interactive screens. The main selection consists of the batch size, AQL, inspection level, and type of inspection as shown in Fig. 9.14. The table values for the sample size and the M-value are obtained on the second screen, as shown in Fig. 9.15. We then provide the sample average, the calculated sample standard deviation, and the lower and upper specification

Enter your process parameters:		
Variability	⊙ Unknown ○ Known	Select "Known" if it is given or you know the variability from historical data. Select "Unknown" if you plan to estimate the variability from the sample.
Batch/lot size (N):	3 to 8 ▼	The number of items in the batch (lot). More info...
AQL:	1.0% ▼	The Acceptable Quality Level. More info... What to do if my AQL is different?
Inspection Level:	IV ▼	Determines the discrimination power of the plan. More info...
Type of inspection:	Normal ▼	Depends on the quality history. More info...
	Submit	

Figure 9.14. Input screen 1 (variable sampling plan)

For a lot of **3** to **8** items, and **AOL = 1.0%**, with inspection level IV, the **Normal** inspection plan is:

Sample **4*** items.
If the estimated percent of non-conforming (defective) items is **1.53% or less** – > accept the lot. Otherwise, reject it.

To estimate the percent of non-conforming items is your process, take a sample of size **4** and enter the value into the next table.

To estimate your process % non-conforming (defectives) enter:		
Sample Average (x̄):	25.6	The average of the 4 measurements
Process Standard Deviation (s):	0.4	The standard deviation of the 4 measurements
Lower Specification Limit:	23.4	The smallest value for your measurement that is considered acceptable. Leave blank if there is no lower limit.
Upper Specification Limit:	26.8	The largest value for your measurement that is considered acceptable. Leave blank if there is no upper limit.
	Submit	

Figure 9.15. Input screen 2 (variable sampling plan)

203

Here is a summary of your process parameters, the sampling plan, and the sample statistics:

Information Summary	
Process Parameters	Lot size = 3 to 8 AQL = 1.0% Lower Spec. Limit (L) = 23.4 Upper Spec. Limit (U) = 26.8 Standard Dev. (s) = 0.4 Variability = unknown
Sampling Plan	Sample size (n) = 4 Non-conforming limit (M) = 1.53% Inspection Type = Normal Inspection Inspection Level = IV
Sample statistics	Sample average ($\bar{x}$) = 26.6

Based on these values, the estimated proportion of non-conforming (defective) items in your process is **0.00%**.

Since this precent is not higher than 1.53%, **the lot should be accepted.**

The following table details the calculations involved in computing the above results, as detailed in MIL-STD-**414**. It is provided for reference only.

Calculation step for unknown variability		
Information needed	Value	Explaination
Sample size code letter	B	See Table A-2 in MIL-STD-414
Upper Quality Index: Q_U	3.00	$(U-\bar{x})v/\sigma$
Lower Quality Index: Q_L	5.50	$(\bar{x}-L)v/\sigma$
Est. of lot % defective above Upper Spec: P_U	0.00%	See Table B-5 in MIL-STD-414
Est of lot % defective below Lower Spec: P_L	0.00%	See Table B-5 in MIL-STD-414
Total Est. % defective in lot	0.00%	0.00% + 0.00%
Max. allowable % defective: M	1.53%	See Table B-3/4 in MIL-STD-414

Figure 9.16. Output screen for variable sampling plan

limits. The summary information and the calculation steps are provided in the ensuing screens, as shown in Fig. 9.16. The final output includes p_L and p_U, which are the percentage nonconforming below the lower specification limit, and above the upper specification limit, respectively. If $p_L + p_U \leq M$, the lot is accepted.

9.4 Summary

The chapter dealt with acceptance sampling for attributes, which is widely used. Steps for the design of a single sampling plan and a double sampling plan have been discussed in detail. The programs *SP1_ATR.xls* and *SP2_ATR.xls* may be used for interactive development of sampling plans. The concepts of AQL and RQL have been extended to sampling problems involving variables. An interactive solution using web-based software has been presented.

The computer programs discussed in this chapter include the following:

SP1_ATR.xls
DodgeRomig.xls
SP2_ATR.xls
SP_VAR.xls

(Note: For using www.sqconline.com, click the link from the book Web site, www.cambridge.org/9780521151221.)

EXERCISE PROBLEMS

9.1. Plot the OC curve for the single sampling plan for $n = 60, c = 2$. What is the probability of acceptance at an AQL of 6%?

9.2. For the data in Problem 9.1, plot the OC curve using a hypergeometric distribution for a finite population of $N = 150$. Determine the probability of acceptance at an AQL of 6%.

9.3. A sampling plan is to be chosen for a producer's risk of 0.06 at AQL = 0.8% and consumer's risk of 0.1 at LQL = 6% nonconforming. Find a single sampling plan that meets these requirements.

9.4. For the double sampling plan $n_1 = 50, c_1 = 1, n_2 = 75, c_2 = 4$, determine the probability of accepting the lot at 5% nonconforming. What is the probability of rejecting 5% nonconforming on the first sample?

9.5. Design a double sampling plan that accepts batches of 1.5% nonconforming with a probability of 0.95 and accepts batches of 7.5% nonconforming with a probability of 0.1. Find the average sample number for the plan if the lot is 4% nonconforming.

9.6. Lots of size 5,000 are submitted for inspection. The AQL is specified as 0.5% and the inspection level is II. Find the single sampling plan using MIL-STD 105E.

9.7. A fabric manufacturer tests yarn for acceptance. The specification limit for the breaking strength is 20 g. The breaking strength is normally distributed with a standard deviation of 2 g. Develop a variable sampling plan that accepts 2.5% nonconforming with a probability of 0.95 and accepts lots with 7% nonconforming with a probability of 0.1.

9.8. Find a two-sided sampling plan using MIL-STD 414 for a lot size of 800, AQL = 2.5%, variability unknown, normal inspection at level III. The sample average is 24.3 and the sample standard deviation is 0.5. The specification limits are 24.1 ± 1.3.

10

Experimental Design

10.1 Introduction

An *experiment* is a test under controlled conditions that is made to demonstrate a known truth, examine the validity of a hypothesis, or determine the efficacy of something previously untried.[1] Scientists and engineers perform experiments. Engineers create new products and devices that are useful to society. Product development involves building a prototype and testing. Manufacturing the product involves the choice of different processes and making decisions to fix the variables at optimum levels. A well-designed experiment is an important step in the development of a robust design or in the improvement of an existing product or process. The foundation for experimental design was laid in the 1920s by Sir Ronald Fisher at the Rothamsted Statistical Laboratory. He did extensive work in the field of agriculture, where he studied the effect of environmental variables (soil qualities, drainage gradients, etc.) on the response variable (crop yields). He formulated the analysis of variance (ANOVA) techniques which form the basis for experimental design.

10.2 Basic Concepts

We first introduce the basic terminology used in formulating, designing, conducting, and analyzing experiments.

Experimental design is used to study the effect of one or more factors on a response variable. A *response variable* is a characteristic of an experimental unit that is measured to judge its performance, and a *factor* is an independent variable which may have an effect on the response variable. Factor values are referred to

---

[1] *American Heritage Dictionary of the English Language*, 4 Ed., Houghton Mifflin Co, Boston, MA, 2006.

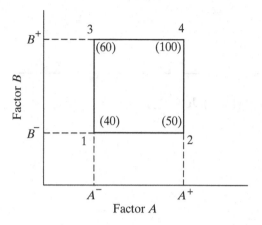

Figure 10.1. Two-factor factorial experiment

as *levels* of the factor. Factors are *controllable* when the levels can be adjusted or *uncontrollable* when the variation may be due to chance causes. As an example, in a turning operation on a lathe, the response variable may be the surface finish, and the controllable factors may be the cutting speed, feed, depth of cut, coolant flow, and the tool geometry. The uncontrollable factors may be the initial surface condition of the workpiece, the material variation, and so on.

The *effect* of a factor is the change in response value produced by a change in the level of that factor. A *treatment* is a certain combination of factor levels whose effect on the response variable is studied. An *interaction* between two factors is said to exist when a change in one factor produces a different change in the response variable at two levels of another factor. A *factorial design* is one in which every level of a factor is paired with every level of another factor in the group for each trial. Consider an experiment with two variables, A and B, each with two levels, A^-, A^+, and B^-, B^+; each replication of the experiment consists of $2 \times 2 = 4$ combinations, as shown in Fig. 10.1. The four combinations are represented by the four corners of the square in the figure. The observations are taken in a random order to remove any bias from the experiment, which is called *randomization*. We give a number to each combination and then generate a set of random numbers equal to the number of combinations. The trials are then conducted in the order suggested by the random numbers. In our 2×2 factorial experiment, the four combinations are 1, 2, 3, and 4. If the random numbers generated are 3, 1, 4, and 2, then the combination corresponding to corner 3 is run first, that for corner 1 next, and so on. The CD includes the program *RandomOrderIntegers.xls*, which gives the required number of nonrepeating integers in random order. The response value for each corner combination is given in Fig. 10.1 in parentheses. The *response curves* are shown in Fig. 10.2. The two response lines illustrate B held at B^- and B^+, respectively. If the response lines are parallel, the conclusion is that there is no interaction between A

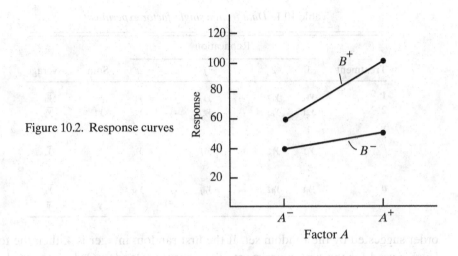

Figure 10.2. Response curves

and B; the plot shows interaction between A and B since the response lines are not parallel. We will systematically develop the relationships for effects and interactions for two or more variables in the 2^k factorial experiments section. In the ANOVA approach used in the next section, we develop relations for mean squared deviations for both effects and interactions.

Another feature of the experimental design is *blocking*, where similar experimental units are grouped together. The variability within a block can be attributed solely to the treatments.

10.3 Factorial Experiments

In our study we concentrate on the *fixed effects model*, where the factor levels are set at specified values. If these levels are selected randomly, the model is called a *random effects model*. In most common experiments, specified levels of factors are tested based on previous experience or study. We discuss the single-factor completely randomized experiment followed by randomized block design and the two-factor factorial experiment. We then proceed to present the 2^k factorial models.

Completely Randomized Single-Factor Experiments

We consider n levels (treatments) of the single factor with each level replicated r times; y_{ij} is the value of the response variable at treatment i and replication j. We consider a *balanced* experiment, where all levels are tested for the same number of replications. The treatments $i = 1, 2, \ldots, n$ are attempted in a randomized order. We choose a set of n random integers and the experiments are conducted in the

Table 10.1. *Data from a single-factor experiment*

Treatment	Replication					Sum	Average
	1	2	...	j	... r		
1	y_{11}	y_{12}	...	y_{1j}	... y_{1r}	$y_{1.}$	$\bar{y}_{1.}$
2	y_{21}	y_{22}	...	y_{2j}	... y_{2r}	$y_{2.}$	$\bar{y}_{2.}$
⋮	⋮						
i	y_{i1}	y_{i2}	...	y_{ij}	... y_{ir}	$y_{i.}$	$\bar{y}_{i.}$
⋮	⋮						
n	y_{n1}	y_{n2}	...	y_{nj}	... y_{nr}	$y_{n.}$	$\bar{y}_{n.}$
						$y_{..}$	$\bar{y}_{..}$

order suggested by the random set. If the first random integer is 4, then the fourth level is used for the first experiment. The program *RandomOrderIntegers.xls* provides natural numbers in a random order. The data are then entered into tabular form as shown in Table 10.1.

Some calculations were performed on the data in the last two columns of Table 10.1. We used the following notation for the calculated data:

$$y_{i.} = \sum_{j=1}^{r} y_{ij} \quad (i = 1, 2, \ldots, n) \quad (10.1)$$

$$\bar{y}_{i.} = \frac{y_{i.}}{r}$$

$$y_{..} = \sum_{i=1}^{p} \sum_{j=1}^{r} y_{ij}$$

$$\bar{y}_{..} = \frac{y_{..}}{rn} \quad (10.2)$$

where y_{ij} is represented by the linear model

$$y_{ij} = \mu + \tau_i + \varepsilon_{ij} \quad (10.3)$$

and where μ is the global mean, and $\mu_i = \mu + \tau_i$ is the mean of the treatment i; ε_{ij} is the independent normally distributed random error $\sim N(0, \sigma_\varepsilon^2)$; and $\tau_1, \tau_2, \ldots, \tau_n$ are parameters attributed to treatments. Without loss of generality, we consider that the mean of n treatment means μ_i ($i = 1, 2, \ldots, n$) is μ. We then have

$$\tau_1 + \tau_2 + \cdots + \tau_n = 0. \quad (10.4)$$

The hypothesis to be tested is

$$H_0: \mu_1 = \mu_2 = \cdots = \mu_n$$
$$H_1: \mu_i \neq \mu_j \text{ (for some } i \text{ and } j\text{).} \quad (10.5)$$

10.3 Factorial Experiments

To test the hypothesis, we use the F-test, which is based on the ratio of variances, or the ratio of the variance of treatment means and the variance of the error. If the null hypothesis is rejected, we need to check which means are significantly different. We first develop relationships for the sum-squared errors by making use of the following steps:

$$\sum_{i=1}^{n}\sum_{j=1}^{r}(y_{ij}-\bar{y}_{..})^2 = \sum_{i=1}^{n}\sum_{j=1}^{r}(y_{ij}-\bar{y}_{i.}+\bar{y}_{i.}-\bar{y}_{..})^2$$

$$= \sum_{i=1}^{n}\sum_{j=1}^{r}[(y_{ij}-\bar{y}_{i.})^2+2(y_{ij}-\bar{y}_{i.})(\bar{y}_{i.}-\bar{y}_{..})+(\bar{y}_{i.}-\bar{y}_{..})^2]$$

$$= \sum_{i=1}^{n}\sum_{j=1}^{r}(y_{ij}-\bar{y}_{i.})^2 + 2\sum_{i=1}^{n}(\bar{y}_{i.}-\bar{y}_{..})\sum_{j=1}^{r}(y_{ij}-\bar{y}_{i.})$$

$$+ r\sum_{i=1}^{n}(\bar{y}_{i.}-\bar{y}_{..})^2.$$

It is clear that the middle term in the last step is zero since $\sum_{j=1}^{r}(y_{ij}-\bar{y}_{i.})=0$. Thus,

$$\sum_{i=1}^{n}\sum_{j=1}^{r}(y_{ij}-\bar{y}_{..})^2 = \sum_{i=1}^{n}\sum_{j=1}^{r}(y_{ij}-\bar{y}_{i.})^2 + r\sum_{i=1}^{n}(\bar{y}_{i.}-\bar{y}_{..})^2. \quad (10.6)$$

The preceding equation is written as

$$\text{SST} = \text{SSE} + \text{SSTR},$$

where SST is the total sum of squares, which is given by

$$\text{SST} = \sum_{i=1}^{n}\sum_{j=1}^{r}(y_{ij}-\bar{y}_{..})^2 = \sum_{i=1}^{n}\sum_{j=1}^{r}y_{ij}^2 - \frac{y_{..}^2}{rn}. \quad (10.7)$$

The sum-squared treatment effects SSTR are given by

$$\text{SSTR} = r\sum_{i=1}^{n}(\bar{y}_{i.}-\bar{y}_{..})^2 = \frac{\sum_{i=1}^{n}y_{i.}^2}{r} - \frac{y_{..}^2}{rn}. \quad (10.8)$$

The sum-squared error (SSE) is then evaluated using

$$\text{SSE} = \text{SST} - \text{SSTR}. \quad (10.9)$$

We note that SST has $rn-1$ degrees of freedom, SSTR has $n-1$ degrees of freedom, and SSE has $rn-1-(n-1)=n(r-1)$ degrees of freedom. The mean

Table 10.2. *ANOVA table single-factor experiment*

Source of variation	Sum of squares	df	Mean square	F
Treatments	SSTR	$n-1$	MSTR	MSTR/MSE
Error	SSE	$n(r-1)$	MSE	
Total	SST	$nr-1$		

squared errors (MSEs) are obtained by dividing the squared errors by the corresponding degrees of freedom. The MSEs are the variances of the errors. The mean squared errors, MSTR and MSE, are calculated using.

$$\text{MSTR} = \frac{\text{SSTR}}{n-1} \tag{10.10}$$

$$\text{MSE} = \frac{\text{SSE}}{n(r-1)}. \tag{10.11}$$

MSE is an unbiased estimate of the variance of the experimental error, σ_ε^2. To check if the treatment error is large in comparison to the experimental error, the ratio of interest is the *F-statistic*:

$$F_{TR} = \frac{\text{MSTR}}{\text{MSE}}. \tag{10.12}$$

F_{TR} has an *F*-distribution with the numerator degrees of freedom, $n-1$, and the denominator degrees of freedom $n(r-1)$. We choose a confidence level of $1-\alpha$ or a level of significance, α, and evaluate $F_{\alpha, n-1, n(r-1)}$. The decision is now made as follows:

$$F_{TR} > F_{\alpha, n-1, n(r-1)} \Rightarrow \text{ reject null hypothesis.} \tag{10.13}$$

Alternatively we use the *p*-method. We evaluate the probability p of the *F*-distribution at F_{TR}: $p = \text{FDIST}(F_{TR}, n-1, n(r-1))$. If $p < \alpha$, we reject the null hypothesis, which implies that at least one mean differs from the others. If $p \geq \alpha$, the difference in the means is not significant enough to reject the null hypothesis. The steps used for analysis of variance (ANOVA) are summarized in Table 10.2.

Difference in Means when H_0 is Rejected – Tukey Intervals

When the null hypothesis is rejected, we need to determine which means differ. The Tukey interval approach is generally recommended for the multiple comparison of means. The studentized range is used in the Tukey approach, which uses the following steps in establishing the interval. If $\overline{Y}_{i.} - \mu_i$ $(i = 1, 2, \ldots, n)$ are treated as random variables, $R = \max(\overline{Y}_{i.} - \mu_i) - \min(\overline{Y}_{i.} - \mu_i)$, and $S = \sqrt{\frac{\text{MSE}}{n}}$, R/S has

10.3 Factorial Experiments

a studentized range distribution $Q_{n,n(r-1)}$ defined by Eq. (4.58) with numerator degrees of freedom n and denominator degrees of freedom $n(r-1)$. For a confidence level of $1-\alpha$ or a level of significance α, the critical value of the studentized range $Q_{\alpha,n,n(r-1)}$ is obtained. Thus, $R \leq Q_{\alpha,n,n(r-1)}S$. We now see that $|(\overline{Y}_{i.} - \mu_i) - (\overline{Y}_{j.} - \mu_j)| \leq R$. Since the absolute value is involved, both positive and negative values of the quantity inside satisfy the inequality. This gives an interval for $\mu_i - \mu_j$. Denoting

$$D = Q_{\alpha,n,n(r-1)}\sqrt{\frac{\text{MSE}}{n}}, \tag{10.14}$$

the Tukey interval for $\mu_i - \mu_j$ is given by

$$\mu_i - \mu_j = (\overline{y}_{i.} - \overline{y}_{j.} - D, \overline{y}_{i.} - \overline{y}_{j.} + D). \tag{10.15}$$

The critical values of the studentized range are given in Appendix Table A.10.

The Tukey interval is evaluated for each pairwise difference of the means. For n treatments, the number of Tukey intervals is $n(n-1)/2$. This method is referred to as the *Tukey honestly significant difference* (HSD).

If the interval contains zero, we conclude that the difference is not significant and the hypothesis that $\mu_i - \mu_j$ is otherwise rejected.

Example 10.1. Four heat treatment cycles were tested in an experiment. Each treatment was tested on five parts. The measured hardness values are shown in the following table:

Tr.\Replic.	1	2	Hardness 3	4	5	Sum	Mean
Tr 1	43	45	44	42	43	217	43.4
Tr 2	42	41	42	40	41	206	41.2
Tr 3	44	45	46	43	44	222	44.4
Tr 4	46	45	43	47	46	227	45.4

Test the hypothesis that the mean hardness from each heat treatment is the same, at a level of significance of 0.05.

Solution. We have $n = 4, r = 5$ (see *ANOVA1B2.xls*, sheet1, on the CD):

$$\frac{y_{..}^2}{rn} = 38019.2$$

$$\text{SST} = \sum_{i=1}^{n}\sum_{j=1}^{r} y_{ij}^2 - \frac{y_{..}^2}{rn} = 70.8,$$

where total degrees of freedom are $rn - 1 = 19$;

$$\text{SSTR} = \frac{\sum_{i=1}^{n} y_{i\cdot}^2}{r} - \frac{y_{\cdot\cdot}^2}{rn} = 48.4,$$

where treatment degrees of freedom are $n - 1 = 3$;

$$\text{MSTR} = \frac{48.4}{3} = 16.133;$$

$$\text{SSE} = \text{SST} - \text{SSTR} = 70.8 - 48.4 = 22.4,$$

where error degrees of freedom are $19 - 3 = 16$;

$$\text{MSE} = \frac{22.4}{16} = 1.4;$$

and

$$F = \frac{\text{MSTR}}{\text{MSE}} = \frac{16.133}{1.4} = 11.524.$$

At a level of significance of 0.05, $F_{0.05,3,16} = 3.239$. Since $F > F_{0.05,3,16}$ (from Appendix Table A.7, also $p = \text{FDIST}(11.524, 3, 16) = 2.84\text{e-}4 < 0.05$), we reject the null hypothesis that the mean hardnesses for the treatments are equal, and we conclude that the difference in means exists.

We now calculate the Tukey intervals for the differences in means. We have four treatments, and the MSE number of degrees of freedom is 16. From the studentized range in Appendix Table A.10, $Q_{0.05,4,16} = 4.05$, which is evaluated as strnginv$(1 - \alpha, n, n(r - 1))$ in the program *ANOVA1B2.xls*. We determine

$$D = Q_{\alpha,n,n(r-1)}\sqrt{\frac{\text{MSE}}{n}} = 4.05\sqrt{\frac{1.4}{4}} = 2.142$$

and the Tukey interval is

$$\mu_1 - \mu_2 = (\bar{y}_{1\cdot} - \bar{y}_{2\cdot} - D, \bar{y}_{1\cdot} - \bar{y}_{2\cdot} + D)$$
$$= (2.2 - 2.142, 2.2 + 2.142) = (0.058, 4.342) \text{ (reject)}.$$

Similarly,

$\mu_1 - \mu_3$ Tukey interval $(-3.142, 1.142)$ (not significant)
$\mu_1 - \mu_4$ Tukey interval $(-4.142, 0.142)$ (not significant)
$\mu_2 - \mu_3$ Tukey interval $(-5.342, -1.058)$ (reject)
$\mu_2 - \mu_4$ Tukey interval $(-6.342, -2.058)$ (reject)
$\mu_3 - \mu_4$ Tukey interval $(-3.142, 1.142)$ (not significant).

The difference is considered to be significant when the interval does not include a zero. The solution shows that treatment 2 is significantly lower.

Table 10.3. *Data from a single-factor block experiment*

Treatment	Block 1	2	...	j	...	r	Sum	Average
1	y_{11}	y_{12}	...	y_{1j}	...	y_{1r}	$y_{1.}$	$\bar{y}_{1.}$
2	y_{21}	y_{22}	...	y_{2j}	...	y_{2r}	$y_{2.}$	$\bar{y}_{2.}$
.	.							
i	y_{i1}	y_{i2}	...	y_{ij}	...	y_{ir}	$y_{i.}$	$\bar{y}_{i.}$
.								
n	y_{n1}	y_{n2}	...	y_{nj}	...	y_{nr}	$y_{n.}$	$\bar{y}_{n.}$
	$y_{.1}$	$y_{.2}$	...	$y_{.j}$	...	$y_{.r}$	$y_{..}$	$\bar{y}_{..}$
	$\bar{y}_{.1}$	$\bar{y}_{.2}$	...	$\bar{y}_{.j}$	...	$\bar{y}_{.r}$		

The program *ANOVA1B2.xls* (sheet1) solves the single-variable full factorial problems. It can handle five or fewer treatments and six or fewer replications. The example problem data have been entered into the program and the solution steps have been clearly identified. The program can be modified for a larger number of treatments and replications.

Randomized Block Experiments

Blocking is done to eliminate the influence of extraneous factors. Consider the example where our idea is to investigate the difference in the materials supplied by three vendors. These materials are processed on four different machines using similar processing conditions. We would like to isolate the influence due to machines. The requirement is that the material from each vendor must have an equal chance of being processed on each of the machines. We divide the experiment into four blocks (representing the four machines). The processing of the material from the three vendors is done in a random order within each block. We consider n treatments and r blocks; y_{ij} is the value of the response variable for treatment i in block j. The data are arranged in Table 10.3.

The linear model for the randomized block design is

$$y_{ij} = \mu + \tau_i + \beta_j + \varepsilon_{ij}, \qquad (10.16)$$

where $\sum_{i=1}^{n} \tau_i = 0$ and $\sum_{j=1}^{r} \beta_j = 0$.

SST is given by

$$\text{SST} = \sum_{i=1}^{n} \sum_{j=1}^{r} y_{ij}^2 - \frac{y_{..}^2}{rn}. \qquad (10.17)$$

SSTR is given by

$$\text{SSTR} = \frac{\sum_{i=1}^{n} y_{i.}^2}{r} - \frac{y_{..}^2}{rn}. \quad (10.18)$$

The block sum-squared effect (SSBL) is given by

$$\text{SSBL} = \frac{\sum_{j=1}^{n} y_{.j}^2}{n} - \frac{y_{..}^2}{rn}. \quad (10.19)$$

We now write $y_{ij} - \bar{y}_{..} = y_{ij} - \bar{y}_{i.} - \bar{y}_{.j} + \bar{y}_{..} + (\bar{y}_{i.} - \bar{y}_{..}) + (\bar{y}_{.j} - \bar{y}_{..})$, which is in line with $y_{ij} - \mu = \varepsilon_{ij} + \tau_i + \beta_j$. We square both sides, apply the summation, and expand as before. On canceling the zero terms, we get

$$\sum_{i=1}^{n}\sum_{j=1}^{r}(y_{ij} - \bar{y}_{..})^2 = \sum_{i=1}^{n}\sum_{j=1}^{r}(y_{ij} - \bar{y}_{i.} - \bar{y}_{.j} + \bar{y}_{..})^2$$

$$+ r\sum_{i=1}^{n}(\bar{y}_{i.} - \bar{y}_{..})^2 + n\sum_{j=1}^{r}(\bar{y}_{.j} - \bar{y}_{..})^2$$

$$\Rightarrow \text{SST} = \text{SSE} + \text{SSTR} + \text{SSBL}.$$

We now have

$$\text{SSE} = \text{SST} - \text{SSTR} - \text{SSBL}. \quad (10.20)$$

The hypothesis to be tested is Eq. (10.5), which is about the equality of the treatment means. The degrees of freedom are SST df $= rn - 1$, SSTR df $= n - 1$, SSBL df $= r - 1$, and SSE df $= rn - 1 - (n - 1) - (r - 1) = (n - 1)(r - 1)$.

The mean squared errors are the variances of the errors. The mean squared effects MSTR and MSBL, and the mean squared error MSE, are calculated using the following:

$$\text{MSTR} = \frac{\text{SSTR}}{n - 1} \quad (10.21)$$

$$\text{MSBL} = \frac{\text{SSBL}}{r - 1} \quad (10.22)$$

$$\text{MSE} = \frac{\text{SSE}}{(n - 1)(r - 1)}. \quad (10.23)$$

The following mean square ratios (variance ratios) are calculated:

$$F_{TR} = \frac{\text{MSTR}}{\text{MSE}} \quad F_{BL} = \frac{\text{MSBL}}{\text{MSE}}. \quad (10.24)$$

10.3 Factorial Experiments

Table 10.4. *ANOVA table single-factor block experiment*

Source of variation	Sum of squares	df	Mean square	F
Treatments	SSTR	$n-1$	MSTR	MSTR/MSE
Blocks	SSBL	$r-1$	MSBL	MSBL/MSE
Error	SSE	$(n-1)(r-1)$	MSE	
Total	SST	$nr-1$		

We now evaluate $p = \text{FDIST}(F_{TR}, n-1, (n-1)(r-1))$. For a level of significance α, the null hypothesis is rejected if $p < \alpha$. We also evaluate $p' = \text{FDIST}(F_{BL}, r-1, (n-1)(r-1))$. If $p' > \alpha$, we conclude that the blocking may not be effective. The steps are summarized in Table 10.4.

If the null hypothesis is rejected, we evaluate the Tukey intervals for differences in means as presented in the fully randomized single-factor experiment.

Example 10.2. A manufacturer planned to study bar stock from three vendors. The pieces are machined on four machines with the same cutting conditions. Material from each vendor is cut on the four machines, and the order of cutting is randomized. The surface finish of the machined surface is the measured response. The data from the tests are in the following table:

		Surface finish				
Tr\Blocks	1	2	3	4	Sum	Mean
Tr 1	17	16	10	12	55	13.75
Tr 2	22	19	15	16	72	18
Tr 3	21	22	18	21	82	20.5
Sum	60	57	43	49	209	17.417
Mean	20.000	19.000	14.333	16.333		

Is there a difference in the mean surface finish for the three vendors at a level of significance of 0.05? (see *ANOVA1B2.xls* (sheet2))

Solution. We have $n = 3, r = 4$, which gives the following:

$$\frac{y_{..}^2}{rn} = 3640.083;$$

$$\text{SST} = \sum_{i=1}^{n} \sum_{j=1}^{r} y_{ij}^2 - \frac{y_{..}^2}{rn} = 164.917,$$

where the total degrees of freedom are $rn - 1 = 11$;

$$\text{SSTR} = \frac{\sum_{i=1}^{n} y_{i.}^2}{r} - \frac{y_{..}^2}{rn} = 93.167,$$

where the treatment degrees of freedom are $n - 1 = 2$;

$$\text{MSTR} = \frac{93.167}{2} = 46.583;$$

$$\text{SSBL} = \frac{\sum_{j=1}^{n} y_{j\cdot}^2}{n} - \frac{y_{\cdot\cdot}^2}{rn} = 59.583,$$

where the blocking degrees of freedom are $r - 1 = 3$;

$$\text{MSBL} = \frac{59.583}{3} = 19.861;$$

$$\text{SSE} = \text{SST} - \text{SSTR} - \text{SSBL} = 164.917 - 93.167 - 59.583 = 12.167,$$

where the error degrees of freedom are $11 - 3 - 2 = 6$;

$$\text{MSE} = \frac{12.167}{6} = 2.028;$$

$$F_{TR} = \frac{\text{MSTR}}{\text{MSE}} = \frac{46.583}{2.028} = 22.97;$$

and

$$F_{BL} = \frac{\text{MSBL}}{\text{MSE}} = \frac{19.861}{2.028} = 9.795.$$

We first check the p-value for F_{BL}:

$$p = \text{FDIST}(F_{BL}, 2, 6) = 0.001 < 0.05$$

from Appendix Table A.7. We conclude that the blocking is effective.

$$p = \text{FDIST}(F_{TR}, 2, 6) = 0.00154 < 0.05$$

from Appendix Table A.7. At a level of significance of 0.05, we reject the null hypothesis that the mean surface finish values for the treatments are equal. We conclude that the difference in means exists.

We now calculate the Tukey intervals for the differences in means.

We have 3 treatments, and the MSE number of degrees of freedom is 6. From the studentized range in Appendix Table A.10, $Q_{0.05,3,6} = 4.342$, and we have the following:

$$D = Q_{\alpha, n, (n-1)(r-1)} \sqrt{\frac{\text{MSE}}{n}} = 4.342 \sqrt{\frac{2.028}{3}} = 3.570.$$

10.3 Factorial Experiments

Table 10.5. *Two-factor factorial experiments*

Factor A	Factor B			Sum	Average
	1	j	b		
1	...	...			
i	...	$y_{ij1}, y_{ij1}, \ldots, y_{ijk}$ sum $= y_{ij.}$ Av. $= \bar{y}_{ij.}$	...	$y_{i..}$	$\bar{y}_{i..}$
a	...	...			
Sum	$y_{.j.}$			$y_{...}$	
Average	$\bar{y}_{.j.}$				$\bar{y}_{...}$

The Tukey interval is

$$\mu_1 - \mu_2 : (\bar{y}_{1.} - \bar{y}_{2.} - D, \bar{y}_{1.} - \bar{y}_{2.} + D)$$
$$= (-4.25 - 3.570, -4.25 + 3.570) = (-7.820, -0.680) \text{ (reject)}.$$

Similarly, for other Tukey intervals,

$\mu_1 - \mu_3$: $(-10.320, -3.180)$ (reject)

$\mu_2 - \mu_3$: $(-6.070, 1.070)$ (not significant).

Therefore, μ_1 is the lowest surface finish and its differences with others are significant. See *ANOVA1B2.xls* (sheet2) for all details.

Two-Factor Factorial Experiment

We now discuss the two-factor fully randomized factorial experiment. Consider factors A and B each having a levels and b levels, respectively. The ab treatments are assigned randomly to the experimental units. The number of replications is r. The model is represented by

$$y_{ijk} = \mu + \tau_i + \beta_j + (\tau\beta)_{ij} + \varepsilon_{ijk} \quad (i = 1, 2, \ldots, a; \ j = 1, 2, \ldots, b; \ k = 1, 2, \ldots, r).$$
(10.25)

The $(\tau\beta)$ term is the interaction parameter for the two variables; y_{ijk} is the response value for variable 1 at level i and for variable 2 at level j, and k is the replication number. The data for a two-factor experiment are arranged in Table 10.5. Only the i, j location is filled. Indices i, j, k ranges are as shown in Eqs. (10.26). We consider an equal number of replications with the following notation:

$$y_{ij.} = \sum_{k=1}^{r} y_{ijk} \quad \bar{y}_{ij.} = \frac{y_{ij.}}{r} \quad y_{i..} = \sum_{j=1}^{b}\sum_{k=1}^{r} y_{ijk} \quad \bar{y}_{i..} = \frac{y_{i..}}{br}$$

$$y_{.j.} = \sum_{i=1}^{a}\sum_{k=1}^{r} y_{ijk} \quad \bar{y}_{.j.} = \frac{y_{.j.}}{ar} \quad y_{...} = \sum_{i=1}^{a}\sum_{j=1}^{b}\sum_{k=1}^{r} y_{ijk} \quad \bar{y}_{...} = \frac{y_{...}}{abr}.$$
(10.26)

In the development of the sum-squared effects, we make use of the following identity:

$$\sum_{i=1}^{a}\sum_{j=1}^{b}\sum_{k=1}^{r}(y_{ijk}-\overline{y}_{...})^2 = \sum_{i=1}^{a}\sum_{j=1}^{b}\sum_{k=1}^{r}(y_{ijk}-\overline{y}_{ij.})^2 + br\sum_{i=1}^{a}(\overline{y}_{i..}-\overline{y}_{...})^2$$

$$+ ar\sum_{j=1}^{b}(\overline{y}_{.j.}-\overline{y}_{...})^2 + r\sum_{i=1}^{a}\sum_{j=1}^{b}(\overline{y}_{ij.}-\overline{y}_{i..}-\overline{y}_{.j.}+\overline{y}_{...})^2$$

$$\Rightarrow \text{SST} = \text{SSE} + \text{SSA} + \text{SSB} + \text{SSAB}. \quad (10.27)$$

The total sum of squares, SST, is given by

$$\text{SST} = \sum_{i=1}^{a}\sum_{j=1}^{b}\sum_{k=1}^{r} y_{ijk}^2 - \frac{y_{...}^2}{abr}. \quad (10.28)$$

The sum of squares for the main effects, SSA and SSB, are given by

$$\text{SSA} = \frac{\sum_{i=1}^{a} y_{i..}^2}{br} - \frac{y_{...}^2}{abr} \quad (10.29)$$

$$\text{SSB} = \frac{\sum_{j=1}^{b} y_{.j.}^2}{ar} - \frac{y_{...}^2}{abr}. \quad (10.30)$$

The interaction effect, SSAB, is given by

$$\text{SSAB} = \frac{\sum_{i=1}^{a}\sum_{j=1}^{b} y_{ij.}^2}{r} - \frac{y_{...}^2}{abr} - \text{SSA} - \text{SSB}. \quad (10.31)$$

The sum-squared error, SSE, is then obtained using

$$\text{SSE} = \text{SST} - \text{SSA} - \text{SSB} - \text{SSAB}. \quad (10.32)$$

The associated degrees of freedom are SST df $= abr - 1$, SSA df $= a - 1$, SSB df $= b - 1$, SSAB df $= (a-1)(b-1)$, and SSE df $= ab(r-1)$.

Table 10.6 provides all the mean square ratios.

The p-values corresponding to F_A, F_B, and F_{AB} are obtained to check if the null hypothesis can be rejected. The computer program *ANOVA1B2.xls* (sheet3) has been prepared to make all the calculations leading to the ANOVA table. It can be used for A (5 levels), B (3 levels), and 9 replications; values of a, b, and r have to be appropriately modified and some formulas need to be copied if the size is larger than the example problem that has been entered.

Table 10.6. *ANOVA table for two-factor factorial experiment*

Source of variation	Sum of squares	df	Mean square	F
Factor A	SSA	$a-1$	$MSA = \dfrac{SSA}{a-1}$	$F_A = \dfrac{MSA}{MSE}$
Factor B	SSB	$b-1$	$MSB = \dfrac{SSB}{b-1}$	$F_B = \dfrac{MSB}{MSE}$
Interaction AB	SSAB	$(a-1)(b-1)$	$MSAB = \dfrac{SSAB}{(a-1)(b-1)}$	$F_{AB} = \dfrac{MSAB}{MSE}$
Error	SSE	$ab(r-1)$	$MSE = \dfrac{SSE}{ab(r-1)}$	
Total	SST	$abr-1$		

Example 10.3. Two variables – tempering temperature and tempering cycle – are investigated in an experiment. Three levels of each variable are tested and the number of replications for each variable level combination is 3. The response variable measured is the toughness of the material. The data are given in the following table:

	Cycle								
Temperature	1			2			3		
1	42	44	41	39	46	41	30	35	33
2	40	45	41	43	46	42	47	44	43
3	36	46	45	40	42	41	31	35	30

Draw your conclusions based on the two way ANOVA. (Hint: Higher toughness is desirable.)

Solution. We have $a = 3$, $b = 3$, and $r = 3$. It is convenient to prepare a table with sums in i, j locations as shown (see *ANOVA1B2.xls*, sheet3):

$A \backslash B$	1	2	3	$y_{i..}$	$\bar{y}_{i..}$
1	127	126	98	351	39.000
2	126	131	134	391	43.444
3	127	123	96	346	38.444
$y_{.j.}$	380	380	328	1088	$= y_{...}$
$\bar{y}_{.j.}$	42.22	42.22	36.44		

Note that the p-values for factors A and B are 0.003 and 0.001, and the value for the AB interaction is 0.005; the effects are significant. The program also gives

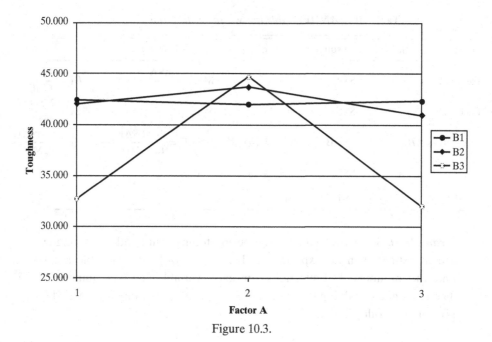

Figure 10.3.

the confidence intervals for each of the mean values. The calculated results are tabulated as follows:

Source of variation	Sum of squares	df	Mean square	F	P
Factor A	135.2	2	57.6	8.04	0.003
Factor B	200.3	2	100.15	11.91	0.001
Interaction AB	180.8	4	45.2	5.38	0.005
Error	151.33	18	8.41		
Total	667.63	26			

The toughness values are plotted against the levels of A as shown in Fig. 10.3. The curves shown are for various levels of B. At level 2 of A, all toughness values are high. However, the values of B at level 3 are low for A at 1 and A at 3; the values for A at level 2 and B at level 2 seem to be a good choice for the tempering operation.

Gage Repeatability and Reproducibility (GR&R) for Measurement Systems

The repeatability and reproducibility problem considered in Chapter 8, Example 8.5, was solved using the range and average method. It can also be solved using the ANOVA approach. The GR&R problem is a two-factor factorial experiment. The two variables are the part(s) and the operator(s). Let A represent the parts, with a total parts, and let B represent the operators, with b total operators, and we have r replications. AB is the interaction. We then get the variance estimates:

$$\sigma_{\text{repeatability}} = 5.15\sqrt{\text{MSE}}$$

$$\sigma_{\text{reproducibility}} = 5.15\sqrt{\frac{\text{MSA} - \text{MSAB}}{br}}$$

$$V_P = 5.15\sqrt{\frac{\text{MSB} - \text{MSAB}}{ar}}$$

$$\sigma_{\text{GR\&R}} = 5.15\sqrt{\sigma_{\text{repeatability}}^2 + \sigma_{\text{reproducibility}}^2 + V_P^2}$$

$$V_T = \sqrt{\sigma_{\text{GR\&R}}^2 + V_P^2}.$$

We note that $\sigma_{\text{reproducibility}}$ is evaluated when MSA > MSAB and is taken as zero otherwise; V_P is similarly taken as zero when MSB is not greater than MSAB.

Example 10.4. Solve the GR&R problem in Example 8.5 using the ANOVA approach.

Solution. The ANOVA approach has been implemented in sheet2 of *GageR&R.xls*:

Repeatability	0.04
Reproducibility	0.00
Interaction	0.11
R&R	0.12
Part Error VP	0.14
Total Error VT	0.18

The values compare well with the range approach presented in Chapter 8.

We now turn our attention to the consideration of two or more variables ($k \geq 2$), each having two levels.

10.4 2^k Factorial Experiments

The experiments discussed in the previous section involved one or two factors or variables at several levels, which enabled one to decide the best choice of treatments. As the number of factors and the number of levels of each factor increase, the number of experiments to be performed increases exponentially. For example, 10 factors at 5 levels each will need 5^{10}, or 9,765,625, experiments. At the development stage engineers face the problem of making a decision about what variables play significant roles. Multifactor experiments with each factor at two levels, high (+1) and low (−1), come in handy. We designate k as the number of factors, with each factor having two levels. A full factorial experiment involves 2^k runs. We call

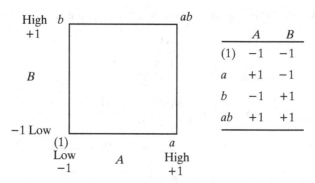

Figure 10.4. 2^2 factorial experiment

these 2^k *factorial experiments*. The number of these two-level experiments increases exponentially as the number of factors increases. Some strategies are employed to reduce the number of these experiments. We first consider full factorial experiments for two variables.

Let us consider the two-level experiment with two variables, A and B. The 2^2 experiment is shown in Fig. 10.4. The low and high levels of A are represented by -1 and $+1$, respectively, and the low and high levels of B are represented in a similar manner by -1 and $+1$. Thus, the four corners of the square are $(-1, -1), (+1, -1), (-1, +1),$ and $(+1, +1)$. These may be seen as two coordinates of a corner. The first coordinate is the level of A, and the second coordinate, the level of B. The figure also shows the four corners as $(1), a, b, ab$. The convention is as follows: if a lower-case letter appears, the corresponding variable is at its high and other variables are at a low; if there is no letter, each variable is at its low. Thus,

$$(1) \to A \text{ low}, \ B \text{ low or } (-1, -1)$$
$$a \to A \text{ high}, \ B \text{ low or } (+1, -1)$$
$$b \to A \text{ low}, \ B \text{ high or } (-1, +1)$$
$$ab \to A \text{ high}, \ B \text{ high or } (+1, +1).$$

The experimental setup table suggested by this arrangement is shown on the right-hand side in Fig. 10.4, and the table is easily prepared. We write A, B in the row at the top of the table and $(1), a, b, ab$ as the column at the left. In each row where a lower-case letter appears, we place $+1$ at the corresponding column location for the variable. In row a, we place $+1$ at column A; in the row b, we place $+1$ at column B; and in the row ab, we place $+1$ at columns A and B; and at all the other locations we place -1. This table shows the settings for the four experiments. However, these four experiments are conducted in a random order.

In the 2^3 factorial experiment, we add the variable C. Now we have eight experiments, represented by the corners of a cube as shown in Fig. 10.5: (1) represents A, B, C at levels $-1, -1, -1$. Also we add the four rows $c, ac, bc,$ and abc as

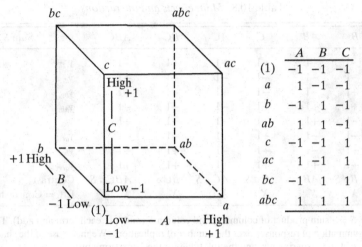

Figure 10.5. 2^3 factorial experiment

shown in the table at the right in Fig. 10.5. Extension of the table to 2^4 factorial experiments is easy, although the four-dimensional hypercube is not easily represented in a diagram. The data for experimental setup and collection are shown in Table 10.7. The table has been entered on sheet1 of the program *2KFactorial.xls* for use in conducting experiments.

We note that each column has an equal number of +1s and −1s. The number of +1s (or −1s) in a column is 2^{k-1}. Once the data are collected in Table 10.7, we need to calculate the factorial effects. In a two-variable case, the factorial effects are the

Table 10.7. *Table for $2^2, 2^3, 2^4$ factorial experiments*

Run		A	B	C	D	Repl. 1	...	r	Sum
1	(1)	−1	−1	−1	−1	$y_{(1)1}$	...	$y_{(1)r}$	$y_{(1).}$
2	a	+1	−1	−1	−1	y_{a1}	...	y_{ar}	$y_{a.}$
3	b	−1	+1	−1	−1	...			$y_{b.}$
4	ab	+1	+1	−1	−1	...			$y_{ab.}$
5	c	−1	−1	+1	−1				$y_{c.}$
6	ac	+1	−1	+1	−1				$y_{ac.}$
7	bc	−1	+1	+1	−1				$y_{bc.}$
8	abc	+1	+1	+1	−1				$y_{abc.}$
9	d	−1	−1	−1	+1				$y_{d.}$
10	ad	+1	−1	−1	+1				$y_{ad.}$
11	bd	−1	+1	−1	+1				$y_{bd.}$
12	abd	+1	+1	−1	+1				$y_{abd.}$
13	cd	−1	−1	+1	+1				$y_{cd.}$
14	acd	+1	−1	+1	+1				$y_{acd.}$
15	bcd	−1	+1	+1	+1				$y_{bcd.}$
16	abcd	+1	+1	+1	+1				$y_{abcd.}$

Table 10.8. *Main effects and interactions*

A	B	AB	C	AC	BC	ABC	Sum S
−1	−1	+1	−1	+1	+1	−1	$y_{(1).}$
+1	−1	−1	−1	−1	+1	+1	$y_{a.}$
−1	+1	−1	−1	+1	−1	+1	$y_{b.}$
+1	+1	+1	−1	−1	−1	−1	$y_{ab.}$
−1	−1	+1	+1	−1	−1	+1	$y_{c.}$
+1	−1	−1	+1	+1	−1	−1	$y_{ac.}$
−1	+1	−1	+1	−1	+1	−1	$y_{bc.}$
+1	+1	+1	+1	+1	+1	+1	$y_{abc.}$
$A \circ S$	$B \circ S$	$AB \circ S$	$C \circ S$	$AC \circ S$	$BC \circ S$	$ABC \circ S$	Contrast
X	X	X	X	X	X	X	Eff. = Contrast/$(r2^{k-1})$

Note: $A \circ S \Rightarrow$ sum product of column A and column S (multiply each term and add). The "." is used for summation of responses over the number of replications. We make use of the shaded part for the 2^2 factorial experiments, and the full table for the 2^3 experiments.

main effects, A and B, and the interaction AB. For the three-variable case, we have three main effects, three two-way interactions AB, AC, and BC, and one three-way interaction, ABC. If we have k variables, the number of main effects is k, the number of two-way interactions is the number of combinations of picking two variables from $k = \binom{k}{2}$, and so on. The number of p-way effects with k variables is $\binom{k}{p} = \frac{k!}{p!(k-p)!}$. In order to determine all effects, we need to prepare the effects table. In the effects table, the columns from Table 10.7 for the main variables are directly transferred. We create a column for each interaction. As an example, for the interaction AB, its column will be the product of columns A and B. Table 10.8 shows the addition of the interactions.

The properties of the columns in Table 10.8 are as follows:

1. In each column, the number of +1s equals the number of −1s.
2. The columns are *orthogonal*; this means that the sum product of any two columns is zero.

The *contrast* of A is obtained by multiplying each term of column A with the corresponding term in the sum column and adding. Since the columns are orthogonal, the contrast is referred to as the *orthogonal contrast*:

$\text{Contrast}_A = -y_{(1).} + y_{a.} - y_{b.} + y_{ab.}$ \hfill (for 2^2)
...
$\text{Contrast}_{AB} = y_{(1).} - y_{a.} - y_{b.} + y_{ab.}$ \hfill (for 2^2)
$\text{Contrast}_A = -y_{(1).} + y_{a.} - y_{b.} + y_{ab.} - y_{c.} + y_{ac.} - y_{bc.} + y_{abc.}$ \hfill (for 2^3) \hfill (10.33)
...
$\text{Contrast}_{AB} = y_{(1).} - y_{a.} - y_{b.} + y_{ab.} + y_{c.} - y_{ac.} - y_{bc.} + y_{abc.}$ \hfill (for 2^3)
...
$\text{Contrast}_X = \text{Sum product}(X \circ S)$ $(X = A, B, AB, \ldots)$

10.4 2^k Factorial Experiments

X is a placeholder for A, B, AB, and so on. The meaning of $(X \circ S)$ as the sum product of S and the column at X is explained clearly in Table 10.8. The function SUMPRODUCT(array1,array2) may be used in Excel to calculate the contrast.

The number of differences in the contrast is the same as the number of plus signs (or minus signs) in the column. The effect is obtained by dividing the contrast by $r2^{k-1}$ ($= 2r$ for $k = 2$, and $4r$ for $k = 3$). Thus, Effect$_A = \frac{\text{Contrast}_A}{r2^{k-1}}$, Effect$_B = \frac{\text{Contrast}_B}{r2^{k-1}}$, ... Using the representation of X for $A, B, \ldots$,

$$\text{Effect}_X = \frac{\text{Contrast}_X}{r2^{k-1}}. \tag{10.34}$$

The effect is the mean at X high minus the mean at X low. This also applies to interactions.

The sum square, SS, is obtained by dividing the square of the Contrast by the number of terms $r2^k$ ($= 4r$ for $k = 2$, and $8r$ for $k = 3$):

$$SS_X = \frac{\text{Contrast}_X^2}{r2^k}. \tag{10.35}$$

Total sum of squares, SST, is obtained as

$$SST = \sum_{i=1}^{2^k} \sum_{j=1}^{r} y_{ij}^2 - \frac{y_{..}^2}{r2^k}. \tag{10.36}$$

The sum-squared, error, SSE, is obtained by subtracting the sum of all SS_X values from SST. Each effect or interaction has one degree of freedom, and the total df is $r2^k - 1$. The ANOVA procedure for 2^k full-factorial experiments is presented in the example.

Example 10.4. In a milling operation, the cutting conditions of feed rate and depth of cut are to be chosen to achieve a good surface finish. Low and high values for the feed rate and depth of cut have been established and a 2^2 factorial experiment has been performed. Variable A is the feed rate and variable B is the depth of cut. Find the main effects and the interaction effects. Discuss what effects are significant.

Run#		A	B	Replications			SUM
1	(1)	−1	−1	19	20	21	60
2	a	1	−1	55	58	61	174
3	b	−1	1	26	25	27	78
4	ab	1	1	61	62	63	186

Solution. The program *2KFactorial.xls* is used for the calculations. (Sheet2 is for 2^2 factorial experiments.) We have $k = 2$ and $r = 3$:

	A	B	AB	S
(1)	−1	−1	1	60
a	1	−1	−1	174
b	−1	1	−1	78
ab	1	1	1	186

$$\text{Contrast}_A = -60 + 174 - 78 + 186 = 222$$

$$\text{Contrast}_B = -60 - 174 + 78 + 186 = 30$$

$$\text{Contrast}_{AB} = 60 - 174 - 78 + 186 = -6.$$

$$\text{Effect}_A = \frac{\text{Contrast}_A}{r2^{k-1}} = \frac{222}{3 \times 2} = 37$$

$$\text{Effect}_B = \frac{\text{Contrast}_B}{r2^{k-1}} = \frac{30}{3 \times 2} = 5$$

$$\text{Effect}_{AB} = \frac{\text{Contrast}_{AB}}{r2^{k-1}} = \frac{-6}{3 \times 2} = -1$$

$$y_{..} = 60 + 174 + 78 + 186 = 498.$$

$$\text{SST} = \sum_{i=1}^{2^k} \sum_{j=1}^{r} y_{ij}^2 - \frac{y_{..}^2}{r2^k} = 24876 - \frac{498^2}{3 \times 4} = 4209.0 \quad (\text{df} = 3 \times 4 - 1 = 11)$$

$$\text{SS}_A = \frac{\text{Contrast}_A^2}{r2^k} = \frac{222^2}{3 \times 4} = 4107 \qquad (\text{df} = 1)$$

$$\text{SS}_B = \frac{\text{Contrast}_B^2}{r2^k} = \frac{30^2}{3 \times 4} = 75 \qquad (\text{df} = 1)$$

$$\text{SS}_{AB} = \frac{\text{Contrast}_{AB}^2}{r2^k} = \frac{(-6)^2}{3 \times 4} = 3 \qquad (\text{df} = 1)$$

$$\text{SSE} = \text{SST} - \text{SS}_A - \text{SS}_B - \text{SS}_{AB} = 24.$$

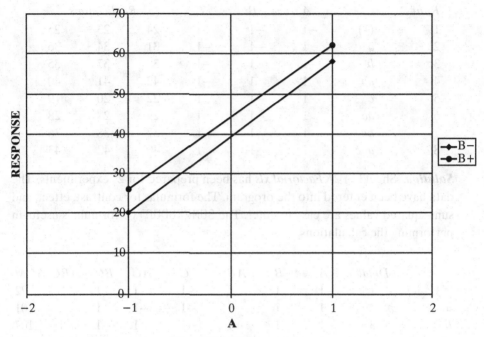

Figure 10.6. Response curves

We make use of the preceding calculations in preparing the ANOVA table:

	SumSq	df	MeanSq	F	P
A	4107	1	4107	1369	3.12E-10
B	75	1	75	25	1.05E-03
AB	3	1	3	1	3.47E-01
SSE	24	8	3		
SST	4209	11			

We note that the effects of A and B are significant. The interaction AB is not significant since the p-value is 0.347. The plot in Fig. 10.6 shows the response (surface finish) values at various levels of A and B. The finish is the best at A low (low feed rate) and B low (low depth of cut). In practice, for a higher material removal rate low feed rate, and high depth of cut may be employed with a small compromise in surface finish. The curves are nearly parallel implying that there is no interaction between A and B.

Example 10.5. Feed (A), depth of cut (B), and the cutting speed (C) are the three variables used in a turning operation and the measured response is the cutting force. Lower cutting force is expected to reduce the dimensional errors. Draw your conclusions by analyzing the following data.

Run#		A	B	C	Replications		
1	(1)	−1	−1	−1	24	23	25
2	a	1	−1	−1	31	34	36
3	b	−1	1	−1	35	37	35
4	ab	1	1	−1	42	41	40
5	c	−1	−1	1	22	20	19
6	ac	1	−1	1	29	27	28
7	bc	−1	1	1	37	38	36
8	abc	1	1	1	40	42	43

Solution. Sheet3 of *2KFactorial.xls* has been prepared for 2^3 experiments. The data have been entered into the program. The formulas for contrast, effect, and sum-squared values are easy to enter. The SUMPRODUCT() formula is useful in performing the calculations.

	Data#	A	B	AB	C	AC	BC	ABC	SUM
(1)	1	−1	−1	1	−1	1	1	−1	72
a	2	1	−1	−1	−1	−1	1	1	101
b	3	−1	1	−1	−1	1	−1	1	107
ab	4	1	1	1	−1	−1	−1	−1	123
c	5	−1	−1	1	1	−1	−1	1	61
ac	6	1	−1	−1	1	1	−1	−1	84
bc	7	−1	1	−1	1	−1	1	−1	111
abc	8	1	1	1	1	1	1	1	125
Contrast		82	148	−22	−22	−8	34	4	$y_{..} =$
Effect		6.83	12.33	−1.83	−1.83	−0.67	2.83	0.33	784

The ANOVA table is summarized here:

	SumSq	df	MeanSq	F	P
A	280.17	1	280.17	137.22	2.91E-09
B	912.67	1	912.67	447.02	4.06E-13
C	20.17	1	20.17	9.88	6.29E-03
AB	20.17	1	20.17	9.88	6.29E-03
AC	2.67	1	2.67	1.31	2.70E-01
BC	48.17	1	48.17	23.59	1.75E-04
ABC	0.67	1	0.67	0.33	5.76E-01
SSE	32.67	16	2.04		
SST	1317.33	23			

We see that A and B (feed and depth of cut) have significant effects since $p <$ 1e-8 for both. The next one is the interaction, BC. The p-value is 0.000175;

10.4 2^k Factorial Experiments

Table 10.9. *Blocking*

A	B	AB	C	AC	BC	ABC	
+1	−1	−1	−1	−1	+1	+1	a
−1	+1	−1	−1	+1	−1	+1	b
−1	−1	+1	+1	−1	−1	+1	c
+1	+1	+1	+1	+1	+1	+1	abc
−1	−1	+1	−1	+1	+1	−1	(1)
+1	+1	+1	−1	−1	−1	−1	ab
+1	−1	−1	+1	+1	−1	−1	ac
−1	+1	−1	+1	−1	+1	−1	bc

the choice may be A low and B low. The BC interaction suggests that high cutting speed (C) may be the correct choice. Plotting the data will help in observing the nature of the response curves. The mean values provided in the spreadsheet may be used for plotting various scenarios.

Blocking and Confounding

In the 2^k full-factorial experiments, the samples were randomly drawn from a homogeneous population. If the experimental units do not come from a homogeneous population, blocking is used. Suppose we have a situation where all eight experimental units of a full 2^k experiment cannot be performed under similar conditions. The situation may be that four units can be performed one day and the other four the next day. Blocking isolates the change in conditions between the two days. In another example, the samples come from vendors, and not all samples can be obtained at the same time. Blocking is used in these situations. Higher order interactions are generally used for blocking. For 2^3 factorial experiments, the choice is ABC. We choose the rows in Table 10.8 where the ABC column has all +1 or −1 values. Table 10.8 is rearranged into Table 10.9 so that all +1s for ABC are at the top. For $k = 3$, the +1s for ABC correspond to a, b, c, abc (when $k = 2$, the −1s for AB correspond to rows a, b). These rows are shown as shaded in Table 10.9. If we perform the upper four experimental units, columns A, B, and C are the same as columns BC, AC, and AB, respectively. Since the ABC column has all +1s, we define $I = ABC$. Then, noting that $A^2 = B^2 = C^2 = I$, we have

$$A = A.I = A.ABC = A^2BC = BC$$

and similarly

$$B = AC \quad \text{and} \quad C = AB.$$

We say that A and BC are *aliases*, B and AC are aliases, and C and AB are aliases. ABC is called the *generator* of the fractional factorial. The blocking effect is *confounded* with the ABC effect.

For the lower half of the table, $A = -BC$, $B = -AC$, and $C = -AB$. The effects of A, B, and C are seen as $A + BC$, $B + AC$, and $C + AB$ for the upper half and $A - BC$, $B - AC$, and $C - AB$ for the lower half. If the second half of the experiments are performed another time, and the data from the two fractions are combined, the main effects and the interactions can be isolated by simple addition and subtraction of the data.

The ideas of confounding are also used in developing fractional factorial experiments. If higher order interaction effects are negligible, we block the highest order interaction. In the 2^3 experiment, if ABC interaction is negligible, we can run the experiment with four runs and BC, AC, and AB confounded with A, B, and C.

We have provided an overview of blocking and confounding in 2^k factorial experiments. A vast literature is available on fractional factorial experiments, which interested readers should explore.

10.5 Summary

Factorial experiments and the related theory have been introduced in this chapter. The example problems have been readily implemented in the computer programs included in the CD. These programs may be used as is for problems of similar size and may be modified for larger problems.

The computer programs discussed in this chapter include the following:

RandomOrderIntegers.xls
ANOVA1B2.xls
GageR&R.xls
2KFactorial.xls.

EXERCISE PROBLEMS

10.1. Three different brands of varnish have been tested to determine the drying time measured in minutes. For each brand, the experiment was replicated four times. The drying times are given as follows:

	Drying time, min			
Varnish 1	23.5	22.4	21.8	20.7
Varnish 2	19.2	18.8	17.9	20.1
Varnish 3	24.5	25.3	26.2	23.8

Test the hypothesis that the mean drying time for each brand of varnish is the same at a level of significance of 0.1.

10.2. The breaking strength of yarn from three different vendors is compared. Yarn samples are taken from the stock in five different rooms where there is a likelihood of variation in humidity. Use a randomized block approach to test the null hypothesis that the breaking strength from the three vendors is the same at a level of significance of 0.05.

	Breaking strength, g				
Vendor 1	16.2	18.3	17.1	18.5	17.8
Vendor 2	18.1	20.4	19.7	21.2	20.3
Vendor 3	21.5	22.2	21.9	22.9	23.1

10.3. A randomized block experiment is performed for four operators on four machines. The machine is used as the blocking variable. The machining time for the same type of parts is compared.

	Machine			
	1	2	3	4
Operator 1	19.1	20.5	19.7	20.5
Operator 2	19.2	20.3	20.5	19.6
Operator 3	19.1	18.2	19.2	18.9
Operator 4	21.4	20.2	21	21.7

Test the null hypothesis that the performance of the four operators is the same.

10.4. Three different soil conditions and four types of fertilizer were tested to study crop yield. Three equal-size plots were prepared for each combination and the scaled yield data are given in the following table:

Soil	Fertilizer											
	1			2			3			4		
1	12	17	14	14	18	16	11	14	15	15	19	24
2	16	18	20	20	23	18	19	17	18	16	20	19
3	21	24	23	23	27	22	20	23	22	17	23	17

Draw your conclusions based on two-way ANOVA.

10.5. Injection pressure (A) and hold pressure (B) are used as the variables in an injection molding operation to study the effect on flash. Flash results in wasted material. Each of the variables is tried at two levels:

	Flash, mm		
(1)	2.5	2.9	2.4
a	7.5	6.8	7.1
b	10.8	11.6	11.1
ab	9.6	10.1	9.9

Find the main effects and the interaction effects. Draw your conclusions.

10.6. A furniture manufacturer tested three variables – type of wood (A), feed rate (B), and depth of cut (C) – to improve the surface finish of the product. Two levels of each variable are tested. Four replicates of scaled surface finish values are given in the following table:

	Surface Finish			
(1)	5	7	7	8
a	11	12	14	15
b	15	13	16	17
ab	11	10	9	10
c	18	22	25	20
ac	29	31	28	30
bc	41	36	39	35
abc	31	26	27	29

Determine the main effects and interaction effects, and discuss what effects are significant.

10.7. The impact of three factors, A, B, and C, is considered in an experiment. The performance levels are given for two replications in the following table:

Treatment	(1)	a	b	ab	c	ac	bc	abc
Repl. 1	5	7	8	11	10	6	10	5
Repl. 2	7	8	9	10	12	8	9	8

Determine the main effect of A, and the effect of interaction AB.

10.8. Show how a 2^4 experiment may be set up in two blocks. Discuss the confounding aspects of this experiment.

10.9. A paper helicopter model is shown in Fig. P10.9; set up a three-factor full-factorial experiment using the parameters paper (regular, thick), wing length (70, 80), and paper clip (small, large) and determine the optimal combination. (Figure from the reference given in Problem 10.10.)

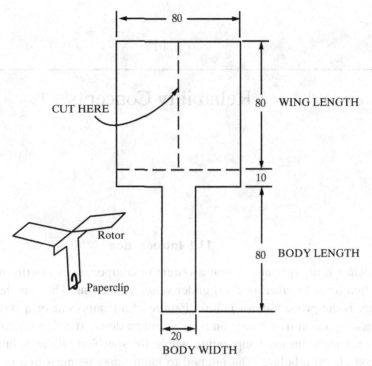

Figure P10.9. Paper helicopter model

10.10.[2] In the paper helicopter experiment in Problem 10.9, the control parameters are paper type, wing length, wing width, body length, wing shape, and number of paper clips attached. The objective of the experiment is to determine the optimal parameters to maximize the time of flight. Set up a 2^{6-2} fractional factorial experiment and evaluate the main effects and interactions.

---

[2] Data from J. Antony and N. Capon, Teaching Experimental Design Techniques to Industrial Engineers, *International Journal of Engineering Education*, v14, n5 (1998) pp. 335–343. (Note: Article may be accessed at www.ijee.dit.ie/articles/Vol14-5/ijee1033.pdf.)

11

Reliability Concepts

11.1 Introduction

Reliability is the probability that a system or component can perform its intended function for a specified interval under stated conditions. The complement of reliability is the probability of failure. Failure of a component or a system does not necessarily mean that it has completely broken down. If a thermostat works but it does not maintain the temperature within the specified tolerance limits, it should be considered a failure. The interval to failure may be measured in time (hours) or in number of cycles. The stated conditions of operation must be clearly defined. Quality and reliability go hand in hand. Customers look not only for good product quality but also for its reliable performance.

11.2 Reliability Functions

We define reliability $R(t)$ as the probability of survival of an item at time t. If the item continues to operate at time t as intended, clearly the reliability function must satisfy $R(0) = 1$ and $R(\infty) = 0$. The reliability function $R(t)$ may be seen as a monotone decreasing function, as shown in Fig. 11.1. Given that the item has survived at time t, the conditional probability of failure in the interval Δt is $\frac{R(t) - R(t + \Delta t)}{R(t)}$. The failure rate $h(t)$ is obtained by dividing this by Δt and taking the limit as Δt tends to zero:

$$h(t) = -\frac{1}{R(t)} \lim_{\Delta t \to 0} \left(\frac{R(t + \Delta t) - R(t)}{\Delta t} \right) = -\frac{1}{R(t)} \frac{dR(t)}{dt} = -\frac{R'(t)}{R(t)}. \quad (11.1)$$

The function $h(t)$ is the *failure rate* at time t and is also referred to as the *hazard function*. The probability of failure, $F(t)$, is the cumulative distribution function and we have the following relation (see Fig. 11.1):

$$R(t) + F(t) = 1. \quad (11.2)$$

11.2 Reliability Functions

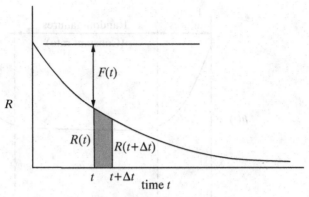

Figure 11.1. Reliability

Integrating the left- and right-hand sides of Eq. (11.1), we get $\int_0^t h(x)dx = -\ln R(x)|_0^t = -\ln R(t)$. This can be written as

$$R(t) = e^{-\int_0^t h(x)dx}. \tag{11.3}$$

From Eqs. (11.2) and (11.3), we have

$$F(t) = 1 - e^{-\int_0^t h(x)dx}. \tag{11.4}$$

The *probability density of failure*, $f(t) = F'(t) = -R'(t)$, is given by

$$f(t) = h(t)e^{-\int_0^t h(x)dx}. \tag{11.5}$$

The *mean time to failure* (MTTF), measured as the expected time to failure after the unit is put into operation, is given by

$$\text{MTTF} = \int_0^\infty tf(t)dt = -\int_0^\infty tR'(t)dt = \int_0^\infty R(t)dt, \tag{11.6}$$

where integration by parts has been used in the last step.

A typical hazard curve has a bathtub shape as shown in Fig. 11.2. The hazard curve is also referred to as the *life-cycle curve*. In the initial stages, the failure rate is high. This early phase may be interpreted as the *debugging phase*, *infant mortality*, or the *break-in period*. In the second phase, the failure rate is constant. The failures are random. This is the *steady life* period. During the final phase, the failure rate increases rapidly, which is called the *rapid breakdown* period. Manufacturers put great effort into completing the break-in or the burn-in period before the product reaches the user. Once the useful life is established, it is good practice to replace the

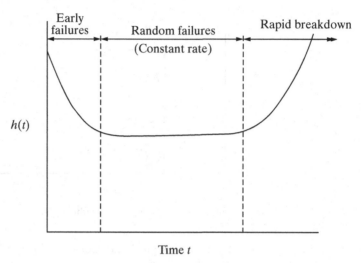

Figure 11.2. Hazard function

part at the onset of the rapid breakdown period. Various reliability distributions are defined based on the type of hazard function $h(t)$.

11.3 Failure Distributions

The most widely used distributions in reliability studies are the exponential distribution, the Weibull distribution, and the lognormal distribution.

Exponential Distribution

In the bathtub model shown in Fig. 11.2, assuming that the burn-in period is completed, we may model the failure rate as a constant value, λ. In the simplest model we assume a constant failure rate for $0 \leq t \leq \infty$. We then have $\int_0^t h(x)dx = \int_0^t \lambda dx = \lambda t$. Equations (11.5) and (11.3) can now be written as

$$f(t) = \lambda e^{-\lambda t} \qquad (11.7)$$

$$R(t) = e^{-\lambda t}, \qquad (11.8)$$

where $f(t)$ given in Eq. (11.7) is the exponential distribution function for the failure model. The model is used extensively in reliability studies.

Using Eq. (11.6), we obtain the MTTF as follows:

$$\text{MTTF} = \frac{1}{\lambda}. \qquad (11.9)$$

11.3 Failure Distributions

If K components fail in a period of time T, the constant failure rate λ may be calculated using

$$\lambda = \frac{K}{T}. \tag{11.10}$$

We discuss more precise ways of calculating the failure rate values in life testing in Chapter 12. Equation (11.10) may be used for failures reported in field studies.

Memoryless Property of Exponential Distribution

We define $R(t|T)$ as the conditional reliability of a new mission of t duration undertaken after the unit has completed a duration of time T. Time t starts from the point of completion of time T. We see that $R(t|T) = R(T+t)/R(T)$. On evaluating the right-hand side of the equation, we find that

$$R(t|T) = e^{-\lambda t}. \tag{11.11}$$

The conditional reliability is not dependent on the previous life history. This property is referred to as the *memoryless* property. This property is unique for the exponential distribution model.

MTBF and MTTR

In addition to MTTF introduced earlier, the *mean time to repair* (MTTR) is the average time it takes to restore a failed unit in consideration. The *mean time between failures* (MTBF) is the average time from one failure to the next. MTBF includes the MTTR. Thus,

$$\text{MTBF} = \text{MTTF} + \text{MTTR}. \tag{11.12}$$

The *availability*, A, is the probability that the system or an item operates satisfactorily at any given time when used under specified conditions:

$$A = \frac{\text{MTTF}}{\text{MTTF} + \text{MTTR}}. \tag{11.13}$$

When MTTR is small, we do not differentiate between MTTF and MTBF.

Example 11.1. A washing machine reported 6 failures during a period of 1,500 hours of operation. The average repair time per failure is 1 hour. Determine the failure rate λ, MTTF, and MTBF.

Solution.

$$\text{MTBF} = \frac{1500}{6} = 250 \text{ hours}$$

$$\text{MTTR} = 250 - 1 = 249 \text{ hours}$$

$$\lambda = \frac{1}{\text{MTTF}} = \frac{1}{249} = 0.00402 \text{ per hour.}$$

Example 11.2. A transistor failure is modeled using a constant failure rate of 0.00008/hour. Find the reliability of the transistor after 6,000 hours of operation. Also determine the MTTF.

Solution. Given $\lambda = 0.00008$ and $R(t) = e^{-\lambda t}$, after 5,000 hours,

$$R(5000) = e^{-(0.00008)(6000)} = e^{-0.48} = 0.6188$$

$$\text{MTTF} = \frac{1}{\lambda} = 12{,}500 \text{ hours.}$$

Example 11.3. If a product reliability of 0.9 is to be achieved after 8,000 hours of operation, determine the failure rate, assuming an exponential distribution.

Solution.

$$0.9 = R(8000) = e^{-8000\lambda}$$

$$\lambda = -\frac{\ln(0.9)}{8000} = 0.00001317/\text{hour}$$

which is 13.17 failures per million hours (10^6 hours).

Weibull Distribution

Swedish engineer Waloddi Weibull (1887–1979) developed the failure model using a hazard function of the type Ct^γ. The Weibull distribution is used extensively in life studies for engineering components. The general model uses three parameters – the *shape parameter* β, the *scale parameter* α, and the *location parameter* γ. The hazard function, $h(t)$, is given by

$$h(t) = \frac{\beta}{\alpha}\left(\frac{t-\gamma}{\alpha}\right)^{\beta-1} \quad (t \geq \gamma). \tag{11.14}$$

The distribution function and the reliability are given by

$$f(t) = \frac{\beta}{\alpha}\left(\frac{t-\gamma}{\alpha}\right)^{\beta-1} e^{-\left(\frac{t-\gamma}{\alpha}\right)^\beta} \quad (t \geq \gamma) \tag{11.15}$$

$$R(t) = e^{-\left(\frac{t-\gamma}{\alpha}\right)^\beta} \quad (t \geq \gamma). \tag{11.16}$$

11.3 Failure Distributions

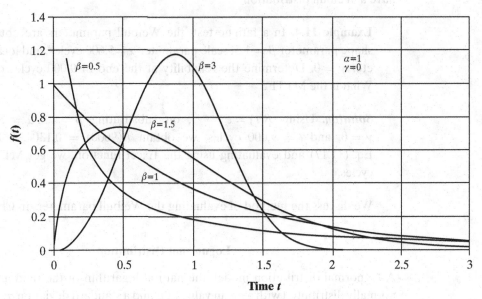

Figure 11.3. Weibull probability density function

We note that, if $\gamma = 0$ and $\beta = 1$, the Weibull distribution is precisely an exponential distribution with $\lambda = 1/\alpha$. Exponential distribution is thus a special case of the Weibull distribution. The Weibull probability distribution function is shown in Fig. 11.3 for $\alpha = 1$, $\gamma = 0$, and various values of β. For $\beta = 3$, the Weibull distribution function is similar to the normal distribution. Typical Weibull shape and scale parameter values are given in Weibull databases (see, for example, http://www.barringer1.com/wdbase.htm).

For the Weibull distribution, the MTTF is given by

$$\text{MTTF} = \gamma + \alpha \Gamma\left(\frac{1}{\beta} + 1\right), \tag{11.17}$$

where $\Gamma(x) = \int_0^\infty e^{-t} t^{x-1} dt$ is the gamma function of x. When x is an integer, $\Gamma(x) = (x-1)!$. In Excel, the gamma function is calculated using exp(Gammaln(x)). The distribution with $\gamma = 0$ is referred to as a two-parameter Weibull distribution.

Conditional Reliability $R(t|T)$ for the Weibull Distribution

The conditional reliability $R(t|T) = R(T+t)/R(T)$, which defines the reliability at $T + t$ starting from completed life T, is given by

$$R(t|T) = e^{-[(\frac{T+t-\gamma}{\alpha})^\beta - (\frac{t-\gamma}{\alpha})^\beta]}. \tag{11.18}$$

This expression is used for the estimation of left-life of engineering components that have a Weibull distribution.

Example 11.4. In a fatigue test, the Weibull parameters are obtained as the shape parameter $\beta = 1.5$, scale parameter $\alpha = 5{,}600$ cycles, and location parameter $\gamma = 0$. Determine the reliability at the end of 9,000 cycles of operation. What is the MTTF?

Solution. Using $R(t) = e^{-(\frac{t-\gamma}{\alpha})^\beta}$, on substituting $\beta = 1.5, \alpha = 5{,}600$ cycles, $\gamma = 0$, and $t = 9{,}000$ cycles, we obtain $R(9000) = 0.1304$. Substituting in Eq. (11.17) and evaluating using the Excel function, we get MTTF = 5055.4 cycles.

We discuss the method of evaluating the Weibull parameters in Chapter 12.

Lognormal Distribution

A lognormal distribution model, the natural logarithm of the random variable, is normally distributed with a mean value of μ and a standard deviation σ; μ and σ may be treated as two parameters of the lognormal distribution. The random variable of interest in reliability is time t. The probability distribution function (pdf) is given by

$$f(t) = \frac{1}{\sqrt{2\pi}\sigma t} e^{-\frac{1}{2}(\frac{\ln t - \mu}{\sigma})^2} \quad (0 \le t < \infty). \tag{11.19}$$

We note that $f(0) \to 0$ as $t \to 0$. A typical curve for the lognormal pdf is shown in Fig. 11.4. The curve is skewed to the right.

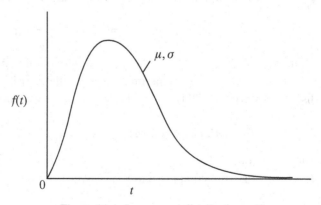

Figure 11.4. Lognormal distribution pdf

The transformation to the standard variable $z(t)$ is carried out using $z(t) = \frac{\ln t - \mu}{\sigma}$, and the probability of failure at time t, $F(t)$, is obtained as $\Phi(z(t))$. The reliability at time t, $R(t)$, is given by

$$R(t) = 1 - \Phi(z(t)). \tag{11.20}$$

Whereas the expected value and the variance of $\ln(t)$ are μ and σ^2, the expected value and variance of t are derived as

$$E(t) = \text{MTTF} = e^{\frac{2\mu + \sigma^2}{2}} \tag{11.21}$$

$$\text{Var}(t) = e^{2\mu + 2\sigma^2} - e^{2\mu + \sigma^2}. \tag{11.22}$$

Example 11.5. The life (hours) of a component is modeled using lognormal distribution. The two parameters for the distribution are given as mean ($\mu = 5.5$) and standard deviation ($\sigma = 1.6$). Determine the reliability of the component at 1,000 hours and its MTTF.

Solution. At time $t = 1000$ hours,

$$z(t) = \frac{\ln t - \mu}{\sigma} = \frac{\ln(1000) - 5.5}{1.6} = \frac{1.408}{1.6} = 0.880.$$

From Appendix Table A.3, $\Phi(0.880) = 0.8106$:

$$R(t) = 1 - \Phi(z(t)) = 1 - 0.8106 = 0.19.$$

From Eq. (11.21), we have

$$\text{MTTF} = e^{\frac{2\mu + \sigma^2}{2}} = e^{\frac{2 \times 5.5 + 1.6^2}{2}} = 880.1 \text{ hours}.$$

11.4 System Reliability

A system is made up of several *elements* or *components*. In analyzing the system reliability, the basic assumption is that the reliability of each of the elements is known. If it is a time-dependent system, we need the failure rate for each of the elements. We then calculate the element reliability using the exponential distribution. A typical element is represented in a block diagram as shown in Fig. 11.5. In a system, these

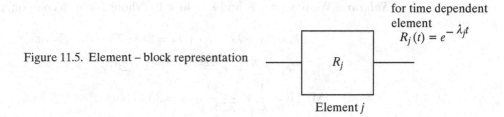

for time dependent element
$R_j(t) = e^{-\lambda_j t}$

Figure 11.5. Element – block representation

Element j

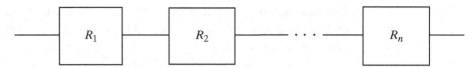

Figure 11.6. Elements in series

elements may appear in series, parallel, or a combination of series and parallel. We first consider the series and parallel systems.

Systems with Elements in Series

A block diagram of a system with n elements connected in series is shown in Fig. 11.6. In a real system they may not be connected end-to-end as shown. If the success of every element in the system results in the success of the system, we represent it as a series connection. The system is said to fail if any one of the components fails. The links in a bicycle chain may be considered elements in series. The system reliability, R_s, of a series system is given by

$$R_s = R_1 R_2 \cdots R_n. \tag{11.23}$$

The system reliability decreases as more elements are added. If it is a time-dependent system, we make use of the failure rates of individual components in calculating the system reliability. Let $\lambda_1, \lambda_2, \ldots, \lambda_n$ be the constant failure rates for an exponential model. Then

$$R_s = e^{-(\lambda_1 + \lambda_1 + \cdots + \lambda_n)t}. \tag{11.24}$$

The equivalent failure rate of the system is $\lambda_s = \lambda_1 + \lambda_1 + \cdots + \lambda_n$.

The MTTF for the system is given by

$$\text{MTTF} = \frac{1}{\lambda_1 + \lambda_2 + \cdots + \lambda_n}. \tag{11.25}$$

Example 11.6. A television camera focus system has 8 components in series. Each component failure has an exponential distribution with a failure rate of 40 per 10^6 hours. Determine the reliability at the end of 5,000 hours of operation. Also calculate the MTTF for the system.

Solution. We have $n = 8$, and $\lambda = 40 \times 10^{-6}$/hour for each component:

$$\lambda_s = \lambda_1 + \lambda_1 + \cdots + \lambda_8 = 8\lambda = 32 \times 10^{-5}/\text{hour}$$

$$R_s = e^{-\lambda_s t} = e^{-32 \times 10^{-5} \times 5000} = 0.2019$$

$$\text{MTTF} = \frac{1}{\lambda_s} = \frac{1}{32 \times 10^{-5}} = 3{,}125 \text{ hours.}$$

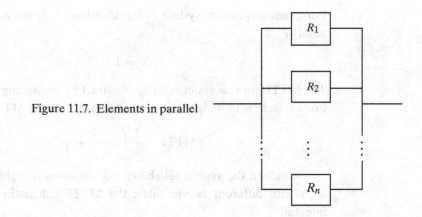

Figure 11.7. Elements in parallel

Example 11.7. If a reliability of 0.95 is desired after 5,000 hours for the television camera in Example 11.6, what should be the failure rate for each component?

Solution. We need $e^{-8\lambda t} = 0.95$. Taking the natural logarithm, we get

$$-8\lambda t = \ln(0.95) = -0.0513$$

$$\lambda = \frac{0.0513}{8 \times 5000} = 1.28 \times 10^{-6}/\text{hour}.$$

Systems with Elements in Parallel

A parallel system is one where the success of any one of the elements in the system results in the system's success. A block diagram with n elements connected in parallel is shown in Fig. 11.7. A system with elements connected in parallel is said to be a redundant system. If a twin engine turbojet can land safely with only one engine, we may model the two engines as a parallel system. A parallel system fails only when all the elements in the system fail. Using the cumulative probabilities of failure,

$$F_s = F_1 F_2 \cdots F_n. \qquad (11.26)$$

The system reliability R_s is obtained by using the following relationships:

$$\begin{aligned} R_s &= 1 - F_s \\ F_i &= 1 - R_i \quad (i = 1, 2, \ldots, n). \end{aligned} \qquad (11.27)$$

Thus,

$$R_s = 1 - (1 - R_1)(1 - R_2) \cdots (1 - R_n). \qquad (11.28)$$

For a time-dependent system with n identical elements, each with exponential reliability $e^{-\lambda t}$,

$$R_s = 1 - (1 - e^{-\lambda t})^n. \tag{11.29}$$

The MTTF for this system can be obtained by evaluating the integral $\int_0^\infty R_s(t)\, dt$. For a system with n identical elements in parallel, the MTTF is given by

$$\text{MTTF} = \frac{1}{\lambda}\left(1 + \frac{1}{2} + \frac{1}{3} + \cdots + \frac{1}{n}\right). \tag{11.30}$$

Although the system reliability calculation is straightforward for parallel systems with different failure rates, the MTTF calculation involves the following integral:

$$\text{MTTF} = \int_0^\infty (1 - (1 - e^{-\lambda_1 t})(1 - e^{-\lambda_2 t}) \cdots (1 - e^{-\lambda_n t}))\, dt. \tag{11.31}$$

For the MTTF of a system with three parallel elements with failure rates λ_1, λ_2, and λ_3, the integral gives the following result:

$$\text{MTTF} = \frac{1}{\lambda_1} + \frac{1}{\lambda_2} - \frac{1}{\lambda_1 + \lambda_2} + \frac{1}{\lambda_3} - \frac{1}{\lambda_2 + \lambda_3} - \frac{1}{\lambda_3 + \lambda_1} + \frac{1}{\lambda_1 + \lambda_2 + \lambda_3}. \tag{11.32}$$

The MTTF for two elements can be obtained by taking the first three terms in Eq. (11.32). The Excel program *ReliabilityOfSystems.xls* provides the calculation of MTTF for up to six parallel elements.

Example 11.8. A system has three components connected in parallel. The reliabilities of the components are 0.92, 0.88, and 0.95, respectively. Determine the system reliability. If these reliabilities are at time 2,000 hours of operation, what is the MTTF of the system? (Note: Assume exponential distribution for the reliabilities.)

Solution.

$$\begin{aligned} R_s &= 1 - (1 - R_1)(1 - R_2)(1 - R_3) \\ &= 1 - (1 - 0.92)(1 - 0.88)(1 - 0.95) \\ &= 0.99952. \end{aligned}$$

In order to calculate the MTTF, we need the failure rates of the elements. We evaluate them using

$$e^{-\lambda_1 t} = 0.92 \Rightarrow \lambda_1 = -\frac{\ln(0.92)}{t} = \frac{0.08338}{2000} = 4.17 \times 10^{-5}/\text{hour}.$$

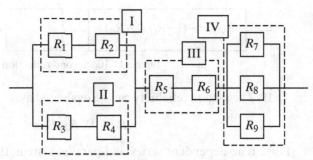

Figure 11.8. Combined system

Similarly, $\lambda_2 = 6.39 \times 10^{-5}$/hour, $\lambda_3 = 2.565 \times 10^{-5}$/hour.

Now we use Eq. (11.32):

$$\text{MTTF} = \frac{1}{\lambda_1} + \frac{1}{\lambda_2} - \frac{1}{\lambda_1 + \lambda_2} + \frac{1}{\lambda_3} - \frac{1}{\lambda_2 + \lambda_3} - \frac{1}{\lambda_3 + \lambda_1} + \frac{1}{\lambda_1 + \lambda_2 + \lambda_3}.$$

On substituting for λ_1, λ_2, and λ_3, we get MTTF = 50,751 hours.

Systems with Series and Parallel Subsystems

For a system with various series and parallel subsystems, each of the series and parallel subunits is identified and the system reliability is then obtained by combining these subunits. Consider the system shown in Fig. 11.8.

The series-and-parallel systems are identified as I, II, III, and IV. In the first reduction, shown in Fig. 11.9, the subsystem reliabilities are evaluated. A second parallel subsystem is identified and evaluated, as shown in the second reduction in

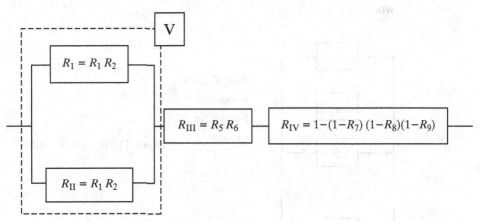

Figure 11.9. First reduction

$$R_V = 1-(1-R_I)(1-R_{II}) \quad R_{III} \quad R_{IV}$$

Figure 11.10. Second reduction

Fig. 11.10. The system reliability can be evaluated as

$$R_s = R_V R_{III} R_{IV}.$$

For a time-dependent series-and-parallel system, the MTTF can be evaluated for simple configurations. As the complexity increases, it is not easy to evaluate the MTTF using the techniques described here. The *Markov model approach* is used for such evaluation. This method is given in books devoted to reliability.

11.5 K-of-N Systems

A *K-of-N system* consists of N identical elements of which K elements ($K > 1$) must operate for the success of the system. A schematic K-of-N system is shown in Fig. 11.11. Examples of such a system are an aircraft with four engines where at least two engines must operate for success; a satellite battery system in which six of ten batteries must operate for the system operation; and a V8 engine where four of eight cylinders must operate for the automobile to run. Let R be the reliability of a component. The probability of its failure is $1 - R$. The requirement that K or more units are operational is equivalent to K or more successes in N trials. Thus, the system reliability R_s follows the binomial distribution, given by

$$R_s = \sum_{j=K}^{N} \binom{N}{j} R^j (1-R)^{N-j}, \qquad (11.33)$$

where $\binom{N}{j} = \frac{N!}{j!(N-j)!}$.

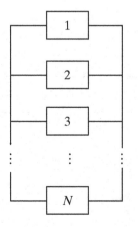

Any K-of-N to operate for system success

Figure 11.11. K-of-N system

11.5 K-of-N Systems

Since $\sum_{j=0}^{N} \binom{N}{j} R^j (1-R)^{N-j} = 1$ for the binomial distribution, if $K \leq N/2$ then the following expression may be used to evaluate system reliability with fewer computations:

$$R_s = 1 - \sum_{j=0}^{K-1} \binom{N}{j} R^j (1-R)^{N-j}. \tag{11.34}$$

Note that if $K = N$ then all N components must operate for the system to function, which is equivalent to a series system. For this case we get the familiar result, $R_s = R^N$, for the series system from Eq. (11.33). When $K = 1$, any one component operating successfully will satisfy the system requirements. Setting $K = 1$ in Eq. (11.33), we get $R_s = 1 - (1-R)^N$, which is the formula for the parallel system.

If the K-of-N system is a time-dependent system, R is given by $R = e^{-\lambda t}$. In order to calculate MTTF we evaluate $\int_0^\infty R_s(t)\,dt$ in Eq. (11.33) to get

$$\text{MTTF} = \frac{1}{\lambda} \sum_{j=K}^{N} \frac{1}{j}. \tag{11.35}$$

K-of-N calculations are provided in sheet2 of the Excel program *ReliabilityOfSystems.xls*. The program includes the function GetR(K,N,Rs), which finds R when K, N, and R_s are given in Eqs. (11.33) or (11.34). This function is useful at the system design stage.

Example 11.9. A system has five components with system success defined as 4 out of 5. The reliability of each component is $R = 0.9$ at time $T = 1$. Determine the system reliability and the MTTF.

Solution. The reliability of a 4-of-5 system is given by

$$R_s = \sum_{j=4}^{5} \binom{5}{j} R^j (1-R)^{N-j} = \binom{5}{4} 0.9^4 (0.1) + \binom{5}{5} (0.9)^5$$

$$= (5)0.9^4(0.1) + (1)0.9^5 = 0.9185,$$

where

$$R = e^{-\lambda} = 0.9$$
$$\lambda = -\ln(0.9) = 0.1054.$$

The MTTF is given by

$$\text{MTTF} = \frac{1}{0.1054}\left(\frac{1}{4} + \frac{1}{5}\right)$$

$$= 4.271.$$

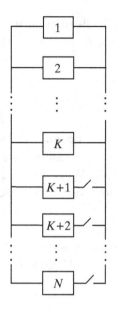

Figure 11.12. *K*-of-*N* standby redundant system

Example 11.10. A 2-of-4 system is to be designed to achieve a system reliability of 0.95. Determine the component reliability.

Solution. R can be obtained by trial and error. The function GetR(2,4,0.95) included in the Excel program *ReliabilityOfSystems.xls* can be used to find it. We find $R = 0.7514$.

11.6 Standby Systems

We now consider systems where some elements are available in standby mode. The system reliability improves when it has standby redundancy. A standby unit is inoperative until needed; it is switched on when a unit in the primary system fails. An example of standby redundancy is the use of a standby generator in a building to ensure the continuity of the power supply. The generator kicks in when the main power supply fails. Fig. 11.12 shows a K-of-N redundant system that starts with $N-K$ standby components. When any one of the components fails, a standby component is switched on.

We develop some relations for the reliability analysis of systems with standby units. The basic assumptions are the following:

1. The sensing and switching are perfect. This implies that no delay occurs. When a unit fails, the standby unit comes into operation instantaneously.
2. The standby unit is always in working order and it does not fail while it is in standby mode.

Let $R = e^{-\lambda t}$ be the reliability of each identical component of the system. Since K units must operate for the system's success, the average number of failures expected at time t is $\lambda K t$. The system reliability R_s is the sum of probabilities of $0, 1, 2, \ldots, N-K$ failures. Using the Poisson distribution, we evaluate this as

$$R_s = e^{-\lambda K t} \left[1 + \lambda K t + \frac{(\lambda K t)^2}{2!} + \frac{(\lambda K t)^3}{3!} + \cdots + \frac{(\lambda K t)^{N-K}}{(N-K)!} \right]. \quad (11.36)$$

In Excel, we can evaluate this using `Poisson(N-K, λKt, TRUE)`, which is the cumulative Poisson probability. Equation (11.36) has been implemented in sheet2 of the program *ReliabilityOfSystems.xls*.

The MTBF for the system is given by

$$\text{MTBF} = \frac{N - K + 1}{\lambda K}. \quad (11.37)$$

If only one component is used as a standby unit, we set $K = N - 1$.

Example 11.11. A system has one basic unit and two standby units. The failure rate for each component is 0.005/hour. Find the system reliability at 400 hours of operation. Also determine the MTTF of the system.

Solution. For this system $K = 1$, $N = 3$, $\lambda = 0.005$/hour, $t = 400$ hours, $\lambda t = 2$. Using Eq. (11.36),

$$R_s = e^{-\lambda t} \left[1 + \lambda t + \frac{(\lambda t)^2}{2!} \right]$$

$$= e^{-2} \left[1 + 2 + \frac{2^2}{2!} \right]$$

$$= 0.6767$$

$$\text{MTTF} = \frac{N}{\lambda} = \frac{3}{0.005} = 600 \text{ hours}.$$

11.7 Summary

In this chapter we developed the concepts of reliability. We first discussed the reliability functions leading to the exponential and Weibull distributions. Reliability and MTTF calculations for series and parallel systems were presented, which provided the necessary background for analyzing K-of-N redundant systems and standby redundant systems. We deal with aspects of reliability testing in Chapter 12.

The computer program discussed in this chapter is *ReliabilityOfSystems.xls*.

EXERCISE PROBLEMS

11.1. A newly developed automobile is undergoing a test on the proving ground. The testing accumulated 4,240 hours and 16 failures occurred during this period. Calculate the MTBF value.

11.2. A computer chip failure rate is 0.00002 per hour. Estimate its reliability after 1,000 hours of operation using a constant failure rate model.

11.3. A pressure transducer failure follows the Weibull distribution. The scale parameter and the shape parameter are estimated as 550 hours and 0.6, respectively. What is the reliability after 600 hours of operation? Determine the MTTF for the transducer. (Note: Use a two-parameter Weibull distribution.)

11.4. The life (in hours) of a component is modeled using a lognormal distribution. The two parameters for the distribution are given as mean, $\mu = 4.8$, and standard deviation, $\sigma = 2.3$. Determine the reliability of the component at 200 hours, and its MTTF.

11.5. A video camera has 20 components in series. Find its reliability if each component has a reliability of 0.999.

11.6. Two components are placed in a parallel configuration. Determine the reliability of the system after 500 hours of operation if the component failure rates are 0.0009/hour and 0.00075/hour. Also calculate the MTBF for the system.

11.7. A personal computer system consists of the components, the processor unit, the monitor, the mouse, and the keyboard. All components must operate for system success. The expected failures per million hours for the components are 120 for the processor unit, 200 for the monitor, 150 for the mouse, and 180 for the keyboard. Determine the system reliability for a mission time of 500 hours and the system MTTF.

11.8. A mechanical system consists of the 4 components shown in Fig. P11.8. The component reliabilities are given in the blocks. Find the system reliability.

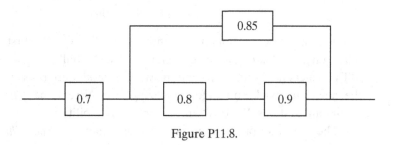

Figure P11.8.

11.9. A mechanical system consists of the 6 components shown in Fig. P11.9. Component reliabilities at 100 hours of operation are given in the blocks. Assume an exponential distribution. Find (a) the system reliability, (b) the system failure rate, and (c) the system MTTF.

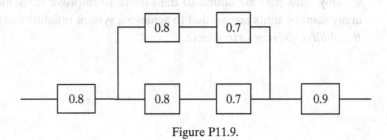

Figure P11.9.

11.10. A mechanical system consists of the 7 components shown Fig. P11.10. Component failure rates (λ/hour) are given in the blocks. Assume an exponential distribution and find the system reliability at 200 hours.

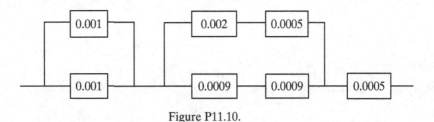

Figure P11.10.

11.11. An aircraft operates on 4 identical turbojet engines, each with a reliability of 0.96 at 500 hours of operation. Determine the system reliability if 2 of the 4 active elements are required for successful operation. Also evaluate the MTTF.

11.12. A power station has 5 generators operating. Four of 5 generators must operate for system success. If each generator has an MTTF of 10,000 hours, determine the system reliability at 500 hours. Also calculate the system MTTF.

11.13. A system has three identical components, two of which must operate for system success. The third component is a standby unit. If each component has a failure rate of 0.0006/hour, find the system reliability for a mission time of 400 hours. Assume perfect sensing and switching. Compare the reliability and MTTF of a two-components system with no standby.

11.14. A system consists of 15 identical units, each with an MTTF of 1,000 hours. Determine the MTTF if all units must operate for system success. If standby units may be added to the system to increase the MTTF, determine how many standby units are needed to achieve a system MTTF of 250 hours. (Note: Use *ReliabilityOf-Systems.xls*, sheet2.)

11.15. A system consists of 6 identical units, each with a failure rate of 0.0001/hour. All 6 units must operate for system success. Determine the system reliability. If standby units may be added to the system to improve reliability, determine how many standby units are needed to achieve a system reliability of 0.9999. (Note: Use *ReliabilityOfSystems.xls*, sheet2.)

12

Reliability Testing

12.1 Introduction

In this chapter we discuss two aspects of life testing. One aspect deals with the analysis of life test data of components to evaluate the parameters α, β, and γ for the Weibull distribution, λ for the exponential distribution, and μ and σ for the lognormal distribution. Once the parameters are evaluated, the reliability can be evaluated at a given life. The second aspect deals with exponent-based life testing. Here we discuss the confidence intervals for the mean life, followed by the acceptance sampling procedures used in life testing.

12.2 Weibull Distribution Parameter Estimation

We develop a procedure for the estimation of the three Weibull parameters, α, β, and γ. This procedure also works for the exponential distribution. If the shape parameter β takes a value close to 1, we conclude that the distribution is exponential. For an exponential distribution, the failure rate is $\lambda = 1/\alpha$. The location parameter γ is generally zero, which defines the two-parameter Weibull distribution. The parameter γ provides an estimate of the earliest time to failure. If $\gamma > 0$, it defines a failure-free operating period from 0 to γ. A negative value of γ may indicate that failures occurred prior to the start of the test, possibly in storage or in transit. In this case γ occurs earlier than the first failure at t_1.

In order to evaluate the parameters, the time-to-failure data for a sample size of N are collected. The data for n failures are arranged in an ascending order of the time to failure: $t_1 \leq t_2 \leq \cdots \leq t_n (n \leq N)$. The data are arranged in a table as shown:

Time to failure, hours or cycles	Failure order
t_1	1
t_2	2
...	...
t_j	j
...	...
t_n	n

We use the median rank MR to evaluate the probability of failure of the jth failure with a confidence level of 0.5. MR is evaluated by solving the binomial expression

$$\sum_{k=j}^{N} \binom{N}{k} \mathrm{MR}^k (1 - \mathrm{MR})^{N-k} = 0.5. \tag{12.1}$$

This has been included as the function MR(j,N) in the Excel program *Weibull3.xls*. Alternatively, we may use the remarkably accurate approximation

$$\mathrm{MR}_j = \frac{j - 0.3}{N + 0.4}, \tag{12.2}$$

which is also referred to as Bernard's approximation for the median rank.

The reliability at the jth failure is evaluated as

$$R(t_j) = 1 - \mathrm{MR}_j. \tag{12.3}$$

For the three-parameter Weibull distribution, we have the relation

$$R(t) = e^{-(\frac{t-\gamma}{\alpha})^\beta}. \tag{12.4}$$

On taking the natural logarithm twice, we get

$$\ln(-\ln R(t)) = \beta \ln(t - \gamma) - \beta \ln \alpha. \tag{12.5}$$

Setting $y = \ln(-\ln R(t))$, $x = \ln(t - \gamma)$, $a = \beta$, and $b = -\beta \ln \alpha$, Eq. (12.5) takes the form

$$y = ax + b, \tag{12.6}$$

which represents the equation of a straight line.

We develop two additional columns in the table for

$$x_j = \ln(t_j - \gamma), \qquad y_j = \ln(-\ln(1 - \mathrm{MR}_j)). \tag{12.7}$$

The table is prepared for $\gamma = 0$.

12.2 Weibull Distribution Parameter Estimation

The squared error with respect to the straight-line fit is given by

$$\varepsilon = \sum_{j=1}^{n} [y_j - (ax_j + b)]^2. \qquad (12.8)$$

For the least squares error, the slope a and the intercept b for the best-fitting straight line are given by

$$a = \frac{\sum_{j=1}^{n} x_j y_j - n\bar{x}\bar{y}}{\sum_{j=1}^{n} x_j^2 - n\bar{x}^2} \qquad b = \bar{y} - a\bar{x}. \qquad (12.9)$$

The shape parameter β and the scale parameter α are then given by $\beta = a$, and $\alpha = e^{-(\frac{b}{\beta})}$.

For $\gamma = 0$, the parameters for the two-parameter Weibull distribution are thus evaluated; the analysis also applies for any given value of γ. We can make a one-dimensional search on γ from, say, $-2t_1$ to t_1 for the least squares error and determine the three parameters γ, α, and β for the data. This approach has been implemented in the program *Weibull3.xls* included in the CD. Sheet1 of the program uses the golden section search for γ. On sheet3, γ can be changed from $-t_1$ to t_1 using the spin button to maximize the coefficient of determination, R^2 (RSQ). The plot changes dynamically as the spin button operates. Sheet2 shows the plot of t vs. $R(t)$ using the parameters evaluated on sheet1.

Example 12.1. The following data were collected in a test for the evaluation of Weibull parameters:

Time to failure, cycles	Failure order
346100	1
434990	2
457270	3
475380	4
553890	5
600020	6
653440	7
679700	8
727075	9
764055	10

The sample size is 10. Determine the Weibull parameters and evaluate the reliability at 800,000 cycles.

Solution.

j	t	MR	$\gamma = 0$ $t - \gamma$	x $\ln(t - \gamma)$	y $\ln(-\ln(1 - R))$
1	346100	0.0670	346100	12.7545	−2.6692
2	434990	0.1623	434990	12.9831	−1.7313
3	457270	0.2586	457270	13.0330	−1.2067
4	475380	0.3551	475380	13.0719	−0.8240
5	553890	0.4517	553890	13.2247	−0.5093
6	600020	0.5483	600020	13.3047	−0.2297
7	653440	0.6449	653440	13.3900	0.0348
8	679700	0.7414	679700	13.4294	0.3020
9	727075	0.8377	727075	13.4968	0.5980
10	764055	0.9330	764055	13.5464	0.9946

From the spreadsheet, we have $\sum_{j=1}^{10} x_j y_j = -66.746, \sum_{j=1}^{10} x_j^2 = 1749.19$, $\bar{x} = 13.223$, and $\bar{y} = -0.524$. On substituting these in Eq. (12.9), we get $a = 4.31$, $b = -57.55$, $\beta = a = 4.31$, and $\alpha = e^{-(\frac{b}{\beta})} = 624{,}638$; MTTF = 568,653.

For $\gamma = 143{,}980$, we get the maximum RSQ and the parameters are $a = 3.09$, $b = -40.39$, $\beta = a = 3.09$, and $\alpha = e^{-(\frac{b}{\beta})} = 475{,}060$; MTTF = 570,817. The Weibull plots for the two cases are shown in Figs. 12.1a and 12.1b.

The survival graph for the three-parameter Weibull is shown in Fig. 12.1c. The reliability at 800,000 cycles is 0.0692.

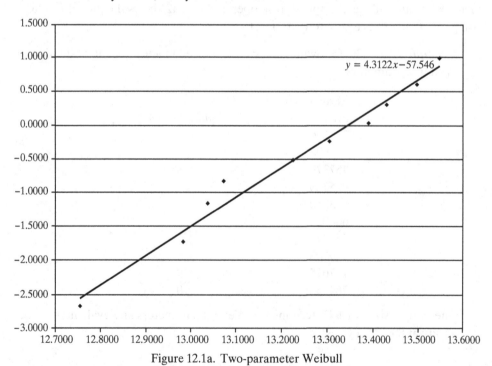

Figure 12.1a. Two-parameter Weibull

12.2 Weibull Distribution Parameter Estimation

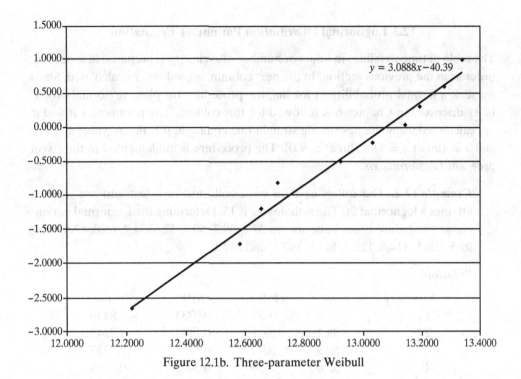

Figure 12.1b. Three-parameter Weibull

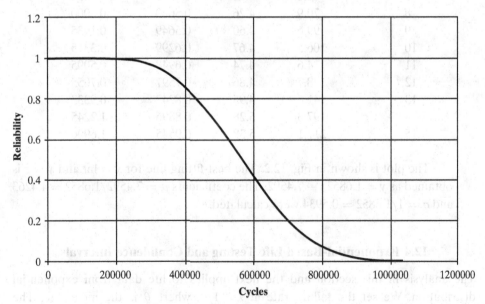

Figure 12.1c. Three-parameter Weibull

12.3 Lognormal Distribution Parameter Estimation

The table of time to failure in hours or number of cycles (t_i) is prepared in ascending order as in the previous section. In the next column, $\ln t_i$ values are calculated. Since we seek a normal probability fit for $\ln t_i$, the procedure for plotting normal probability discussed in Chapter 6 is followed for this column. The parameters μ and σ are calculated using the best-fitting straight line equation; μ is the $\ln t$ value at $z = 0$ and σ is ($\ln t$ at $z = 1$) − ($\ln t$ at $z = 0$). The procedure is implemented in the Excel program *Lognormal.xls*.

Example 12.2. The following data were collected in a test and are used to attempt a lognormal fit. The sample size is 15. Determine the lognormal parameters. The failure times in hours are 37.2, 39.2, 50.3, 52.6, 54.2, 66.0, 67.6, 70.9, 99.6, 106.5, 114.6, 128.7, 141.6, 197.3, and 217.1.

Solution.

Number i	t_i	$\ln(t_i)$	MR	Z-value
1	37.2	3.62	0.0333	−1.8339
2	39.2	3.67	0.1104	−1.2245
3	50.3	3.92	0.1753	−0.9333
4	52.6	3.96	0.2403	−0.7055
5	54.2	3.99	0.3052	−0.5095
6	66.0	4.19	0.3701	−0.3315
7	67.6	4.21	0.4351	−0.1635
8	70.9	4.26	0.5000	0.0000
9	99.6	4.60	0.5649	0.1635
10	106.5	4.67	0.6299	0.3315
11	114.6	4.74	0.6948	0.5095
12	128.7	4.86	0.7597	0.7055
13	141.6	4.95	0.8247	0.9333
14	197.3	5.28	0.8896	1.2245
15	217.1	5.38	0.9545	1.6906

The plot is shown in Fig. 12.2. The best-fitting line for $x = \ln t$ and $y = z$ is obtained as $y = 1.6852x - 7.4592$. The coefficients $\mu = 7.4592/1.6852 = 4.4263$ and $\sigma = 1/1.6852 = 0.5934$ were calculated.

12.4 Exponential-Based Life Testing and Confidence Intervals

The analysis in this section and the next applies to life data from exponential distribution. We set the failure rate at $\lambda = 1/\theta$, where θ is the mean life. The

12.4 Exponential-Based Life Testing and Confidence Intervals

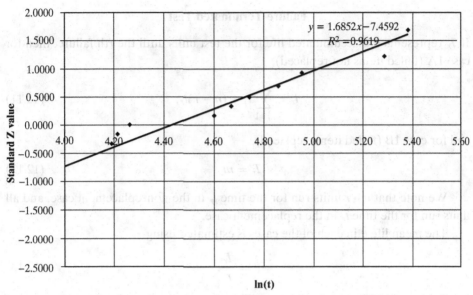

Figure 12.2. Lognormal plot

failure probability density function is given by

$$f(t) = \frac{1}{\theta} e^{-\frac{t}{\theta}}. \tag{12.10}$$

The exponential distribution is widely used to model constant failure rate processes. If we substitute $u = 2t/\theta$, then the distribution is $f(u) = (1/2)\exp(-u/2)$, which is a chi-squared distribution with two degrees of freedom. Thus, we say $2t/\theta$ has a chi-squared distribution with df = 2.

Let n be the number of units placed in the life test, and let t_j denote the time when the jth unit fails. Let k be the most recent failure. The failure times are arranged in the order $t_1 \leq t_2 \leq \cdots \leq t_k$. A preassigned number of failures, r, is decided at the test design stage. We have several choices of testing:

1. To stop the testing when the number of failures, r, is reached (*failure terminated test*) or
2. To stop the test when a preassigned time T_0 is reached. If the rth failure occurs before this time, the lot is rejected (*time terminated test*).

In each of the preceding two cases, we have one of the following choices:

A. The failed unit is not replaced (*testing without replacement*).
B. The failed unit is immediately replaced (*testing with replacement*).

Failure-Terminated Test

If T_r represents the accumulated life for the test units until the rth failure, then for case 1A (failed items not replaced),

$$T_r = \sum_{j=1}^{r} t_j + (n-r)t_r, \tag{12.11}$$

and for case 1B (failed items replaced),

$$T_r = nt_r. \tag{12.12}$$

We note that $n - r$ units run for the time t_r in the nonreplacement case, and all units run for the time t_r in the replacement case.

The mean life $\hat{\theta}$ in each of the cases is estimated using

$$\hat{\theta} = \frac{T_r}{r}. \tag{12.13}$$

It is known that $2T_r/\theta (= 2r\hat{\theta}/\theta)$ has a chi-squared distribution with $2r$ degrees of freedom (see the comment following Eq. (12.10)). The $1 - \alpha$ confidence interval for the mean life is given by

$$\frac{2T_r}{\chi^2_{\alpha/2, 2r}} \leq \theta \leq \frac{2T_r}{\chi^2_{1-\alpha/2, 2r}}. \tag{12.14}$$

Example 12.3. A failure-terminated test was performed on 15 gyro units without replacement. The test was terminated after 5 failures. The failure times in hours are 642, 674, 705, 722, and 732. Determine the 95% confidence interval for the mean life using the exponential time-to-failure model.

Solution. We have $n = 15$ and $r = 5$. From Eq. (12.11), we get

$$T_r = 642 + 674 + 705 + 722 + 732 + (15 - 5)\,732 = 10{,}795 \text{ hours}$$

$$\hat{\theta} = \frac{10795}{5} = 2{,}159 \text{ hours.}$$

From Appendix Table A.5, we have $\chi^2_{0.025, 10} = 20.483$, $\chi^2_{0.975, 10} = 3.247$. Substituting in Eq. (12.14), we get

$$\frac{2 \times 10795}{20.483} \leq \theta \leq \frac{2 \times 10795}{3.247}$$

$$1054 \leq \theta \leq 6649.$$

The calculations are similar for the test with replacement, except that the T_r calculation is done using Eq. (12.12).

Time-Terminated Test

Let T_0 be the preassigned time for the life test conducted on a random sample of n units. Let x be the number of failures observed during T_0. Let the preassigned number of failures be r. If r failures occur before the time T_0, the lot is rejected. Analysis for further decision is only performed on the tests where $x < r$. For case 2A (failed items not replaced),

$$T_x = \sum_{j=1}^{x} t_j + (n - x)T_0 \tag{12.15}$$

and for case 2B (failed items replaced),

$$T_x = nT_0. \tag{12.16}$$

The mean life is estimated using

$$\hat{\theta} = \frac{T_x}{x}. \tag{12.17}$$

The $1 - \alpha$ confidence interval for the mean life is given by

$$\frac{2T_x}{\chi^2_{\alpha/2, 2x+2}} \leq \theta \leq \frac{2T_x}{\chi^2_{1-\alpha/2, 2x+2}}. \tag{12.18}$$

This interval is established under the assumption that the $(x + 1)$th failure is about to occur when the test is time terminated.

12.5 Sampling Procedures for Life Testing (Exponential-Based)

This presentation on the sampling procedures is based on the *Quality Control and Reliability Handbook* H-108, and a fundamental paper on life testing by Epstein and Sobel.[1] Let $\hat{\theta}$ be the mean life from a failure-terminated or time-terminated life test. Epstein and Sobel state: "the Neyman-Pearson theory tells us that a simple test for $\theta = \theta_0$ against $\theta < \theta_0$ with Type I error $= \alpha$ is given by an acceptance region of the form"

$$\hat{\theta} > \theta_0 \frac{\chi^2_{1-\alpha, 2r}}{2r}. \tag{12.19}$$

---

[1] B. Epstein and M. Sobel, Life Testing, *Journal of the American Statistical Association*, v48, n263 (1953) pp. 486–502.

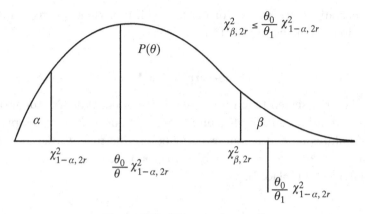

Figure 12.3. OC curve for $r = 5$

The operating characteristic (OC) curve is developed using the procedure specified in Eq. (12.19). Denoting $P(\theta)$ as the probability of accepting $\theta = \theta_0$ when θ is the true value (Type II error),

$$P(\theta) = \Pr\left(\hat{\theta} > \theta_0 \frac{\chi^2_{1-\alpha,2r}}{2r}\right) = \Pr\left(\frac{2r\hat{\theta}}{\theta} > \frac{\theta_0}{\theta}\chi^2_{1-\alpha,2r}\right) = \Pr\left(\chi^2_{2r} > \frac{\theta_0}{\theta}\chi^2_{1-\alpha,2r}\right),$$

(12.20)

and Eq. (12.20) can be interpreted as finding P for a given θ satisfying

$$\chi^2_{P,2r} = \frac{\theta_0}{\theta}\chi^2_{1-\alpha,2r}.$$

(12.21)

The interpretation of Eqs. (12.20) and (12.21) is shown in the chi-squared distribution curve in Fig. 12.3. The probability of acceptance is the area to the right of $\frac{\theta_0}{\theta}\chi^2_{1-\alpha,2r}$, for $\theta_0/\theta \geq 0$. We also have the parameter r, which appears from df = $2r$. If the consumer's risk point is defined as (θ_1, β), we need to choose the smallest integer value of r that satisfies

$$\chi^2_{\beta,2r} \leq \frac{\theta_0}{\theta_1}\chi^2_{1-\alpha,2r}.$$

(12.22)

The final expression in Eq. (12.21) is calculated in Excel using **chidist** $(\frac{\theta_0}{\theta}\chi^2_{1-\alpha,2r}, 2r)$ for given values of α, r, and various values of θ. The plot θ/θ_0 vs. $P(\theta)$ is the OC curve. The OC curve plot has been introduced in the program *LifeTestOC.xls*. The OC curve for $r = 5$, and $1 - \alpha = 0.9$ is shown in Fig. 12.4. The values of r, the number of units tested, n, and α can be varied using the spin button. The OC curve passes through the point $(\theta/\theta_0 = 1, P(\theta) = 1 - \alpha)$. The OC curve must keep to the right of the point $(\theta_1/\theta_0, P(\theta_1) = \beta)$, defining the consumer's risk. If the curve can be manipulated to keep it to the right of the point at β, it gives the right choice for r and other parameters. MIL-HDBK-108 gives twenty pages of OC

12.5 Sampling Procedures for Life Testing (Exponential-Based)

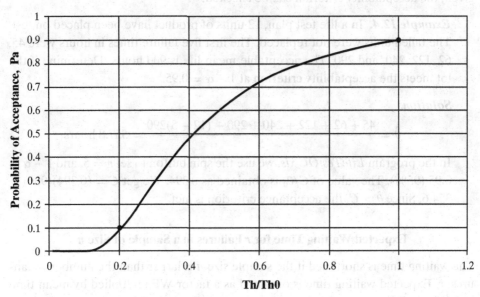

Figure 12.4. OC curve $r = 3$

curves for various values of r and $1 - \alpha$. The program can be used to generate all these curves.

Lot Acceptance Procedure for Termination upon Preassigned Number of Failures

The estimate of the lot mean life $\hat{\theta}$ is calculated as given in Eq. (12.13). T_r is calculated using Eq. (12.11) for case 1A, where the failed items are not replaced, and Eq. (12.12) for case 1B, where the failed items are replaced. Turning our attention to Eq. (12.19), we define the constant C:

$$\frac{C}{\theta_0} = \frac{\chi^2_{1-\alpha,2r}}{2r}. \tag{12.23}$$

The lot is accepted if

$$\hat{\theta} \geq C \tag{12.24}$$

and rejected otherwise.

Values of C/θ_0 for various code letters are given in a table in the handbook. The value r is changed so that the OC curve is just to the right of the consumer's point in the program *LifeTestOC.xls*. The value of C/θ_0 is obtained in cell C4. If the acceptable mean life θ_0 and the corresponding probability of acceptance, $1 - \alpha$, are given, the acceptability criterion can be checked.

Example 12.4. In a life test plan, 12 units of product have been placed on test. The failed units were not replaced. The first five failure times in hours were 45, 62, 122, 240, and 290. The acceptable mean life is 900 hours. Determine if the lot meets the acceptability criterion at $1 - \alpha = 0.95$.

Solution.
$$\hat{\theta} = \frac{45 + 62 + 122 + 240 + 290 + (12 - 5)290}{5} = 557.8 \text{ hours}.$$

In the program *LifeTestOC.xls*, we use the spin button to set $r = 5$, and $1 - \alpha = 0.95$ (95%). The value of C/θ_0 is obtained as 0.394. We get $C = (0.394)(900) = 354.6$. Since $\hat{\theta} > C$, the acceptance criterion is met.

Expected Waiting Time for *r* Failures in a Sample of Size *n*

The waiting time is shortened if the sample size n is larger than the number of failures, r. Expected waiting time is obtained as a factor WF multiplied by mean time to failure θ_0. The expected waiting time factor (WF) is given by

$$\text{WF} = \frac{1}{n} + \frac{1}{n-1} + \cdots + \frac{1}{n-r+1} \tag{12.25}$$

without replacement, and by

$$\text{WF} = \frac{r}{n} \tag{12.26}$$

with replacement.

The waiting time is given by

$$\text{Waiting Time} = \text{WF}\,\theta_0. \tag{12.27}$$

The function WF(r,n) has been introduced in the program *LifeTestOC.xls* to calculate the ratio WF in Eq. (12.25).

Example 12.5. Determine the waiting time in terms of θ_0 for a test without replacement when the sample size is 12 and the number of failures for the test plan is 5. Compare it with the waiting time for the test with replacement.

Solution.
$$\text{WF} = \frac{1}{12} + \frac{1}{11} + \frac{1}{10} + \frac{1}{9} + \frac{1}{8} = 0.5104;$$

the waiting time $= 0.5104\theta_0$. For the test with replacement,

$$\text{WF} = \frac{5}{12} = 0.4167, \text{ and waiting time} = 0.4167\theta_0.$$

12.5 Sampling Procedures for Life Testing (Exponential-Based)

Example 12.6. The number of failures, r, and the sample size n are to be chosen to complete the failure-terminated test without replacement in an average time of 300 hours. The mean life $\theta_0 = 1{,}000$ hours and $1 - \alpha = 0.9$, $\beta = 0.1$ at $\theta_1/\theta_0 = 0.2$. Choose the parameters.

Solution. We enter $1 - \alpha = 0.9$, $\beta = 0.1$ at $\theta_1/\theta_0 = 0.2$ on sheet1 of the program *LifeTestOC.xls*. After manipulating the curve by changing r, we find that $r = 3$ satisfies the producer's and consumer's points. Since the test is to be completed in an average time of 300 hours, the waiting factor needs to be WF $= 300/1000 = 0.3$. By changing the n-value, we find that $n = 11$ gives WF $= 0.302$, and $n = 12$ gives WF $= 0.2742$. We choose $n = 12$.

Lot Acceptance Procedure for Termination upon Preassigned Time

In designing an acceptance procedure based on preassigned time or r failures, the known quantities are the mean life θ_0, the probability of acceptance, $1 - \alpha$, at the mean life, and the probability of acceptance, β, at a given value of $\theta_1(< \theta_0)$. Using these data we obtain r by manipulating the spin buttons on the OC curve in the program *LifeTestOC.xls*. This choice of r is similar to that in the failure-terminated test.

Once the value of r is established, we need to determine the termination time T for assignment. For the failure-terminated test, we established C/θ_0, and we accept the lot if the estimated life $\hat{\theta}$ exceeds C. Using the concept of the waiting factor, the following formula is used to calculate the termination time T in the H108 handbook:

$$\frac{T}{\theta_0} = \text{WF} \, \frac{C}{\theta_0}. \qquad (12.28)$$

WF is found using Eq. (12.25) for the test without replacement and Eq. (12.26) for the test with replacement. Equation (12.28) has been implemented in the program *LifeTestOC.xls*. For the time-terminated test, the value of T/θ_0 can be found in cell C9 for the nonreplacement case, and C10 for the replacement case. T can be calculated once this ratio is found. The handbook gives T/θ_0 ratios in several tables for $n = 2r, 3r, 4r, \ldots$. We can obtain these ratios for any value of n using the program. We illustrate the design of a time-terminated life test plan by means of an example.

Example 12.7. Design a time-terminated life test sampling plan without replacement which will accept a lot having an acceptable mean life of 1,000 hours with probability 0.95 (Type I error, 0.05) and the probability of acceptance of 0.1 for accepting a mean life of 300 hours (Type II error, 0.1). Due to the constraints in the test lab, the maximum sample size is restricted to 12.

Solution. We have $\theta_0 = 1{,}000$ h, $1 - \alpha = 0.95$, and $\beta = 0.1$ at $\theta_1/\theta_0 = 300/1000 = 0.3$. Entering these values in sheet1 of the program *LifeTestOC.xls*, and changing r using the spin button, we find that $r = 7$ satisfies the producer's and consumer's points.

We use the maximum sample size that is permitted, $n = 12$. (n is changed to 12 using the spin button.) For the nonreplacement case, the value of T/θ_0 is found to be 0.3848 (from cell C9). The termination time T is now calculated:

$$T = \theta_0(T/\theta_0) = 1000 \times 0.3848 = 385 \text{ hours.}$$

The tables from the handbook only give a choice of $n = 2r = 14$ for the minimum. If this is possible in the lab, we get T/θ_0 as 0.3092.

Sequential Life Test Sampling Plans

In the sequential test plan, the life testing goes on in a continuous manner until a decision can be reached. Let j be the number of failed product units observed in time t_j. Let n be the sample size. Consider the kth failure, which takes place at the current time t. We calculate $V(t)$, the total length of time survived by all units of the product in the test:

$$V(t) = \sum_{j=1}^{k} t_j + (n - k)t \text{ (failed items not replaced)} \qquad (12.29)$$

$$V(t) = nt \text{ (failed items replaced).} \qquad (12.30)$$

We establish the acceptance time line

$$h_0 + ks, \qquad (12.31)$$

and the rejection time line

$$h_1 + ks, \qquad (12.32)$$

as shown in Fig. 12.5. The acceptability criterion is

$$h_1 + ks < V(t) < h_0 + ks. \qquad (12.33)$$

12.5 Sampling Procedures for Life Testing (Exponential-Based)

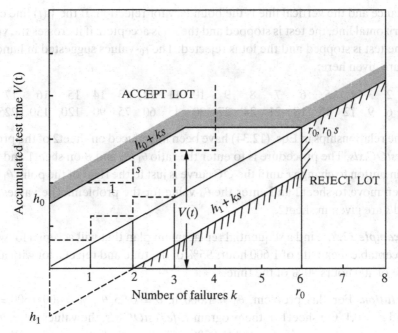

Figure 12.5. Sequential life test

The military handbook H108 uses the formulas developed by Epstein.[2] The constants $h_0, h_1,$ and s are calculated using

$$\frac{h_0}{\theta_0} = \frac{-\ln B}{\theta_0/\theta_1 - 1} \qquad \frac{h_1}{\theta_0} = \frac{-\ln A}{\theta_0/\theta_1 - 1} \qquad \frac{s}{\theta_0} = \frac{-\ln(\theta_0/\theta_1)}{\theta_0/\theta_1 - 1},$$
$$A = (1 - \beta)/\alpha \qquad B = \beta/(1 - \alpha) \tag{12.34}$$

where α is the probability of a Type I error at θ_0, and β is the probability of a Type II error at θ_1. A and B used in the preceding calculations are approximations of the exact expressions. The calculated values obtained using these expressions closely match the published handbook values.

For a known value of acceptable mean life $\theta_0, h_0, h_1,$ and s are calculated using Eqs. (12.34). The sequential life test boundaries are then drawn as shown in Fig. 12.5. $V(t)$ defines a point at each failure k. A horizontal dotted line is drawn to $k + 1$. The $V(t)$ line is a stepped line. If $V(t)$ continues to be inside $h_1 + ks, h_0 + ks$, the test can go on and on. In practice we set a limit of r_0 ($r_0 \gg r$) and, from the point $(r_0, r_0 s)$, a horizontal line is drawn to intersect with $h_0 + ks$ and a vertical line is drawn to intersect with $h_1 + ks$. The horizontal line represents the boundary for

---

[2] B. Epstein, Statistical Life Test Acceptance Procedures, *Technometrics*, v2, n4 (1960) pp. 435–446.

acceptance and the vertical line is the boundary for rejection. If the $V(t)$ line crosses the horizontal line, the test is stopped and the lot is accepted; if it crosses the vertical line, the test is stopped and the lot is rejected. The r_0-values suggested in handbook H108 are given here:

r	1	2	3	4	5	6	7	8	9	10	11	12	13	14	15	16	17	18
r_0	3	6	9	12	15	18	21	24	27	30	45	60	75	90	120	150	225	300

The relationships in Eqs. (12.34) have been introduced on sheet2 of the program *LifeTestOC.xls*. The procedure is to enter the ratio θ_1/θ_0 and β on sheet1 and to use the spin button to change r until the OC curve is just to the right of the point $\theta_1/\theta_0, \beta$. We then move to sheet2 and enter the θ_0-value for the problem. The values of h_0, h_1, and s are given in sheet2.

Example 12.8. Find a sequential replacement plan that will accept a lot with an acceptable mean life of 1,600 hours 95% of the time and reject a lot with a mean life of 400 hours 90% of the time.

Solution. For this problem, $\theta_0 = 1{,}600$ h, $\alpha = 0.05$, $\theta_1/\theta_0 = 400/1600 = 0.25$, and $\beta = 0.1$. On sheet1 of the program *LifeTestOC.xls*, the values for θ_1/θ_0 and β are entered; $1 - \alpha$ is changed to 95% using the spin button, and r is then varied until the curve moves to the right of the consumer risk point. We obtain $r = 6$. The value of θ_0 is entered on sheet2. The sequential test plan parameters are obtained as $h_0 = 1412.9$, $h_1 = -1813.98$, and $s = 795.03$. The sequential life test sheet can be prepared using these data.

The handbook also gives parameters that enable us to calculate the expected time to reach a decision in the sequential test.

12.6 Summary

In this chapter we developed Weibull parameter estimation concepts. The program *Weibull3.xls* calculates the three parameters using life test data. We have also discussed the life test plans based on the military handbook, H108. The theory of using an OC curve for life tests has been presented in detail. The program *LifeTestOC.xls* gives the OC curve for any specified values of Type I and Type II errors. It provides all the necessary parameters for both failure-terminated and time-terminated tests. The program also provides the parameters for the sequential life test.

The computer programs discussed in this chapter include the following:

Weibull3.xls
Lognormal.xls
LifeTestOC.xls.

EXERCISE PROBLEMS

12.1. A sample of 10 units was subjected to a life test. The failure times in hours are obtained as 6, 12, 23, 31, 42, 60, 71, 99, 126, and 187. Determine the Weibull parameters and check if an exponential distribution can be used to model the failures. Estimate the MTTF and predict the reliability for a 15-hour mission.

12.2. A sample of 50 units was subjected to a life test. The failure times in hours are obtained as 8, 21, 32, 46, 59, 70, 84, 98, 117, and 126. Determine the Weibull parameters and check if an exponential distribution can be used to model the failures. Estimate the MTTF and predict the reliability for a 60-hour mission.

12.3. A sample of 50 units was subjected to a life test. The failure times in hours are obtained as 23, 40, 55, 67, 76, 92, 104, 110, 120, and 124. Determine the Weibull parameters. Estimate the MTTF and plot the survival graph.

12.4. A sample of 50 units was subjected to a life test. The failure times in hours are obtained as 4.8, 5.8, 8.4, 21.3, 34.4, 44.0, 44.1, 61.4, 86.3, and 209.3. Determine the lognormal parameters using the best-fit approach. Estimate the MTTF.

12.5. A random sample of 25 light-emitting diodes (LEDs) is subjected to a life test. The test is terminated after 5 failures. Failed items are not replaced. The failure times in hours are 245, 270, 285, 305, and 362. Estimate the MTTF for the LEDs, assuming exponential distribution, and find a 90% confidence interval for the mean life.

12.6. A life test is performed on brake pedals by an automobile manufacturer. The number of cycles to failure is the criterion used. A random sample of 40 units is tested and failed units are immediately replaced. The test is stopped after 10,000 cycles; 4 units failed at 1,550, 4,320, 6,540, and 7,930 cycles. Estimate the mean time to failure and find a 95% confidence interval for the MTTF.

12.7. A life testing plan is terminated at the sixth failure. The test is conducted without replacement using a random sample of 40 units. The failure times in hours are 340, 456, 484, 630, 710, and 740. The producer's risk is 0.1, the probability of rejecting an acceptable mean life of 8,000 hours. Determine if the lot is acceptable with these criteria. Plot the OC curve and find the probability of accepting a mean life of 2,000 hours.

12.8. A failure-terminated life testing plan is to be designed with a producer's risk of $\alpha = 0.05$ at an acceptable life of $\theta_0 = 1,200$ hours, and consumer's risk $\beta = 0.1$ at $0.2\theta_0$. The test is conducted without replacement. Choose the least number of failures, r, satisfying the criteria and find the sample size so that the expected test completion time is less than $0.3\theta_0$. (Note: Use the program *LifeTestOC.xls*.)

12.9. Design a time-terminated life test sampling plan without replacement. The plan should accept a lot that has an acceptable mean life of 1,500 hours with a probability of 0.9 and should accept a mean life of 300 hours with probability of 0.1. Choose the sample size so that the time for terminating the test can be set at 150 hours. Plot the OC curve.

12.10. Design a sequential replacement plan that will accept a lot with an acceptable mean life of 2,000 hours 90% of the time and reject a lot with a mean life of 300 hours 92% of the time.

Appendix

Table A.1. *Cumulative binomial probabilities*

x	n	p	Term	Cumulative
3	9	0.2	0.176	0.914

									p						
n	x	0.01	0.05	0.1	0.2	0.25	0.3	0.35	0.4	0.45	0.5	0.6	0.7	0.8	0.9
2	0	0.980	0.903	0.810	0.640	0.563	0.490	0.423	0.360	0.303	0.250	0.160	0.090	0.040	0.010
3	0	0.970	0.857	0.729	0.512	0.422	0.343	0.275	0.216	0.166	0.125	0.064	0.027	0.008	0.001
	1	1.000	0.993	0.972	0.896	0.844	0.784	0.718	0.648	0.575	0.500	0.352	0.216	0.104	0.028
4	0	0.961	0.815	0.656	0.410	0.316	0.240	0.179	0.130	0.092	0.063	0.026	0.008	0.002	0.000
	1	0.999	0.986	0.948	0.819	0.738	0.652	0.563	0.475	0.391	0.313	0.179	0.084	0.027	0.004
	2	1.000	1.000	0.996	0.973	0.949	0.916	0.874	0.821	0.759	0.688	0.525	0.348	0.181	0.052
5	0	0.951	0.774	0.590	0.328	0.237	0.168	0.116	0.078	0.050	0.031	0.010	0.002	0.000	0.000
	1	0.999	0.977	0.919	0.737	0.633	0.528	0.428	0.337	0.256	0.188	0.087	0.031	0.007	0.000
	2	1.000	0.999	0.991	0.942	0.896	0.837	0.765	0.683	0.593	0.500	0.317	0.163	0.058	0.009
	3	1.000	1.000	1.000	0.993	0.984	0.969	0.946	0.913	0.869	0.813	0.663	0.472	0.263	0.081
6	0	0.941	0.735	0.531	0.262	0.178	0.118	0.075	0.047	0.028	0.016	0.004	0.001	0.000	0.000
	1	0.999	0.967	0.886	0.655	0.534	0.420	0.319	0.233	0.164	0.109	0.041	0.011	0.002	0.000
	2	1.000	0.998	0.984	0.901	0.831	0.744	0.647	0.544	0.442	0.344	0.179	0.070	0.017	0.001
	3	1.000	1.000	0.999	0.983	0.962	0.930	0.883	0.821	0.745	0.656	0.456	0.256	0.099	0.016
7	0	0.932	0.698	0.478	0.210	0.133	0.082	0.049	0.028	0.015	0.008	0.002	0.000	0.000	0.000
	1	0.998	0.956	0.850	0.577	0.445	0.329	0.234	0.159	0.102	0.063	0.019	0.004	0.000	0.000
	2	1.000	0.996	0.974	0.852	0.756	0.647	0.532	0.420	0.316	0.227	0.096	0.029	0.005	0.000
	3	1.000	1.000	0.997	0.967	0.929	0.874	0.800	0.710	0.608	0.500	0.290	0.126	0.033	0.003
	4	1.000	1.000	1.000	0.995	0.987	0.971	0.944	0.904	0.847	0.773	0.580	0.353	0.148	0.026
8	0	0.923	0.663	0.430	0.168	0.100	0.058	0.032	0.017	0.008	0.004	0.001	0.000	0.000	0.000
	1	0.997	0.943	0.813	0.503	0.367	0.255	0.169	0.106	0.063	0.035	0.009	0.001	0.000	0.000
	2	1.000	0.994	0.962	0.797	0.679	0.552	0.428	0.315	0.220	0.145	0.050	0.011	0.001	0.000
	3	1.000	1.000	0.995	0.944	0.886	0.806	0.706	0.594	0.477	0.363	0.174	0.058	0.010	0.000
	4	1.000	1.000	1.000	0.990	0.973	0.942	0.894	0.826	0.740	0.637	0.406	0.194	0.056	0.005
9	0	0.914	0.630	0.387	0.134	0.075	0.040	0.021	0.010	0.005	0.002	0.000	0.000	0.000	0.000
	1	0.997	0.929	0.775	0.436	0.300	0.196	0.121	0.071	0.039	0.020	0.004	0.000	0.000	0.000
	2	1.000	0.992	0.947	0.738	0.601	0.463	0.337	0.232	0.150	0.090	0.025	0.004	0.000	0.000
	3	1.000	0.999	0.992	0.914	0.834	0.730	0.609	0.483	0.361	0.254	0.099	0.025	0.003	0.000
	4	1.000	1.000	0.999	0.980	0.951	0.901	0.828	0.733	0.621	0.500	0.267	0.099	0.020	0.001
	5	1.000	1.000	1.000	0.997	0.990	0.975	0.946	0.901	0.834	0.746	0.517	0.270	0.086	0.008
10	0	0.904	0.599	0.349	0.107	0.056	0.028	0.013	0.006	0.003	0.001	0.000	0.000	0.000	0.000
	1	0.996	0.914	0.736	0.376	0.244	0.149	0.086	0.046	0.023	0.011	0.002	0.000	0.000	0.000
	2	1.000	0.988	0.930	0.678	0.526	0.383	0.262	0.167	0.100	0.055	0.012	0.002	0.000	0.000
	3	1.000	0.999	0.987	0.879	0.776	0.650	0.514	0.382	0.266	0.172	0.055	0.011	0.001	0.000
	4	1.000	1.000	0.998	0.967	0.922	0.850	0.751	0.633	0.504	0.377	0.166	0.047	0.006	0.000
	5	1.000	1.000	1.000	0.994	0.980	0.953	0.905	0.834	0.738	0.623	0.367	0.150	0.033	0.002

(continued)

Table A.1 (continued)

x	n	p	Term	Cumulative
4	13	0.4	0.184	0.353

n	x	0.01	0.05	0.1	0.2	0.25	0.3	0.35	0.4	0.45	0.5	0.6	0.7	0.8	0.9
11	0	0.895	0.569	0.314	0.086	0.042	0.020	0.009	0.004	0.001	0.000	0.000	0.000	0.000	0.000
	1	0.995	0.898	0.697	0.322	0.197	0.113	0.061	0.030	0.014	0.006	0.001	0.000	0.000	0.000
	2	1.000	0.985	0.910	0.617	0.455	0.313	0.200	0.119	0.065	0.033	0.006	0.001	0.000	0.000
	3	1.000	0.998	0.981	0.839	0.713	0.570	0.426	0.296	0.191	0.113	0.029	0.004	0.000	0.000
	4	1.000	1.000	0.997	0.950	0.885	0.790	0.668	0.533	0.397	0.274	0.099	0.022	0.002	0.000
	5	1.000	1.000	1.000	0.988	0.966	0.922	0.851	0.753	0.633	0.500	0.247	0.078	0.012	0.000
	6	1.000	1.000	1.000	0.998	0.992	0.978	0.950	0.901	0.826	0.726	0.467	0.210	0.050	0.003
12	0	0.886	0.540	0.282	0.069	0.032	0.014	0.006	0.002	0.001	0.000	0.000	0.000	0.000	0.000
	1	0.994	0.882	0.659	0.275	0.158	0.085	0.042	0.020	0.008	0.003	0.000	0.000	0.000	0.000
	2	1.000	0.980	0.889	0.558	0.391	0.253	0.151	0.083	0.042	0.019	0.003	0.000	0.000	0.000
	3	1.000	0.998	0.974	0.795	0.649	0.493	0.347	0.225	0.134	0.073	0.015	0.002	0.000	0.000
	4	1.000	1.000	0.996	0.927	0.842	0.724	0.583	0.438	0.304	0.194	0.057	0.009	0.001	0.000
	5	1.000	1.000	0.999	0.981	0.946	0.882	0.787	0.665	0.527	0.387	0.158	0.039	0.004	0.000
	6	1.000	1.000	1.000	0.996	0.986	0.961	0.915	0.842	0.739	0.613	0.335	0.118	0.019	0.001
13	0	0.878	0.513	0.254	0.055	0.024	0.010	0.004	0.001	0.000	0.000	0.000	0.000	0.000	0.000
	1	0.993	0.865	0.621	0.234	0.127	0.064	0.030	0.013	0.005	0.002	0.000	0.000	0.000	0.000
	2	1.000	0.975	0.866	0.502	0.333	0.202	0.113	0.058	0.027	0.011	0.001	0.000	0.000	0.000
	3	1.000	0.997	0.966	0.747	0.584	0.421	0.278	0.169	0.093	0.046	0.008	0.001	0.000	0.000
	4	1.000	1.000	0.994	0.901	0.794	0.654	0.501	0.353	0.228	0.133	0.032	0.004	0.000	0.000
	5	1.000	1.000	0.999	0.970	0.920	0.835	0.716	0.574	0.427	0.291	0.098	0.018	0.001	0.000
	6	1.000	1.000	1.000	0.993	0.976	0.938	0.871	0.771	0.644	0.500	0.229	0.062	0.007	0.000
	7	1.000	1.000	1.000	0.999	0.994	0.982	0.954	0.902	0.821	0.709	0.426	0.165	0.030	0.001
14	0	0.869	0.488	0.229	0.044	0.018	0.007	0.002	0.001	0.000	0.000	0.000	0.000	0.000	0.000
	1	0.992	0.847	0.585	0.198	0.101	0.047	0.021	0.008	0.003	0.001	0.000	0.000	0.000	0.000
	2	1.000	0.970	0.842	0.448	0.281	0.161	0.084	0.040	0.017	0.006	0.001	0.000	0.000	0.000
	3	1.000	0.996	0.956	0.698	0.521	0.355	0.220	0.124	0.063	0.029	0.004	0.000	0.000	0.000
	4	1.000	1.000	0.991	0.870	0.742	0.584	0.423	0.279	0.167	0.090	0.018	0.002	0.000	0.000
	5	1.000	1.000	0.999	0.956	0.888	0.781	0.641	0.486	0.337	0.212	0.058	0.008	0.000	0.000
	6	1.000	1.000	1.000	0.988	0.962	0.907	0.816	0.692	0.546	0.395	0.150	0.031	0.002	0.000
	7	1.000	1.000	1.000	0.998	0.990	0.969	0.925	0.850	0.741	0.605	0.308	0.093	0.012	0.000
15	0	0.860	0.463	0.206	0.035	0.013	0.005	0.002	0.000	0.000	0.000	0.000	0.000	0.000	0.000
	1	0.990	0.829	0.549	0.167	0.080	0.035	0.014	0.005	0.002	0.000	0.000	0.000	0.000	0.000
	2	1.000	0.964	0.816	0.398	0.236	0.127	0.062	0.027	0.011	0.004	0.000	0.000	0.000	0.000
	3	1.000	0.995	0.944	0.648	0.461	0.297	0.173	0.091	0.042	0.018	0.002	0.000	0.000	0.000
	4	1.000	0.999	0.987	0.836	0.686	0.515	0.352	0.217	0.120	0.059	0.009	0.001	0.000	0.000
	5	1.000	1.000	0.998	0.939	0.852	0.722	0.564	0.403	0.261	0.151	0.034	0.004	0.000	0.000
	6	1.000	1.000	1.000	0.982	0.943	0.869	0.755	0.610	0.452	0.304	0.095	0.015	0.001	0.000
	7	1.000	1.000	1.000	0.996	0.983	0.950	0.887	0.787	0.654	0.500	0.213	0.050	0.004	0.000
	8	1.000	1.000	1.000	0.999	0.996	0.985	0.958	0.905	0.818	0.696	0.390	0.131	0.018	0.000

Table A.1 (continued)

x	n	p	Term	Cumulative
3	17	0.2	0.239	0.549

							p							
n	x	0.01	0.05	0.1	0.2	0.25	0.3	0.35	0.4	0.45	0.5	0.6	0.7	0.8	0.9
16	0	0.851	0.440	0.185	0.028	0.010	0.003	0.001	0.000	0.000	0.000	0.000	0.000	0.000	0.000
	1	0.989	0.811	0.515	0.141	0.063	0.026	0.010	0.003	0.001	0.000	0.000	0.000	0.000	0.000
	2	0.999	0.957	0.789	0.352	0.197	0.099	0.045	0.018	0.007	0.002	0.000	0.000	0.000	0.000
	3	1.000	0.993	0.932	0.598	0.405	0.246	0.134	0.065	0.028	0.011	0.001	0.000	0.000	0.000
	4	1.000	0.999	0.983	0.798	0.630	0.450	0.289	0.167	0.085	0.038	0.005	0.000	0.000	0.000
	5	1.000	1.000	0.997	0.918	0.810	0.660	0.490	0.329	0.198	0.105	0.019	0.002	0.000	0.000
	6	1.000	1.000	0.999	0.973	0.920	0.825	0.688	0.527	0.366	0.227	0.058	0.007	0.000	0.000
	7	1.000	1.000	1.000	0.993	0.973	0.926	0.841	0.716	0.563	0.402	0.142	0.026	0.001	0.000
	8	1.000	1.000	1.000	0.999	0.993	0.974	0.933	0.858	0.744	0.598	0.284	0.074	0.007	0.000
17	0	0.843	0.418	0.167	0.023	0.008	0.002	0.001	0.000	0.000	0.000	0.000	0.000	0.000	0.000
	1	0.988	0.792	0.482	0.118	0.050	0.019	0.007	0.002	0.001	0.000	0.000	0.000	0.000	0.000
	2	0.999	0.950	0.762	0.310	0.164	0.077	0.033	0.012	0.004	0.001	0.000	0.000	0.000	0.000
	3	1.000	0.991	0.917	0.549	0.353	0.202	0.103	0.046	0.018	0.006	0.000	0.000	0.000	0.000
	4	1.000	0.999	0.978	0.758	0.574	0.389	0.235	0.126	0.060	0.025	0.003	0.000	0.000	0.000
	5	1.000	1.000	0.995	0.894	0.765	0.597	0.420	0.264	0.147	0.072	0.011	0.001	0.000	0.000
	6	1.000	1.000	0.999	0.962	0.893	0.775	0.619	0.448	0.290	0.166	0.035	0.003	0.000	0.000
	7	1.000	1.000	1.000	0.989	0.960	0.895	0.787	0.641	0.474	0.315	0.092	0.013	0.000	0.000
	8	1.000	1.000	1.000	0.997	0.988	0.960	0.901	0.801	0.663	0.500	0.199	0.040	0.003	0.000
	9	1.000	1.000	1.000	1.000	0.997	0.987	0.962	0.908	0.817	0.685	0.359	0.105	0.011	0.000
18	0	0.835	0.397	0.150	0.018	0.006	0.002	0.000	0.000	0.000	0.000	0.000	0.000	0.000	0.000
	1	0.986	0.774	0.450	0.099	0.039	0.014	0.005	0.001	0.000	0.000	0.000	0.000	0.000	0.000
	2	0.999	0.942	0.734	0.271	0.135	0.060	0.024	0.008	0.003	0.001	0.000	0.000	0.000	0.000
	3	1.000	0.989	0.902	0.501	0.306	0.165	0.078	0.033	0.012	0.004	0.000	0.000	0.000	0.000
	4	1.000	0.998	0.972	0.716	0.519	0.333	0.189	0.094	0.041	0.015	0.001	0.000	0.000	0.000
	5	1.000	1.000	0.994	0.867	0.717	0.534	0.355	0.209	0.108	0.048	0.006	0.000	0.000	0.000
	6	1.000	1.000	0.999	0.949	0.861	0.722	0.549	0.374	0.226	0.119	0.020	0.001	0.000	0.000
	7	1.000	1.000	1.000	0.984	0.943	0.859	0.728	0.563	0.391	0.240	0.058	0.006	0.000	0.000
	8	1.000	1.000	1.000	0.996	0.981	0.940	0.861	0.737	0.578	0.407	0.135	0.021	0.001	0.000
	9	1.000	1.000	1.000	0.999	0.995	0.979	0.940	0.865	0.747	0.593	0.263	0.060	0.004	0.000
19	0	0.826	0.377	0.135	0.014	0.004	0.001	0.000	0.000	0.000	0.000	0.000	0.000	0.000	0.000
	1	0.985	0.755	0.420	0.083	0.031	0.010	0.003	0.001	0.000	0.000	0.000	0.000	0.000	0.000
	2	0.999	0.933	0.705	0.237	0.111	0.046	0.017	0.005	0.002	0.000	0.000	0.000	0.000	0.000
	3	1.000	0.987	0.885	0.455	0.263	0.133	0.059	0.023	0.008	0.002	0.000	0.000	0.000	0.000
	4	1.000	0.998	0.965	0.673	0.465	0.282	0.150	0.070	0.028	0.010	0.001	0.000	0.000	0.000
	5	1.000	1.000	0.991	0.837	0.668	0.474	0.297	0.163	0.078	0.032	0.003	0.000	0.000	0.000
	6	1.000	1.000	0.998	0.932	0.825	0.666	0.481	0.308	0.173	0.084	0.012	0.001	0.000	0.000
	7	1.000	1.000	1.000	0.977	0.923	0.818	0.666	0.488	0.317	0.180	0.035	0.003	0.000	0.000
	8	1.000	1.000	1.000	0.993	0.971	0.916	0.815	0.667	0.494	0.324	0.088	0.011	0.000	0.000
	9	1.000	1.000	1.000	0.998	0.991	0.967	0.913	0.814	0.671	0.500	0.186	0.033	0.002	0.000
	10	1.000	1.000	1.000	1.000	0.998	0.989	0.965	0.912	0.816	0.676	0.333	0.084	0.007	0.000

(continued)

Table A.1 *(continued)*

x	n	p	Term	Cumulative
3	21	0.3	0.058	0.086

n	x	0.01	0.05	0.1	0.2	0.25	0.3	0.35	0.4	0.45	0.5	0.6	0.7	0.8	0.9
20	0	0.818	0.358	0.122	0.012	0.003	0.001	0.000	0.000	0.000	0.000	0.000	0.000	0.000	0.000
	1	0.983	0.736	0.392	0.069	0.024	0.008	0.002	0.001	0.000	0.000	0.000	0.000	0.000	0.000
	2	0.999	0.925	0.677	0.206	0.091	0.035	0.012	0.004	0.001	0.000	0.000	0.000	0.000	0.000
	3	1.000	0.984	0.867	0.411	0.225	0.107	0.044	0.016	0.005	0.001	0.000	0.000	0.000	0.000
	4	1.000	0.997	0.957	0.630	0.415	0.238	0.118	0.051	0.019	0.006	0.000	0.000	0.000	0.000
	5	1.000	1.000	0.989	0.804	0.617	0.416	0.245	0.126	0.055	0.021	0.002	0.000	0.000	0.000
	6	1.000	1.000	0.998	0.913	0.786	0.608	0.417	0.250	0.130	0.058	0.006	0.000	0.000	0.000
	7	1.000	1.000	1.000	0.968	0.898	0.772	0.601	0.416	0.252	0.132	0.021	0.001	0.000	0.000
	8	1.000	1.000	1.000	0.990	0.959	0.887	0.762	0.596	0.414	0.252	0.057	0.005	0.000	0.000
	9	1.000	1.000	1.000	0.997	0.986	0.952	0.878	0.755	0.591	0.412	0.128	0.017	0.001	0.000
	10	1.000	1.000	1.000	0.999	0.996	0.983	0.947	0.872	0.751	0.588	0.245	0.048	0.003	0.000
21	0	0.810	0.341	0.109	0.009	0.002	0.001	0.000	0.000	0.000	0.000	0.000	0.000	0.000	0.000
	1	0.981	0.717	0.365	0.058	0.019	0.006	0.001	0.000	0.000	0.000	0.000	0.000	0.000	0.000
	2	0.999	0.915	0.648	0.179	0.075	0.027	0.009	0.002	0.001	0.000	0.000	0.000	0.000	0.000
	3	1.000	0.981	0.848	0.370	0.192	0.086	0.033	0.011	0.003	0.001	0.000	0.000	0.000	0.000
	4	1.000	0.997	0.948	0.586	0.367	0.198	0.092	0.037	0.013	0.004	0.000	0.000	0.000	0.000
	5	1.000	1.000	0.986	0.769	0.567	0.363	0.201	0.096	0.039	0.013	0.001	0.000	0.000	0.000
	6	1.000	1.000	0.997	0.891	0.744	0.551	0.357	0.200	0.096	0.039	0.004	0.000	0.000	0.000
	7	1.000	1.000	0.999	0.957	0.870	0.723	0.536	0.350	0.197	0.095	0.012	0.001	0.000	0.000
	8	1.000	1.000	1.000	0.986	0.944	0.852	0.706	0.524	0.341	0.192	0.035	0.002	0.000	0.000
	9	1.000	1.000	1.000	0.996	0.979	0.932	0.838	0.691	0.512	0.332	0.085	0.009	0.000	0.000
	10	1.000	1.000	1.000	0.999	0.994	0.974	0.923	0.826	0.679	0.500	0.174	0.026	0.001	0.000
	11	1.000	1.000	1.000	1.000	0.998	0.991	0.969	0.915	0.816	0.668	0.309	0.068	0.004	0.000
22	0	0.802	0.324	0.098	0.007	0.002	0.000	0.000	0.000	0.000	0.000	0.000	0.000	0.000	0.000
	1	0.980	0.698	0.339	0.048	0.015	0.004	0.001	0.000	0.000	0.000	0.000	0.000	0.000	0.000
	2	0.999	0.905	0.620	0.154	0.061	0.021	0.006	0.002	0.000	0.000	0.000	0.000	0.000	0.000
	3	1.000	0.978	0.828	0.332	0.162	0.068	0.025	0.008	0.002	0.000	0.000	0.000	0.000	0.000
	4	1.000	0.996	0.938	0.543	0.323	0.165	0.072	0.027	0.008	0.002	0.000	0.000	0.000	0.000
	5	1.000	0.999	0.982	0.733	0.517	0.313	0.163	0.072	0.027	0.008	0.000	0.000	0.000	0.000
	6	1.000	1.000	0.996	0.867	0.699	0.494	0.302	0.158	0.071	0.026	0.002	0.000	0.000	0.000
	7	1.000	1.000	0.999	0.944	0.838	0.671	0.474	0.290	0.152	0.067	0.007	0.000	0.000	0.000
	8	1.000	1.000	1.000	0.980	0.925	0.814	0.647	0.454	0.276	0.143	0.021	0.001	0.000	0.000
	9	1.000	1.000	1.000	0.994	0.970	0.908	0.792	0.624	0.435	0.262	0.055	0.004	0.000	0.000
	10	1.000	1.000	1.000	0.998	0.990	0.961	0.893	0.772	0.604	0.416	0.121	0.014	0.000	0.000
	11	1.000	1.000	1.000	1.000	0.997	0.986	0.953	0.879	0.754	0.584	0.228	0.039	0.002	0.000

p = probability of success.

Table A.2. *Cumulative Poisson distribution*

							x	λ	**Term**	**Cumulative**
							3	1.5	0.1255	0.9344

						x					
λ	0	1	2	3	4	5	6	7	8	9	10	11

λ	0	1	2	3	4	5	6	7	8	9	10	11
0.01	0.9900	1	1	1	1	1	1	1	1	1	1	1
0.05	0.9512	0.9988	1	1	1	1	1	1	1	1	1	1
0.1	0.9048	0.9953	0.9998	1	1	1	1	1	1	1	1	1
0.2	0.8187	0.9825	0.9989	0.9999	1	1	1	1	1	1	1	1
0.3	0.7408	0.9631	0.9964	0.9997	1	1	1	1	1	1	1	1
0.4	0.6703	0.9384	0.9921	0.9992	0.9999	1	1	1	1	1	1	1
0.5	0.6065	0.9098	0.9856	0.9982	0.9998	1	1	1	1	1	1	1
0.6	0.5488	0.8781	0.9769	0.9966	0.9996	1	1	1	1	1	1	1
0.7	0.4966	0.8442	0.9659	0.9942	0.9992	0.9999	1	1	1	1	1	1
0.8	0.4493	0.8088	0.9526	0.9909	0.9986	0.9998	1	1	1	1	1	1
0.9	0.4066	0.7725	0.9371	0.9865	0.9977	0.9997	1	1	1	1	1	1
1	0.3679	0.7358	0.9197	0.9810	0.9963	0.9994	0.9999	1.0000	1	1	1	1
1.1	0.3329	0.6990	0.9004	0.9743	0.9946	0.9990	0.9999	1.0000	1	1	1	1
1.2	0.3012	0.6626	0.8795	0.9662	0.9923	0.9985	0.9997	1.0000	1	1	1	1
1.3	0.2725	0.6268	0.8571	0.9569	0.9893	0.9978	0.9996	0.9999	1	1	1	1
1.4	0.2466	0.5918	0.8335	0.9463	0.9857	0.9968	0.9994	0.9999	1	1	1	1
1.5	0.2231	0.5578	0.8088	0.9344	0.9814	0.9955	0.9991	0.9998	1	1	1	1
1.6	0.2019	0.5249	0.7834	0.9212	0.9763	0.9940	0.9987	0.9997	1	1	1	1
1.7	0.1827	0.4932	0.7572	0.9068	0.9704	0.9920	0.9981	0.9996	0.9999	1	1	1
1.8	0.1653	0.4628	0.7306	0.8913	0.9636	0.9896	0.9974	0.9994	0.9999	1	1	1
1.9	0.1496	0.4337	0.7037	0.8747	0.9559	0.9868	0.9966	0.9992	0.9998	1	1	1
2	0.1353	0.4060	0.6767	0.8571	0.9473	0.9834	0.9955	0.9989	0.9998	1	1	1
2.2	0.1108	0.3546	0.6227	0.8194	0.9275	0.9751	0.9925	0.9980	0.9995	0.9999	1	1
2.4	0.0907	0.3084	0.5697	0.7787	0.9041	0.9643	0.9884	0.9967	0.9991	0.9998	1	1
2.6	0.0743	0.2674	0.5184	0.7360	0.8774	0.9510	0.9828	0.9947	0.9985	0.9996	0.9999	1
2.8	0.0608	0.2311	0.4695	0.6919	0.8477	0.9349	0.9756	0.9919	0.9976	0.9993	0.9998	1
3	0.0498	0.1991	0.4232	0.6472	0.8153	0.9161	0.9665	0.9881	0.9962	0.9989	0.9997	0.9999
3.2	0.0408	0.1712	0.3799	0.6025	0.7806	0.8946	0.9554	0.9832	0.9943	0.9982	0.9995	0.9999
3.4	0.0334	0.1468	0.3397	0.5584	0.7442	0.8705	0.9421	0.9769	0.9917	0.9973	0.9992	0.9998
3.6	0.0273	0.1257	0.3027	0.5152	0.7064	0.8441	0.9267	0.9692	0.9883	0.9960	0.9987	0.9996
3.8	0.0224	0.1074	0.2689	0.4735	0.6678	0.8156	0.9091	0.9599	0.9840	0.9942	0.9981	0.9994
4	0.0183	0.0916	0.2381	0.4335	0.6288	0.7851	0.8893	0.9489	0.9786	0.9919	0.9972	0.9991
4.2	0.0150	0.0780	0.2102	0.3954	0.5898	0.7531	0.8675	0.9361	0.9721	0.9889	0.9959	0.9986
4.4	0.0123	0.0663	0.1851	0.3594	0.5512	0.7199	0.8436	0.9214	0.9642	0.9851	0.9943	0.9980
4.6	0.0101	0.0563	0.1626	0.3257	0.5132	0.6858	0.8180	0.9049	0.9549	0.9805	0.9922	0.9971
4.8	0.0082	0.0477	0.1425	0.2942	0.4763	0.6510	0.7908	0.8867	0.9442	0.9749	0.9896	0.9960
5	0.0067	0.0404	0.1247	0.2650	0.4405	0.6160	0.7622	0.8666	0.9319	0.9682	0.9863	0.9945
5.2	0.0055	0.0342	0.1088	0.2381	0.4061	0.5809	0.7324	0.8449	0.9181	0.9603	0.9823	0.9927
5.4	0.0045	0.0289	0.0948	0.2133	0.3733	0.5461	0.7017	0.8217	0.9027	0.9512	0.9775	0.9904
5.6	0.0037	0.0244	0.0824	0.1906	0.3422	0.5119	0.6703	0.7970	0.8857	0.9409	0.9718	0.9875
5.8	0.0030	0.0206	0.0715	0.1700	0.3127	0.4783	0.6384	0.7710	0.8672	0.9292	0.9651	0.9841
6	0.0025	0.0174	0.0620	0.1512	0.2851	0.4457	0.6063	0.7440	0.8472	0.9161	0.9574	0.9799
6.2	0.0020	0.0146	0.0536	0.1342	0.2592	0.4141	0.5742	0.7160	0.8259	0.9016	0.9486	0.9750
6.4	0.0017	0.0123	0.0463	0.1189	0.2351	0.3837	0.5423	0.6873	0.8033	0.8858	0.9386	0.9693
6.6	0.0014	0.0103	0.0400	0.1052	0.2127	0.3547	0.5108	0.6581	0.7796	0.8686	0.9274	0.9627

(continued)

Table A.2 *(continued)*

x	λ	Term	Cumulative
14	6.4	0.0037	0.9974

λ \ x	12	13	14	15	16	17	18	19	20	21	22	23
0.01	1	1	1	1	1	1	1	1	1	1	1	1
0.05	1	1	1	1	1	1	1	1	1	1	1	1
0.1	1	1	1	1	1	1	1	1	1	1	1	1
0.2	1	1	1	1	1	1	1	1	1	1	1	1
0.3	1	1	1	1	1	1	1	1	1	1	1	1
0.4	1	1	1	1	1	1	1	1	1	1	1	1
0.5	1	1	1	1	1	1	1	1	1	1	1	1
0.6	1	1	1	1	1	1	1	1	1	1	1	1
0.7	1	1	1	1	1	1	1	1	1	1	1	1
0.8	1	1	1	1	1	1	1	1	1	1	1	1
0.9	1	1	1	1	1	1	1	1	1	1	1	1
1	1	1	1	1	1	1	1	1	1	1	1	1
1.1	1	1	1	1	1	1	1	1	1	1	1	1
1.2	1	1	1	1	1	1	1	1	1	1	1	1
1.3	1	1	1	1	1	1	1	1	1	1	1	1
1.4	1	1	1	1	1	1	1	1	1	1	1	1
1.5	1	1	1	1	1	1	1	1	1	1	1	1
1.6	1	1	1	1	1	1	1	1	1	1	1	1
1.7	1	1	1	1	1	1	1	1	1	1	1	1
1.8	1	1	1	1	1	1	1	1	1	1	1	1
1.9	1	1	1	1	1	1	1	1	1	1	1	1
2	1	1	1	1	1	1	1	1	1	1	1	1
2.2	1	1	1	1	1	1	1	1	1	1	1	1
2.4	1	1	1	1	1	1	1	1	1	1	1	1
2.6	1	1	1	1	1	1	1	1	1	1	1	1
2.8	1	1	1	1	1	1	1	1	1	1	1	1
3	1	1	1	1	1	1	1	1	1	1	1	1
3.2	1	1	1	1	1	1	1	1	1	1	1	1
3.4	0.9999	1	1	1	1	1	1	1	1	1	1	1
3.6	0.9999	1	1	1	1	1	1	1	1	1	1	1
3.8	0.9998	1	1	1	1	1	1	1	1	1	1	1
4	0.9997	0.9999	1	1	1	1	1	1	1	1	1	1
4.2	0.9996	0.9999	1	1	1	1	1	1	1	1	1	1
4.4	0.9993	0.9998	0.9999	1	1	1	1	1	1	1	1	1
4.6	0.9990	0.9997	0.9999	1	1	1	1	1	1	1	1	1
4.8	0.9986	0.9995	0.9999	1	1	1	1	1	1	1	1	1
5	0.9980	0.9993	0.9998	0.9999	1	1	1	1	1	1	1	1
5.2	0.9972	0.9990	0.9997	0.9999	1	1	1	1	1	1	1	1
5.4	0.9962	0.9986	0.9995	0.9998	0.9999	1	1	1	1	1	1	1
5.6	0.9949	0.9980	0.9993	0.9998	0.9999	1	1	1	1	1	1	1
5.8	0.9932	0.9973	0.9990	0.9996	0.9999	1	1	1	1	1	1	1
6	0.9912	0.9964	0.9986	0.9995	0.9998	0.9999	1	1	1	1	1	1
6.2	0.9887	0.9952	0.9981	0.9993	0.9997	0.9999	1	1	1	1	1	1
6.4	0.9857	0.9937	0.9974	0.9990	0.9996	0.9999	1	1	1	1	1	1
6.6	0.9821	0.9920	0.9966	0.9986	0.9995	0.9998	0.9999	1	1	1	1	1

λ = mean; x = number of events.

Table A.3. *Cumulative standard normal distribution*
Φ(z)

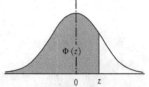

z	0.09	0.08	0.07	0.06	0.05	0.04	0.03	0.02	0.01	0
−3.9	0.0000	0.0000	0.0000	0.0000	0.0000	0.0000	0.0000	0.0000	0.0000	0.0000
−3.8	0.0001	0.0001	0.0001	0.0001	0.0001	0.0001	0.0001	0.0001	0.0001	0.0001
−3.7	0.0001	0.0001	0.0001	0.0001	0.0001	0.0001	0.0001	0.0001	0.0001	0.0001
−3.6	0.0001	0.0001	0.0001	0.0001	0.0001	0.0001	0.0001	0.0001	0.0002	0.0002
−3.5	0.0002	0.0002	0.0002	0.0002	0.0002	0.0002	0.0002	0.0002	0.0002	0.0002
−3.4	0.0002	0.0003	0.0003	0.0003	0.0003	0.0003	0.0003	0.0003	0.0003	0.0003
−3.3	0.0003	0.0004	0.0004	0.0004	0.0004	0.0004	0.0004	0.0005	0.0005	0.0005
−3.2	0.0005	0.0005	0.0005	0.0006	0.0006	0.0006	0.0006	0.0006	0.0007	0.0007
−3.1	0.0007	0.0007	0.0008	0.0008	0.0008	0.0008	0.0009	0.0009	0.0009	0.0010
−3	0.0010	0.0010	0.0011	0.0011	0.0011	0.0012	0.0012	0.0013	0.0013	0.0013
−2.9	0.0014	0.0014	0.0015	0.0015	0.0016	0.0016	0.0017	0.0018	0.0018	0.0019
−2.8	0.0019	0.0020	0.0021	0.0021	0.0022	0.0023	0.0023	0.0024	0.0025	0.0026
−2.7	0.0026	0.0027	0.0028	0.0029	0.0030	0.0031	0.0032	0.0033	0.0034	0.0035
−2.6	0.0036	0.0037	0.0038	0.0039	0.0040	0.0041	0.0043	0.0044	0.0045	0.0047
−2.5	0.0048	0.0049	0.0051	0.0052	0.0054	0.0055	0.0057	0.0059	0.0060	0.0062
−2.4	0.0064	0.0066	0.0068	0.0069	0.0071	0.0073	0.0075	0.0078	0.0080	0.0082
−2.3	0.0084	0.0087	0.0089	0.0091	0.0094	0.0096	0.0099	0.0102	0.0104	0.0107
−2.2	0.0110	0.0113	0.0116	0.0119	0.0122	0.0125	0.0129	0.0132	0.0136	0.0139
−2.1	0.0143	0.0146	0.0150	0.0154	0.0158	0.0162	0.0166	0.0170	0.0174	0.0179
−2	0.0183	0.0188	0.0192	0.0197	0.0202	0.0207	0.0212	0.0217	0.0222	0.0228
−1.9	0.0233	0.0239	0.0244	0.0250	0.0256	0.0262	0.0268	0.0274	0.0281	0.0287
−1.8	0.0294	0.0301	0.0307	0.0314	0.0322	0.0329	0.0336	0.0344	0.0351	0.0359
−1.7	0.0367	0.0375	0.0384	0.0392	0.0401	0.0409	0.0418	0.0427	0.0436	0.0446
−1.6	0.0455	0.0465	0.0475	0.0485	0.0495	0.0505	0.0516	0.0526	0.0537	0.0548
−1.5	0.0559	0.0571	0.0582	0.0594	0.0606	0.0618	0.0630	0.0643	0.0655	0.0668
−1.4	0.0681	0.0694	0.0708	0.0721	0.0735	0.0749	0.0764	0.0778	0.0793	0.0808
−1.3	0.0823	0.0838	0.0853	0.0869	0.0885	0.0901	0.0918	0.0934	0.0951	0.0968
−1.2	0.0985	0.1003	0.1020	0.1038	0.1056	0.1075	0.1093	0.1112	0.1131	0.1151
−1.1	0.1170	0.1190	0.1210	0.1230	0.1251	0.1271	0.1292	0.1314	0.1335	0.1357
−1	0.1379	0.1401	0.1423	0.1446	0.1469	0.1492	0.1515	0.1539	0.1562	0.1587
−0.9	0.1611	0.1635	0.1660	0.1685	0.1711	0.1736	0.1762	0.1788	0.1814	0.1841
−0.8	0.1867	0.1894	0.1922	0.1949	0.1977	0.2005	0.2033	0.2061	0.2090	0.2119
−0.7	0.2148	0.2177	0.2206	0.2236	0.2266	0.2296	0.2327	0.2358	0.2389	0.2420
−0.6	0.2451	0.2483	0.2514	0.2546	0.2578	0.2611	0.2643	0.2676	0.2709	0.2743
−0.5	0.2776	0.2810	0.2843	0.2877	0.2912	0.2946	0.2981	0.3015	0.3050	0.3085
−0.4	0.3121	0.3156	0.3192	0.3228	0.3264	0.3300	0.3336	0.3372	0.3409	0.3446
−0.3	0.3483	0.3520	0.3557	0.3594	0.3632	0.3669	0.3707	0.3745	0.3783	0.3821
−0.2	0.3859	0.3897	0.3936	0.3974	0.4013	0.4052	0.4090	0.4129	0.4168	0.4207
−0.1	0.4247	0.4286	0.4325	0.4364	0.4404	0.4443	0.4483	0.4522	0.4562	0.4602
0	0.4641	0.4681	0.4721	0.4761	0.4801	0.4840	0.4880	0.4920	0.4960	0.5000

| z | Area to left | (See sheet2 for z > 0) | | | | | | | | *(continued)* |
| −2 | 0.02275 | | | | | | | | | |

Table A.3 *(continued)*

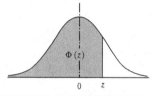

Φ(z)

z	0	0.01	0.02	0.03	0.04	0.05	0.06	0.07	0.08	0.09
0	0.5000	0.5040	0.5080	0.5120	0.5160	0.5199	0.5239	0.5279	0.5319	0.5359
0.1	0.5398	0.5438	0.5478	0.5517	0.5557	0.5596	0.5636	0.5675	0.5714	0.5753
0.2	0.5793	0.5832	0.5871	0.5910	0.5948	0.5987	0.6026	0.6064	0.6103	0.6141
0.3	0.6179	0.6217	0.6255	0.6293	0.6331	0.6368	0.6406	0.6443	0.6480	0.6517
0.4	0.6554	0.6591	0.6628	0.6664	0.6700	0.6736	0.6772	0.6808	0.6844	0.6879
0.5	0.6915	0.6950	0.6985	0.7019	0.7054	0.7088	0.7123	0.7157	0.7190	0.7224
0.6	0.7257	0.7291	0.7324	0.7357	0.7389	0.7422	0.7454	0.7486	0.7517	0.7549
0.7	0.7580	0.7611	0.7642	0.7673	0.7704	0.7734	0.7764	0.7794	0.7823	0.7852
0.8	0.7881	0.7910	0.7939	0.7967	0.7995	0.8023	0.8051	0.8078	0.8106	0.8133
0.9	0.8159	0.8186	0.8212	0.8238	0.8264	0.8289	0.8315	0.8340	0.8365	0.8389
1	0.8413	0.8438	0.8461	0.8485	0.8508	0.8531	0.8554	0.8577	0.8599	0.8621
1.1	0.8643	0.8665	0.8686	0.8708	0.8729	0.8749	0.8770	0.8790	0.8810	0.8830
1.2	0.8849	0.8869	0.8888	0.8907	0.8925	0.8944	0.8962	0.8980	0.8997	0.9015
1.3	0.9032	0.9049	0.9066	0.9082	0.9099	0.9115	0.9131	0.9147	0.9162	0.9177
1.4	0.9192	0.9207	0.9222	0.9236	0.9251	0.9265	0.9279	0.9292	0.9306	0.9319
1.5	0.9332	0.9345	0.9357	0.9370	0.9382	0.9394	0.9406	0.9418	0.9429	0.9441
1.6	0.9452	0.9463	0.9474	0.9484	0.9495	0.9505	0.9515	0.9525	0.9535	0.9545
1.7	0.9554	0.9564	0.9573	0.9582	0.9591	0.9599	0.9608	0.9616	0.9625	0.9633
1.8	0.9641	0.9649	0.9656	0.9664	0.9671	0.9678	0.9686	0.9693	0.9699	0.9706
1.9	0.9713	0.9719	0.9726	0.9732	0.9738	0.9744	0.9750	0.9756	0.9761	0.9767
2	0.9772	0.9778	0.9783	0.9788	0.9793	0.9798	0.9803	0.9808	0.9812	0.9817
2.1	0.9821	0.9826	0.9830	0.9834	0.9838	0.9842	0.9846	0.9850	0.9854	0.9857
2.2	0.9861	0.9864	0.9868	0.9871	0.9875	0.9878	0.9881	0.9884	0.9887	0.9890
2.3	0.9893	0.9896	0.9898	0.9901	0.9904	0.9906	0.9909	0.9911	0.9913	0.9916
2.4	0.9918	0.9920	0.9922	0.9925	0.9927	0.9929	0.9931	0.9932	0.9934	0.9936
2.5	0.9938	0.9940	0.9941	0.9943	0.9945	0.9946	0.9948	0.9949	0.9951	0.9952
2.6	0.9953	0.9955	0.9956	0.9957	0.9959	0.9960	0.9961	0.9962	0.9963	0.9964
2.7	0.9965	0.9966	0.9967	0.9968	0.9969	0.9970	0.9971	0.9972	0.9973	0.9974
2.8	0.9974	0.9975	0.9976	0.9977	0.9977	0.9978	0.9979	0.9979	0.9980	0.9981
2.9	0.9981	0.9982	0.9982	0.9983	0.9984	0.9984	0.9985	0.9985	0.9986	0.9986
3	0.9987	0.9987	0.9987	0.9988	0.9988	0.9989	0.9989	0.9989	0.9990	0.9990
3.1	0.9990	0.9991	0.9991	0.9991	0.9992	0.9992	0.9992	0.9992	0.9993	0.9993
3.2	0.9993	0.9993	0.9994	0.9994	0.9994	0.9994	0.9994	0.9995	0.9995	0.9995
3.3	0.9995	0.9995	0.9995	0.9996	0.9996	0.9996	0.9996	0.9996	0.9996	0.9997
3.4	0.9997	0.9997	0.9997	0.9997	0.9997	0.9997	0.9997	0.9997	0.9997	0.9998
3.5	0.9998	0.9998	0.9998	0.9998	0.9998	0.9998	0.9998	0.9998	0.9998	0.9998
3.6	0.9998	0.9998	0.9999	0.9999	0.9999	0.9999	0.9999	0.9999	0.9999	0.9999
3.7	0.9999	0.9999	0.9999	0.9999	0.9999	0.9999	0.9999	0.9999	0.9999	0.9999
3.8	0.9999	0.9999	0.9999	0.9999	0.9999	0.9999	0.9999	0.9999	0.9999	0.9999
3.9	1.0000	1.0000	1.0000	1.0000	1.0000	1.0000	1.0000	1.0000	1.0000	1.0000

z   Left Tail Area   (See sheet1 for $z < 0$)
1.73   0.95818

Table A.4. *Standard normal inverse*

Rt Area	z-value
0.15	1.0364

z-Values for given right-tail area

Rt area	0	0.0001	0.0002	0.0003	0.0004	0.0005	0.0006	0.0007	0.0008	0.0009
0.000	inf	3.7190	3.5401	3.4316	3.3528	3.2905	3.2389	3.1947	3.1559	3.1214
0.001	3.0902	3.0618	3.0357	3.0115	2.9889	2.9677	2.9478	2.9290	2.9112	2.8943
0.002	2.8782	2.8627	2.8480	2.8338	2.8202	2.8070	2.7944	2.7822	2.7703	2.7589
0.003	2.7478	2.7370	2.7266	2.7164	2.7065	2.6968	2.6874	2.6783	2.6693	2.6606
0.004	2.6521	2.6437	2.6356	2.6276	2.6197	2.6121	2.6045	2.5972	2.5899	2.5828
0.005	2.5758	2.5690	2.5622	2.5556	2.5491	2.5427	2.5364	2.5302	2.5241	2.5181
0.006	2.5121	2.5063	2.5006	2.4949	2.4893	2.4838	2.4783	2.4730	2.4677	2.4624
0.007	2.4573	2.4522	2.4471	2.4422	2.4372	2.4324	2.4276	2.4228	2.4181	2.4135
0.008	2.4089	2.4044	2.3999	2.3954	2.3911	2.3867	2.3824	2.3781	2.3739	2.3698
0.009	2.3656	2.3615	2.3575	2.3535	2.3495	2.3455	2.3416	2.3378	2.3339	2.3301

Rt area	0	0.001	0.002	0.003	0.004	0.005	0.006	0.007	0.008	0.009
0.01	2.3263	2.2904	2.2571	2.2262	2.1973	2.1701	2.1444	2.1201	2.0969	2.0749
0.02	2.0537	2.0335	2.0141	1.9954	1.9774	1.9600	1.9431	1.9268	1.9110	1.8957
0.03	1.8808	1.8663	1.8522	1.8384	1.8250	1.8119	1.7991	1.7866	1.7744	1.7624
0.04	1.7507	1.7392	1.7279	1.7169	1.7060	1.6954	1.6849	1.6747	1.6646	1.6546
0.05	1.6449	1.6352	1.6258	1.6164	1.6072	1.5982	1.5893	1.5805	1.5718	1.5632
0.06	1.5548	1.5464	1.5382	1.5301	1.5220	1.5141	1.5063	1.4985	1.4909	1.4833
0.07	1.4758	1.4684	1.4611	1.4538	1.4466	1.4395	1.4325	1.4255	1.4187	1.4118
0.08	1.4051	1.3984	1.3917	1.3852	1.3787	1.3722	1.3658	1.3595	1.3532	1.3469
0.09	1.3408	1.3346	1.3285	1.3225	1.3165	1.3106	1.3047	1.2988	1.2930	1.2873
0.10	1.2816	1.2759	1.2702	1.2646	1.2591	1.2536	1.2481	1.2426	1.2372	1.2319
0.11	1.2265	1.2212	1.2160	1.2107	1.2055	1.2004	1.1952	1.1901	1.1850	1.1800
0.12	1.1750	1.1700	1.1650	1.1601	1.1552	1.1503	1.1455	1.1407	1.1359	1.1311
0.13	1.1264	1.1217	1.1170	1.1123	1.1077	1.1031	1.0985	1.0939	1.0893	1.0848
0.14	1.0803	1.0758	1.0714	1.0669	1.0625	1.0581	1.0537	1.0494	1.0450	1.0407
0.15	1.0364	1.0322	1.0279	1.0237	1.0194	1.0152	1.0110	1.0069	1.0027	0.9986
0.16	0.9945	0.9904	0.9863	0.9822	0.9782	0.9741	0.9701	0.9661	0.9621	0.9581
0.17	0.9542	0.9502	0.9463	0.9424	0.9385	0.9346	0.9307	0.9269	0.9230	0.9192
0.18	0.9154	0.9116	0.9078	0.9040	0.9002	0.8965	0.8927	0.8890	0.8853	0.8816
0.19	0.8779	0.8742	0.8705	0.8669	0.8633	0.8596	0.8560	0.8524	0.8488	0.8452
0.20	0.8416	0.8381	0.8345	0.8310	0.8274	0.8239	0.8204	0.8169	0.8134	0.8099
0.21	0.8064	0.8030	0.7995	0.7961	0.7926	0.7892	0.7858	0.7824	0.7790	0.7756
0.22	0.7722	0.7688	0.7655	0.7621	0.7588	0.7554	0.7521	0.7488	0.7454	0.7421
0.23	0.7388	0.7356	0.7323	0.7290	0.7257	0.7225	0.7192	0.7160	0.7128	0.7095
0.24	0.7063	0.7031	0.6999	0.6967	0.6935	0.6903	0.6871	0.6840	0.6808	0.6776

(continued)

Table A.4 *(continued)*

L Area	z-Value
0.451	0.1231

z-Values for given right-tail area

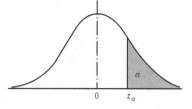

L area	0	0.001	0.002	0.003	0.004	0.005	0.006	0.007	0.008	0.009
0.25	0.6745	0.6713	0.6682	0.6651	0.6620	0.6588	0.6557	0.6526	0.6495	0.6464
0.26	0.6433	0.6403	0.6372	0.6341	0.6311	0.6280	0.6250	0.6219	0.6189	0.6158
0.27	0.6128	0.6098	0.6068	0.6038	0.6008	0.5978	0.5948	0.5918	0.5888	0.5858
0.28	0.5828	0.5799	0.5769	0.5740	0.5710	0.5681	0.5651	0.5622	0.5592	0.5563
0.29	0.5534	0.5505	0.5476	0.5446	0.5417	0.5388	0.5359	0.5330	0.5302	0.5273
0.30	0.5244	0.5215	0.5187	0.5158	0.5129	0.5101	0.5072	0.5044	0.5015	0.4987
0.31	0.4959	0.4930	0.4902	0.4874	0.4845	0.4817	0.4789	0.4761	0.4733	0.4705
0.32	0.4677	0.4649	0.4621	0.4593	0.4565	0.4538	0.4510	0.4482	0.4454	0.4427
0.33	0.4399	0.4372	0.4344	0.4316	0.4289	0.4261	0.4234	0.4207	0.4179	0.4152
0.34	0.4125	0.4097	0.4070	0.4043	0.4016	0.3989	0.3961	0.3934	0.3907	0.3880
0.35	0.3853	0.3826	0.3799	0.3772	0.3745	0.3719	0.3692	0.3665	0.3638	0.3611
0.36	0.3585	0.3558	0.3531	0.3505	0.3478	0.3451	0.3425	0.3398	0.3372	0.3345
0.37	0.3319	0.3292	0.3266	0.3239	0.3213	0.3186	0.3160	0.3134	0.3107	0.3081
0.38	0.3055	0.3029	0.3002	0.2976	0.2950	0.2924	0.2898	0.2871	0.2845	0.2819
0.39	0.2793	0.2767	0.2741	0.2715	0.2689	0.2663	0.2637	0.2611	0.2585	0.2559
0.40	0.2533	0.2508	0.2482	0.2456	0.2430	0.2404	0.2378	0.2353	0.2327	0.2301
0.41	0.2275	0.2250	0.2224	0.2198	0.2173	0.2147	0.2121	0.2096	0.2070	0.2045
0.42	0.2019	0.1993	0.1968	0.1942	0.1917	0.1891	0.1866	0.1840	0.1815	0.1789
0.43	0.1764	0.1738	0.1713	0.1687	0.1662	0.1637	0.1611	0.1586	0.1560	0.1535
0.44	0.1510	0.1484	0.1459	0.1434	0.1408	0.1383	0.1358	0.1332	0.1307	0.1282
0.45	0.1257	0.1231	0.1206	0.1181	0.1156	0.1130	0.1105	0.1080	0.1055	0.1030
0.46	0.1004	0.0979	0.0954	0.0929	0.0904	0.0878	0.0853	0.0828	0.0803	0.0778
0.47	0.0753	0.0728	0.0702	0.0677	0.0652	0.0627	0.0602	0.0577	0.0552	0.0527
0.48	0.0502	0.0476	0.0451	0.0426	0.0401	0.0376	0.0351	0.0326	0.0301	0.0276
0.49	0.0251	0.0226	0.0201	0.0175	0.0150	0.0125	0.0100	0.0075	0.0050	0.0025
0.50	0.0000									

Table A.5. *Chi-square distribution*

df = 12
Rt Area χ^2
0.1   18.549
0.406   12.5

χ^2 Values for given right-tail area

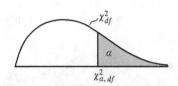

df	0.999	0.995	0.99	0.975	0.95	0.9	0.1	0.05	0.025	0.01	0.005	0.001
1	0.000	0.000	0.000	0.001	0.004	0.016	2.706	3.841	5.024	6.635	7.879	10.828
2	0.002	0.010	0.020	0.051	0.103	0.211	4.605	5.991	7.378	9.210	10.597	13.816
3	0.024	0.072	0.115	0.216	0.352	0.584	6.251	7.815	9.348	11.345	12.838	16.266
4	0.091	0.207	0.297	0.484	0.711	1.064	7.779	9.488	11.143	13.277	14.860	18.467
5	0.210	0.412	0.554	0.831	1.145	1.610	9.236	11.070	12.833	15.086	16.750	20.515
6	0.381	0.676	0.872	1.237	1.635	2.204	10.645	12.592	14.449	16.812	18.548	22.458
7	0.598	0.989	1.239	1.690	2.167	2.833	12.017	14.067	16.013	18.475	20.278	24.322
8	0.857	1.344	1.646	2.180	2.733	3.490	13.362	15.507	17.535	20.090	21.955	26.124
9	1.152	1.735	2.088	2.700	3.325	4.168	14.684	16.919	19.023	21.666	23.589	27.877
10	1.479	2.156	2.558	3.247	3.940	4.865	15.987	18.307	20.483	23.209	25.188	29.588
11	1.834	2.603	3.053	3.816	4.575	5.578	17.275	19.675	21.920	24.725	26.757	31.264
12	2.214	3.074	3.571	4.404	5.226	6.304	18.549	21.026	23.337	26.217	28.300	32.909
13	2.617	3.565	4.107	5.009	5.892	7.042	19.812	22.362	24.736	27.688	29.819	34.528
14	3.041	4.075	4.660	5.629	6.571	7.790	21.064	23.685	26.119	29.141	31.319	36.123
15	3.483	4.601	5.229	6.262	7.261	8.547	22.307	24.996	27.488	30.578	32.801	37.697
16	3.942	5.142	5.812	6.908	7.962	9.312	23.542	26.296	28.845	32.000	34.267	39.252
17	4.416	5.697	6.408	7.564	8.672	10.085	24.769	27.587	30.191	33.409	35.718	40.790
18	4.905	6.265	7.015	8.231	9.390	10.865	25.989	28.869	31.526	34.805	37.156	42.312
19	5.407	6.844	7.633	8.907	10.117	11.651	27.204	30.144	32.852	36.191	38.582	43.820
20	5.921	7.434	8.260	9.591	10.851	12.443	28.412	31.410	34.170	37.566	39.997	45.315
21	6.447	8.034	8.897	10.283	11.591	13.240	29.615	32.671	35.479	38.932	41.401	46.797
22	6.983	8.643	9.542	10.982	12.338	14.041	30.813	33.924	36.781	40.289	42.796	48.268
23	7.529	9.260	10.196	11.689	13.091	14.848	32.007	35.172	38.076	41.638	44.181	49.728
24	8.085	9.886	10.856	12.401	13.848	15.659	33.196	36.415	39.364	42.980	45.559	51.179
25	8.649	10.520	11.524	13.120	14.611	16.473	34.382	37.652	40.646	44.314	46.928	52.620
26	9.222	11.160	12.198	13.844	15.379	17.292	35.563	38.885	41.923	45.642	48.290	54.052
27	9.803	11.808	12.879	14.573	16.151	18.114	36.741	40.113	43.195	46.963	49.645	55.476
28	10.391	12.461	13.565	15.308	16.928	18.939	37.916	41.337	44.461	48.278	50.993	56.892
29	10.986	13.121	14.256	16.047	17.708	19.768	39.087	42.557	45.722	49.588	52.336	58.301
30	11.588	13.787	14.953	16.791	18.493	20.599	40.256	43.773	46.979	50.892	53.672	59.703
35	14.688	17.192	18.509	20.569	22.465	24.797	46.059	49.802	53.203	57.342	60.275	66.619
40	17.916	20.707	22.164	24.433	26.509	29.051	51.805	55.758	59.342	63.691	66.766	73.402
45	21.251	24.311	25.901	28.366	30.612	33.350	57.505	61.656	65.410	69.957	73.166	80.077
50	24.674	27.991	29.707	32.357	34.764	37.689	63.167	67.505	71.420	76.154	79.490	86.661
55	28.173	31.735	33.570	36.398	38.958	42.060	68.796	73.311	77.380	82.292	85.749	93.168
60	31.738	35.534	37.485	40.482	43.188	46.459	74.397	79.082	83.298	88.379	91.952	99.607
70	39.036	43.275	45.442	48.758	51.739	55.329	85.527	90.531	95.023	100.425	104.215	112.317
80	46.520	51.172	53.540	57.153	60.391	64.278	96.578	101.879	106.629	112.329	116.321	124.839
90	54.155	59.196	61.754	65.647	69.126	73.291	107.565	113.145	118.136	124.116	128.299	137.208
100	61.918	67.328	70.065	74.222	77.929	82.358	118.498	124.342	129.561	135.807	140.169	149.449

Table A.6. *Student's t-distribution*

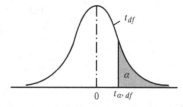

df = 12

Rt Area	t
0.1	1.3562
0.1685	1

t-Values for given right-tail area

df	0.3	0.25	0.2	0.1	0.05	0.025	0.01	0.005	0.001
1	0.7265	1.0000	1.3764	3.0777	6.3138	12.7062	31.8205	63.6567	318.3088
2	0.6172	0.8165	1.0607	1.8856	2.9200	4.3027	6.9646	9.9248	22.3271
3	0.5844	0.7649	0.9785	1.6377	2.3534	3.1824	4.5407	5.8409	10.2145
4	0.5686	0.7407	0.9410	1.5332	2.1318	2.7764	3.7469	4.6041	7.1732
5	0.5594	0.7267	0.9195	1.4759	2.0150	2.5706	3.3649	4.0321	5.8934
6	0.5534	0.7176	0.9057	1.4398	1.9432	2.4469	3.1427	3.7074	5.2076
7	0.5491	0.7111	0.8960	1.4149	1.8946	2.3646	2.9980	3.4995	4.7853
8	0.5459	0.7064	0.8889	1.3968	1.8595	2.3060	2.8965	3.3554	4.5008
9	0.5435	0.7027	0.8834	1.3830	1.8331	2.2622	2.8214	3.2498	4.2968
10	0.5415	0.6998	0.8791	1.3722	1.8125	2.2281	2.7638	3.1693	4.1437
11	0.5399	0.6974	0.8755	1.3634	1.7959	2.2010	2.7181	3.1058	4.0247
12	0.5386	0.6955	0.8726	1.3562	1.7823	2.1788	2.6810	3.0545	3.9296
13	0.5375	0.6938	0.8702	1.3502	1.7709	2.1604	2.6503	3.0123	3.8520
14	0.5366	0.6924	0.8681	1.3450	1.7613	2.1448	2.6245	2.9768	3.7874
15	0.5357	0.6912	0.8662	1.3406	1.7531	2.1314	2.6025	2.9467	3.7328
16	0.5350	0.6901	0.8647	1.3368	1.7459	2.1199	2.5835	2.9208	3.6862
17	0.5344	0.6892	0.8633	1.3334	1.7396	2.1098	2.5669	2.8982	3.6458
18	0.5338	0.6884	0.8620	1.3304	1.7341	2.1009	2.5524	2.8784	3.6105
19	0.5333	0.6876	0.8610	1.3277	1.7291	2.0930	2.5395	2.8609	3.5794
20	0.5329	0.6870	0.8600	1.3253	1.7247	2.0860	2.5280	2.8453	3.5518
21	0.5325	0.6864	0.8591	1.3232	1.7207	2.0796	2.5176	2.8314	3.5272
22	0.5321	0.6858	0.8583	1.3212	1.7171	2.0739	2.5083	2.8188	3.5050
23	0.5317	0.6853	0.8575	1.3195	1.7139	2.0687	2.4999	2.8073	3.4850
24	0.5314	0.6848	0.8569	1.3178	1.7109	2.0639	2.4922	2.7969	3.4668
25	0.5312	0.6844	0.8562	1.3163	1.7081	2.0595	2.4851	2.7874	3.4502
26	0.5309	0.6840	0.8557	1.3150	1.7056	2.0555	2.4786	2.7787	3.4350
27	0.5306	0.6837	0.8551	1.3137	1.7033	2.0518	2.4727	2.7707	3.4210
28	0.5304	0.6834	0.8546	1.3125	1.7011	2.0484	2.4671	2.7633	3.4082
29	0.5302	0.6830	0.8542	1.3114	1.6991	2.0452	2.4620	2.7564	3.3962
30	0.5300	0.6828	0.8538	1.3104	1.6973	2.0423	2.4573	2.7500	3.3852
40	0.5286	0.6807	0.8507	1.3031	1.6839	2.0211	2.4233	2.7045	3.3069
50	0.5278	0.6794	0.8489	1.2987	1.6759	2.0086	2.4033	2.6778	3.2614
60	0.5272	0.6786	0.8477	1.2958	1.6706	2.0003	2.3901	2.6603	3.2317
70	0.5268	0.6780	0.8468	1.2938	1.6669	1.9944	2.3808	2.6479	3.2108
80	0.5265	0.6776	0.8461	1.2922	1.6641	1.9901	2.3739	2.6387	3.1953
90	0.5263	0.6772	0.8456	1.2910	1.6620	1.9867	2.3685	2.6316	3.1833
100	0.5261	0.6770	0.8452	1.2901	1.6602	1.9840	2.3642	2.6259	3.1737
∞	0.5244	0.6745	0.8416	1.2816	1.6449	1.9600	2.3263	2.5758	3.0902

Table A.7. *F-Distribution*

df1	2	Rt Area	F
df2	3	0.1	5.4624
		0.0894	6

F for right-tail area = 0.1

Denom df2	\multicolumn{14}{c}{Numerator df1}
	1	2	3	4	5	6	7	8	9	10	16	25	40	64	100
---	---	---	---	---	---	---	---	---	---	---	---	---	---	---	---
1	39.86	49.50	53.59	55.83	57.24	58.20	58.91	59.44	59.86	60.19	61.35	62.05	62.53	62.83	63.01
2	8.53	9.00	9.16	9.24	9.29	9.33	9.35	9.37	9.38	9.39	9.43	9.45	9.47	9.48	9.48
3	5.54	5.46	5.39	5.34	5.31	5.28	5.27	5.25	5.24	5.23	5.20	5.17	5.16	5.15	5.14
4	4.54	4.32	4.19	4.11	4.05	4.01	3.98	3.95	3.94	3.92	3.86	3.83	3.80	3.79	3.78
5	4.06	3.78	3.62	3.52	3.45	3.40	3.37	3.34	3.32	3.30	3.23	3.19	3.16	3.14	3.13
6	3.78	3.46	3.29	3.18	3.11	3.05	3.01	2.98	2.96	2.94	2.86	2.81	2.78	2.76	2.75
7	3.59	3.26	3.07	2.96	2.88	2.83	2.78	2.75	2.72	2.70	2.62	2.57	2.54	2.51	2.50
8	3.46	3.11	2.92	2.81	2.73	2.67	2.62	2.59	2.56	2.54	2.45	2.40	2.36	2.34	2.32
9	3.36	3.01	2.81	2.69	2.61	2.55	2.51	2.47	2.44	2.42	2.33	2.27	2.23	2.21	2.19
10	3.29	2.92	2.73	2.61	2.52	2.46	2.41	2.38	2.35	2.32	2.23	2.17	2.13	2.10	2.09
11	3.23	2.86	2.66	2.54	2.45	2.39	2.34	2.30	2.27	2.25	2.16	2.10	2.05	2.02	2.01
12	3.18	2.81	2.61	2.48	2.39	2.33	2.28	2.24	2.21	2.19	2.09	2.03	1.99	1.96	1.94
13	3.14	2.76	2.56	2.43	2.35	2.28	2.23	2.20	2.16	2.14	2.04	1.98	1.93	1.90	1.88
14	3.10	2.73	2.52	2.39	2.31	2.24	2.19	2.15	2.12	2.10	2.00	1.93	1.89	1.85	1.83
15	3.07	2.70	2.49	2.36	2.27	2.21	2.16	2.12	2.09	2.06	1.96	1.89	1.85	1.81	1.79
16	3.05	2.67	2.46	2.33	2.24	2.18	2.13	2.09	2.06	2.03	1.93	1.86	1.81	1.78	1.76
17	3.03	2.64	2.44	2.31	2.22	2.15	2.10	2.06	2.03	2.00	1.90	1.83	1.78	1.75	1.73
18	3.01	2.62	2.42	2.29	2.20	2.13	2.08	2.04	2.00	1.98	1.87	1.80	1.75	1.72	1.70
19	2.99	2.61	2.40	2.27	2.18	2.11	2.06	2.02	1.98	1.96	1.85	1.78	1.73	1.69	1.67
20	2.97	2.59	2.38	2.25	2.16	2.09	2.04	2.00	1.96	1.94	1.83	1.76	1.71	1.67	1.65
25	2.92	2.53	2.32	2.18	2.09	2.02	1.97	1.93	1.89	1.87	1.76	1.68	1.63	1.59	1.56
30	2.88	2.49	2.28	2.14	2.05	1.98	1.93	1.88	1.85	1.82	1.71	1.63	1.57	1.53	1.51
40	2.84	2.44	2.23	2.09	2.00	1.93	1.87	1.83	1.79	1.76	1.65	1.57	1.51	1.46	1.43
60	2.79	2.39	2.18	2.04	1.95	1.87	1.82	1.77	1.74	1.71	1.59	1.50	1.44	1.39	1.36
80	2.77	2.37	2.15	2.02	1.92	1.85	1.79	1.75	1.71	1.68	1.56	1.47	1.40	1.35	1.32
100	2.76	2.36	2.14	2.00	1.91	1.83	1.78	1.73	1.69	1.66	1.54	1.45	1.38	1.33	1.29
120	2.75	2.35	2.13	1.99	1.90	1.82	1.77	1.72	1.68	1.65	1.53	1.44	1.37	1.31	1.28

(continued)

Table A.7 (continued)

df1	2	Rt Area	F
df2	3	0.1	5.4624
		0.0894	6

F for right-tail area = 0.05

Denom df2	Numerator df1													
	1	2	3	4	5	6	7	8	9	10	16	25	40	64	100
1	161.45	199.50	215.71	224.58	230.16	233.99	236.77	238.88	240.54	241.88	246.46	249.26	251.14	252.33	253.04
2	18.51	19.00	19.16	19.25	19.30	19.33	19.35	19.37	19.38	19.40	19.43	19.46	19.47	19.48	19.49
3	10.13	9.55	9.28	9.12	9.01	8.94	8.89	8.85	8.81	8.79	8.69	8.63	8.59	8.57	8.55
4	7.71	6.94	6.59	6.39	6.26	6.16	6.09	6.04	6.00	5.96	5.84	5.77	5.72	5.68	5.66
5	6.61	5.79	5.41	5.19	5.05	4.95	4.88	4.82	4.77	4.74	4.60	4.52	4.46	4.43	4.41
6	5.99	5.14	4.76	4.53	4.39	4.28	4.21	4.15	4.10	4.06	3.92	3.83	3.77	3.74	3.71
7	5.59	4.74	4.35	4.12	3.97	3.87	3.79	3.73	3.68	3.64	3.49	3.40	3.34	3.30	3.27
8	5.32	4.46	4.07	3.84	3.69	3.58	3.50	3.44	3.39	3.35	3.20	3.11	3.04	3.00	2.97
9	5.12	4.26	3.86	3.63	3.48	3.37	3.29	3.23	3.18	3.14	2.99	2.89	2.83	2.78	2.76
10	4.96	4.10	3.71	3.48	3.33	3.22	3.14	3.07	3.02	2.98	2.83	2.73	2.66	2.62	2.59
11	4.84	3.98	3.59	3.36	3.20	3.09	3.01	2.95	2.90	2.85	2.70	2.60	2.53	2.48	2.46
12	4.75	3.89	3.49	3.26	3.11	3.00	2.91	2.85	2.80	2.75	2.60	2.50	2.43	2.38	2.35
13	4.67	3.81	3.41	3.18	3.03	2.92	2.83	2.77	2.71	2.67	2.51	2.41	2.34	2.29	2.26
14	4.60	3.74	3.34	3.11	2.96	2.85	2.76	2.70	2.65	2.60	2.44	2.34	2.27	2.22	2.19
15	4.54	3.68	3.29	3.06	2.90	2.79	2.71	2.64	2.59	2.54	2.38	2.28	2.20	2.15	2.12
16	4.49	3.63	3.24	3.01	2.85	2.74	2.66	2.59	2.54	2.49	2.33	2.23	2.15	2.10	2.07
17	4.45	3.59	3.20	2.96	2.81	2.70	2.61	2.55	2.49	2.45	2.29	2.18	2.10	2.05	2.02
18	4.41	3.55	3.16	2.93	2.77	2.66	2.58	2.51	2.46	2.41	2.25	2.14	2.06	2.01	1.98
19	4.38	3.52	3.13	2.90	2.74	2.63	2.54	2.48	2.42	2.38	2.21	2.11	2.03	1.97	1.94
20	4.35	3.49	3.10	2.87	2.71	2.60	2.51	2.45	2.39	2.35	2.18	2.07	1.99	1.94	1.91
25	4.24	3.39	2.99	2.76	2.60	2.49	2.40	2.34	2.28	2.24	2.07	1.96	1.87	1.82	1.78
30	4.17	3.32	2.92	2.69	2.53	2.42	2.33	2.27	2.21	2.16	1.99	1.88	1.79	1.73	1.70
40	4.08	3.23	2.84	2.61	2.45	2.34	2.25	2.18	2.12	2.08	1.90	1.78	1.69	1.63	1.59
60	4.00	3.15	2.76	2.53	2.37	2.25	2.17	2.10	2.04	1.99	1.82	1.69	1.59	1.53	1.48
80	3.96	3.11	2.72	2.49	2.33	2.21	2.13	2.06	2.00	1.95	1.77	1.64	1.54	1.47	1.43
100	3.94	3.09	2.70	2.46	2.31	2.19	2.10	2.03	1.97	1.93	1.75	1.62	1.52	1.44	1.39
120	3.92	3.07	2.68	2.45	2.29	2.18	2.09	2.02	1.96	1.91	1.73	1.60	1.50	1.42	1.37

Table A.7 *(continued)*

df1	2	Rt Area	F
df2	3	0.1	5.4624
		0.0894	6

F for right-tail area = 0.025

Denom df2	Numerator df1														
	1	2	3	4	5	6	7	8	9	10	16	25	40	64	100
1	647.79	799.50	864.16	899.58	921.85	937.11	948.22	956.66	963.28	968.63	986.92	998.08	1005.60	1010.33	1013.17
2	38.51	39.00	39.17	39.25	39.30	39.33	39.36	39.37	39.39	39.40	39.44	39.46	39.47	39.48	39.49
3	17.44	16.04	15.44	15.10	14.88	14.73	14.62	14.54	14.47	14.42	14.23	14.12	14.04	13.99	13.96
4	12.22	10.65	9.98	9.60	9.36	9.20	9.07	8.98	8.90	8.84	8.63	8.50	8.41	8.35	8.32
5	10.01	8.43	7.76	7.39	7.15	6.98	6.85	6.76	6.68	6.62	6.40	6.27	6.18	6.12	6.08
6	8.81	7.26	6.60	6.23	5.99	5.82	5.70	5.60	5.52	5.46	5.24	5.11	5.01	4.95	4.92
7	8.07	6.54	5.89	5.52	5.29	5.12	4.99	4.90	4.82	4.76	4.54	4.40	4.31	4.25	4.21
8	7.57	6.06	5.42	5.05	4.82	4.65	4.53	4.43	4.36	4.30	4.08	3.94	3.84	3.78	3.74
9	7.21	5.71	5.08	4.72	4.48	4.32	4.20	4.10	4.03	3.96	3.74	3.60	3.51	3.44	3.40
10	6.94	5.46	4.83	4.47	4.24	4.07	3.95	3.85	3.78	3.72	3.50	3.35	3.26	3.19	3.15
11	6.72	5.26	4.63	4.28	4.04	3.88	3.76	3.66	3.59	3.53	3.30	3.16	3.06	3.00	2.96
12	6.55	5.10	4.47	4.12	3.89	3.73	3.61	3.51	3.44	3.37	3.15	3.01	2.91	2.84	2.80
13	6.41	4.97	4.35	4.00	3.77	3.60	3.48	3.39	3.31	3.25	3.03	2.88	2.78	2.71	2.67
14	6.30	4.86	4.24	3.89	3.66	3.50	3.38	3.29	3.21	3.15	2.92	2.78	2.67	2.61	2.56
15	6.20	4.77	4.15	3.80	3.58	3.41	3.29	3.20	3.12	3.06	2.84	2.69	2.59	2.52	2.47
16	6.12	4.69	4.08	3.73	3.50	3.34	3.22	3.12	3.05	2.99	2.76	2.61	2.51	2.44	2.40
17	6.04	4.62	4.01	3.66	3.44	3.28	3.16	3.06	2.98	2.92	2.70	2.55	2.44	2.37	2.33
18	5.98	4.56	3.95	3.61	3.38	3.22	3.10	3.01	2.93	2.87	2.64	2.49	2.38	2.31	2.27
19	5.92	4.51	3.90	3.56	3.33	3.17	3.05	2.96	2.88	2.82	2.59	2.44	2.33	2.26	2.22
20	5.87	4.46	3.86	3.51	3.29	3.13	3.01	2.91	2.84	2.77	2.55	2.40	2.29	2.22	2.17
25	5.69	4.29	3.69	3.35	3.13	2.97	2.85	2.75	2.68	2.61	2.38	2.23	2.12	2.04	2.00
30	5.57	4.18	3.59	3.25	3.03	2.87	2.75	2.65	2.57	2.51	2.28	2.12	2.01	1.93	1.88
40	5.42	4.05	3.46	3.13	2.90	2.74	2.62	2.53	2.45	2.39	2.15	1.99	1.88	1.79	1.74
60	5.29	3.93	3.34	3.01	2.79	2.63	2.51	2.41	2.33	2.27	2.03	1.87	1.74	1.66	1.60
80	5.22	3.86	3.28	2.95	2.73	2.57	2.45	2.35	2.28	2.21	1.97	1.81	1.68	1.59	1.53
100	5.18	3.83	3.25	2.92	2.70	2.54	2.42	2.32	2.24	2.18	1.94	1.77	1.64	1.55	1.48
120	5.15	3.80	3.23	2.89	2.67	2.52	2.39	2.30	2.22	2.16	1.92	1.75	1.61	1.52	1.45

(continued)

Table A.7 (continued)

df1	2	Rt Area	F
df2	3	0.1	5.4624
		0.0894	6

F for right-tail area = 0.01

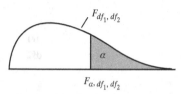

Denom df2	Numerator df1													
	1	2	3	4	5	6	7	8	9	10	16	25	40	64	100
1	4052.2	4999.5	5403.4	5624.6	5763.6	5859.0	5928.4	5981.1	6022.5	6055.8	6170.1	6239.8	6286.8	6316.3	6334.1
2	98.5	99.0	99.2	99.2	99.3	99.3	99.4	99.4	99.4	99.4	99.4	99.5	99.5	99.5	99.5
3	34.1	30.8	29.5	28.7	28.2	27.9	27.7	27.5	27.3	27.2	26.8	26.6	26.4	26.3	26.2
4	21.2	18.0	16.7	16.0	15.5	15.2	15.0	14.8	14.7	14.5	14.2	13.9	13.7	13.6	13.6
5	16.3	13.3	12.1	11.4	11.0	10.7	10.5	10.3	10.2	10.1	9.7	9.4	9.3	9.2	9.1
6	13.7	10.9	9.8	9.1	8.7	8.5	8.3	8.1	8.0	7.9	7.5	7.3	7.1	7.0	7.0
7	12.2	9.5	8.5	7.8	7.5	7.2	7.0	6.8	6.7	6.6	6.3	6.1	5.9	5.8	5.8
8	11.3	8.6	7.6	7.0	6.6	6.4	6.2	6.0	5.9	5.8	5.5	5.3	5.1	5.0	5.0
9	10.6	8.0	7.0	6.4	6.1	5.8	5.6	5.5	5.4	5.3	4.9	4.7	4.6	4.5	4.4
10	10.0	7.6	6.6	6.0	5.6	5.4	5.2	5.1	4.9	4.8	4.5	4.3	4.2	4.1	4.0
11	9.6	7.2	6.2	5.7	5.3	5.1	4.9	4.7	4.6	4.5	4.2	4.0	3.9	3.8	3.7
12	9.3	6.9	6.0	5.4	5.1	4.8	4.6	4.5	4.4	4.3	4.0	3.8	3.6	3.5	3.5
13	9.1	6.7	5.7	5.2	4.9	4.6	4.4	4.3	4.2	4.1	3.8	3.6	3.4	3.3	3.3
14	8.9	6.5	5.6	5.0	4.7	4.5	4.3	4.1	4.0	3.9	3.6	3.4	3.3	3.2	3.1
15	8.7	6.4	5.4	4.9	4.6	4.3	4.1	4.0	3.9	3.8	3.5	3.3	3.1	3.0	3.0
16	8.5	6.2	5.3	4.8	4.4	4.2	4.0	3.9	3.8	3.7	3.4	3.2	3.0	2.9	2.9
17	8.4	6.1	5.2	4.7	4.3	4.1	3.9	3.8	3.7	3.6	3.3	3.1	2.9	2.8	2.8
18	8.3	6.0	5.1	4.6	4.2	4.0	3.8	3.7	3.6	3.5	3.2	3.0	2.8	2.7	2.7
19	8.2	5.9	5.0	4.5	4.2	3.9	3.8	3.6	3.5	3.4	3.1	2.9	2.8	2.7	2.6
20	8.1	5.8	4.9	4.4	4.1	3.9	3.7	3.6	3.5	3.4	3.1	2.8	2.7	2.6	2.5
25	7.8	5.6	4.7	4.2	3.9	3.6	3.5	3.3	3.2	3.1	2.8	2.6	2.5	2.4	2.3
30	7.6	5.4	4.5	4.0	3.7	3.5	3.3	3.2	3.1	3.0	2.7	2.5	2.3	2.2	2.1
40	7.3	5.2	4.3	3.8	3.5	3.3	3.1	3.0	2.9	2.8	2.5	2.3	2.1	2.0	1.9
60	7.1	5.0	4.1	3.6	3.3	3.1	3.0	2.8	2.7	2.6	2.3	2.1	1.9	1.8	1.7
80	7.0	4.9	4.0	3.6	3.3	3.0	2.9	2.7	2.6	2.6	2.2	2.0	1.8	1.7	1.7
100	6.9	4.8	4.0	3.5	3.2	3.0	2.8	2.7	2.6	2.5	2.2	2.0	1.8	1.7	1.6
120	6.9	4.8	3.9	3.5	3.2	3.0	2.8	2.7	2.6	2.5	2.2	1.9	1.8	1.6	1.6

Table A.7 *(continued)*

df1	2	Rt Area	F
df2	3	0.1	5.4624
		0.0894	6

F for right-tail area = 0.005

Denom	Numerator df1														
df2	1	2	3	4	5	6	7	8	9	10	16	25	40	64	100
1	16211	19999	21615	22500	23056	23437	23715	23925	24091	24224	24681	24960	25148	25266	25337
2	199	199	199	199	199	199	199	199	199	199	199	199	199	199	199
3	56	50	47	46	45	45	44	44	44	44	43	43	42	42	42
4	31	26	24	23	22	22	22	21	21	21	20	20	20	20	19
5	23	18	17	16	15	15	14	14	14	14	13	13	13	12	12
6	19	15	13	12	11	11	11	11	10	10	10	9	9	9	9
7	16	12	11	10	10	9	9	9	9	8	8	8	7	7	7
8	15	11	10	9	8	8	8	7	7	7	7	6	6	6	6
9	14	10	9	8	7	7	7	7	7	6	6	6	6	5	5
10	13	9	8	7	7	7	6	6	6	6	5	5	5	5	5
11	12	9	8	7	6	6	6	6	6	5	5	5	5	4	4
12	12	9	7	7	6	6	6	5	5	5	5	4	4	4	4
13	11	8	7	6	6	5	5	5	5	5	4	4	4	4	4
14	11	8	7	6	6	5	5	5	5	5	4	4	4	4	4
15	11	8	6	6	5	5	5	5	5	4	4	4	4	3	3
16	11	8	6	6	5	5	5	5	4	4	4	4	3	3	3
17	10	7	6	5	5	5	5	4	4	4	4	3	3	3	3
18	10	7	6	5	5	5	4	4	4	4	3	3	3	3	3
19	10	7	6	5	5	5	4	4	4	4	3	3	3	3	3
20	10	7	6	5	5	4	4	4	4	4	3	3	3	3	3
25	9	7	5	5	4	4	4	4	4	4	3	3	3	3	3
30	9	6	5	5	4	4	4	4	3	3	3	3	3	2	2
40	9	6	5	4	4	4	4	3	3	3	3	2	2	2	2
60	8	6	5	4	4	3	3	3	3	3	3	2	2	2	2
80	8	6	5	4	4	3	3	3	3	3	2	2	2	2	2
100	8	6	5	4	4	3	3	3	3	3	2	2	2	2	2
120	8	6	4	4	4	3	3	3	3	3	2	2	2	2	2

Table A.8. *Factors for control charts*

n	X-Bar A	A2	A3	d2	D1	R-Chart D2	D3	D4	d3	c4	B3	s-Chart B4	B5	B6
2	2.121	1.88	2.659	1.128	0	3.686	0	3.267	0.853	0.798	0	3.267	0	2.606
3	1.732	1.023	1.954	1.693	0	4.358	0	2.575	0.888	0.886	0	2.568	0	2.276
4	1.5	0.729	1.628	2.059	0	4.698	0	2.282	0.88	0.921	0	2.266	0	2.088
5	1.342	0.577	1.427	2.326	0	4.918	0	2.114	0.864	0.94	0	2.089	0	1.964
6	1.225	0.483	1.287	2.534	0	5.079	0	2.004	0.848	0.952	0.03	1.97	0.029	1.874
7	1.134	0.419	1.182	2.704	0.205	5.204	0.076	1.924	0.833	0.959	0.118	1.882	0.113	1.806
8	1.061	0.373	1.099	2.847	0.388	5.307	0.136	1.864	0.82	0.965	0.185	1.815	0.179	1.751
9	1	0.337	1.032	2.97	0.547	5.394	0.184	1.816	0.808	0.969	0.239	1.761	0.232	1.707
10	0.949	0.308	0.975	3.078	0.686	5.469	0.223	1.777	0.797	0.973	0.284	1.716	0.276	1.669
11	0.905	0.285	0.927	3.173	0.811	5.535	0.256	1.744	0.787	0.975	0.321	1.679	0.313	1.637
12	0.866	0.266	0.886	3.258	0.923	5.594	0.283	1.717	0.778	0.978	0.354	1.646	0.346	1.61
13	0.832	0.249	0.85	3.336	1.025	5.647	0.307	1.693	0.77	0.979	0.382	1.618	0.374	1.585
14	0.802	0.235	0.817	3.407	1.118	5.696	0.328	1.672	0.763	0.981	0.406	1.594	0.399	1.563
15	0.775	0.223	0.789	3.472	1.203	5.74	0.347	1.653	0.756	0.982	0.428	1.572	0.421	1.544
16	0.75	0.212	0.763	3.532	1.282	5.782	0.363	1.637	0.75	0.983	0.448	1.552	0.44	1.526
17	0.728	0.203	0.739	3.588	1.356	5.82	0.378	1.622	0.744	0.985	0.466	1.534	0.458	1.511
18	0.707	0.194	0.718	3.64	1.424	5.856	0.391	1.609	0.739	0.985	0.482	1.518	0.475	1.496
19	0.688	0.187	0.698	3.689	1.489	5.889	0.404	1.596	0.733	0.986	0.497	1.503	0.49	1.483
20	0.671	0.18	0.68	3.735	1.549	5.921	0.415	1.585	0.729	0.987	0.51	1.49	0.504	1.47
21	0.655	0.173	0.663	3.778	1.606	5.951	0.425	1.575	0.724	0.988	0.523	1.477	0.516	1.459
22	0.64	0.167	0.647	3.819	1.66	5.979	0.435	1.565	0.72	0.988	0.534	1.466	0.528	1.448
23	0.626	0.162	0.633	3.858	1.711	6.006	0.443	1.557	0.716	0.989	0.545	1.455	0.539	1.438
24	0.612	0.157	0.619	3.895	1.759	6.032	0.452	1.548	0.712	0.989	0.555	1.445	0.549	1.429
25	0.6	0.153	0.606	3.931	1.805	6.056	0.459	1.541	0.708	0.99	0.565	1.435	0.559	1.42
30	0.548	0.134	0.552	4.086	2.008	6.164	0.491	1.509	0.693	0.991	0.604	1.396	0.599	1.384
40	0.474	0.11	0.477	4.322	2.314	6.329	0.535	1.465	0.669	0.994	0.659	1.341	0.655	1.332
50	0.424	0.094	0.426	4.498	2.542	6.455	0.565	1.435	0.652	0.995	0.696	1.304	0.693	1.297
60	0.387	0.083	0.389	4.639	2.722	6.555	0.587	1.413	0.639	0.996	0.723	1.277	0.72	1.271
80	0.335	0.069	0.336	4.854	2.995	6.712	0.617	1.383	0.619	0.997	0.761	1.239	0.759	1.235
100	0.3	0.06	0.301	5.015	3.2	6.831	0.638	1.362	0.605	0.997	0.787	1.213	0.785	1.21

$\overline{X}-R$ Chart CL at $\overline{\overline{X}}$ $\text{UCL}_{\overline{X}} = \overline{\overline{X}} + A_2 \overline{R}$ $\text{LCL}_{\overline{X}} = \overline{\overline{X}} - A_2 \overline{R}$
CL at $\overline{R}$ $\text{UCL}_R = D_4 \overline{R}$ $\text{LCL}_R = D_3 \overline{R}$

$\overline{X}-s$ Chart CL at $\overline{\overline{X}}$ $\text{UCL}_{\overline{X}} = \overline{\overline{X}} + A_3 \overline{s}$ $\text{LCL}_{\overline{X}} = \overline{\overline{X}} - A_3 \overline{s}$
CL at $\overline{s}$ $\text{UCL}_s = B_4 \overline{s}$ $\text{LCL}_s = B_3 \overline{s}$

Target μ_0, σ_0: $\overline{X}$-Chart $\text{UCL}_{\overline{X}} = \mu_0 + A\sigma_0$ $\text{LCL}_{\overline{X}} = \mu_0 - A\sigma_0$
R-Chart CL at $d_2\sigma_0$ $\text{UCL}_R = D_2\sigma_0$ $\text{LCL}_R = D_1\sigma_0$
s-Chart CL at $c_4\sigma_0$ $\text{UCL}_s = B_6\sigma_0$ $\text{LCL}_s = B_5\sigma_0$

Table A.9. *Tolerance intervals: Sheet1. Factor k_1 for one-sided tolerance limits*

	g	0.95			g	0.99	
n \ p	0.9	0.95	0.99	n \ p	0.9	0.95	0.99
3	6.155	7.656	10.553	3	13.995	17.370	23.896
4	4.162	5.144	7.042	4	7.380	9.083	12.387
5	3.407	4.203	5.741	5	5.362	6.578	8.939
6	3.006	3.708	5.062	6	4.411	5.406	7.335
7	2.755	3.399	4.642	7	3.859	4.728	6.412
8	2.582	3.187	4.354	8	3.497	4.285	5.812
9	2.454	3.031	4.143	9	3.240	3.972	5.389
10	2.355	2.911	3.981	10	3.048	3.738	5.074
11	2.275	2.815	3.852	11	2.898	3.556	4.829
12	2.210	2.736	3.747	12	2.777	3.410	4.633
13	2.155	2.671	3.659	13	2.677	3.290	4.472
14	2.109	2.614	3.585	14	2.593	3.189	4.337
15	2.068	2.566	3.520	15	2.521	3.102	4.222
16	2.033	2.524	3.464	16	2.459	3.028	4.123
17	2.002	2.486	3.414	17	2.405	2.963	4.037
18	1.974	2.453	3.370	18	2.357	2.905	3.960
19	1.949	2.423	3.331	19	2.314	2.854	3.892
20	1.926	2.396	3.295	20	2.276	2.808	3.832
22	1.886	2.349	3.233	22	2.209	2.729	3.727
24	1.853	2.309	3.181	24	2.154	2.662	3.640
26	1.824	2.275	3.136	26	2.106	2.606	3.566
28	1.799	2.246	3.098	28	2.065	2.558	3.502
30	1.777	2.220	3.064	30	2.030	2.515	3.447
35	1.732	2.167	2.995	35	1.957	2.430	3.334
40	1.697	2.125	2.941	40	1.902	2.364	3.249
45	1.669	2.092	2.898	45	1.857	2.312	3.180
50	1.646	2.065	2.862	50	1.821	2.269	3.125
60	1.609	2.022	2.807	60	1.764	2.202	3.038
70	1.581	1.990	2.765	70	1.722	2.153	2.974
80	1.559	1.964	2.733	80	1.688	2.114	2.924
90	1.542	1.944	2.706	90	1.661	2.082	2.883
100	1.527	1.927	2.684	100	1.639	2.056	2.850
150	1.478	1.870	2.611	150	1.566	1.971	2.740
200	1.450	1.837	2.570	200	1.524	1.923	2.679
250	1.431	1.815	2.542	250	1.496	1.891	2.638
300	1.417	1.800	2.522	300	1.476	1.868	2.608
400	1.398	1.778	2.494	400	1.448	1.836	2.567
500	1.385	1.763	2.475	500	1.430	1.814	2.540
1000	1.354	1.727	2.430	1000	1.385	1.762	2.474

Table A.9. *Tolerance intervals: Sheet2. Factor k_2 for two-sided tolerance limits*
(See sheet1 for one-sided tolerance limits)

n \ p	g 0.9	0.95 0.95	0.99	n \ p	g 0.9	0.99 0.95	0.99
3	8.380	9.916	12.861	3	18.930	22.401	29.055
4	5.369	6.370	8.299	4	9.398	11.150	14.527
5	4.275	5.079	6.634	5	6.612	7.855	10.260
6	3.712	4.414	5.775	6	5.337	6.345	8.301
7	3.369	4.007	5.248	7	4.613	5.488	7.187
8	3.136	3.732	4.891	8	4.147	4.936	6.468
9	2.967	3.532	4.631	9	3.822	4.550	5.966
10	2.839	3.379	4.433	10	3.582	4.265	5.594
11	2.737	3.259	4.277	11	3.397	4.045	5.308
12	2.655	3.162	4.150	12	3.250	3.870	5.079
13	2.587	3.081	4.044	13	3.130	3.727	4.893
14	2.529	3.012	3.955	14	3.029	3.608	4.737
15	2.480	2.954	3.878	15	2.945	3.507	4.605
16	2.437	2.903	3.812	16	2.872	3.421	4.492
17	2.400	2.858	3.754	17	2.808	3.345	4.393
18	2.366	2.819	3.702	18	2.753	3.279	4.307
19	2.337	2.784	3.656	19	2.703	3.221	4.230
20	2.310	2.752	3.615	20	2.659	3.168	4.161
22	2.264	2.697	3.543	22	2.584	3.078	4.044
24	2.225	2.651	3.483	24	2.522	3.004	3.947
26	2.193	2.612	3.432	26	2.469	2.941	3.865
28	2.164	2.579	3.388	28	2.424	2.888	3.794
30	2.140	2.549	3.350	30	2.385	2.841	3.733
35	2.090	2.490	3.272	35	2.306	2.748	3.611
40	2.052	2.445	3.212	40	2.247	2.677	3.518
45	2.021	2.408	3.165	45	2.200	2.621	3.444
50	1.996	2.379	3.126	50	2.162	2.576	3.385
60	1.958	2.333	3.066	60	2.103	2.506	3.293
70	1.929	2.299	3.021	70	2.060	2.454	3.225
80	1.907	2.272	2.986	80	2.026	2.414	3.172
90	1.889	2.251	2.958	90	1.999	2.382	3.130
100	1.874	2.233	2.934	100	1.977	2.355	3.096
150	1.825	2.175	2.859	150	1.905	2.270	2.983
200	1.798	2.143	2.816	200	1.865	2.222	2.921
250	1.780	2.121	2.788	250	1.839	2.191	2.880
300	1.767	2.106	2.767	300	1.820	2.169	2.850
400	1.749	2.084	2.739	400	1.794	2.138	2.809
500	1.737	2.070	2.721	500	1.777	2.117	2.783
1000	1.709	2.036	2.676	1000	1.736	2.068	2.718

Table A.10. *Studentized range*

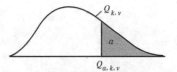

$1-\alpha=0.95$ $v \backslash k$	2	3	4	5	6	7	8	9	10
5	3.64	4.61	5.22	5.68	6.04	6.33	6.58	6.80	7.00
6	3.46	4.34	4.90	5.31	5.63	5.90	6.12	6.32	6.50
7	3.35	4.17	4.68	5.06	5.36	5.61	5.82	6.00	6.16
8	3.26	4.04	4.53	4.89	5.17	5.40	5.60	5.77	5.92
9	3.20	3.95	4.42	4.76	5.03	5.25	5.43	5.60	5.74
10	3.15	3.88	4.33	4.66	4.91	5.13	5.31	5.46	5.60
11	3.11	3.82	4.26	4.58	4.82	5.03	5.20	5.35	5.49
15	3.02	3.68	4.08	4.37	4.60	4.78	4.94	5.08	5.20
20	2.95	3.58	3.96	4.23	4.45	4.62	4.77	4.90	5.01
30	2.89	3.49	3.85	4.10	4.30	4.47	4.60	4.72	4.83
40	2.86	3.44	3.79	4.04	4.23	4.39	4.52	4.64	4.74

$1-\alpha=0.99$ $v \backslash k$	2	3	4	5	6	7	8	9	10
5	5.73	7.00	7.83	8.44	8.93	9.34	9.69	10.00	10.27
6	5.27	6.35	7.05	7.57	7.99	8.34	8.63	8.89	9.12
7	4.97	5.93	6.56	7.02	7.39	7.69	7.95	8.18	8.38
8	4.76	5.65	6.22	6.64	6.97	7.25	7.49	7.69	7.88
9	4.61	5.44	5.97	6.36	6.67	6.92	7.14	7.34	7.51
10	4.49	5.28	5.78	6.15	6.44	6.68	6.88	7.06	7.22
11	4.40	5.16	5.63	5.98	6.25	6.48	6.68	6.85	7.00
15	4.18	4.84	5.26	5.56	5.80	6.00	6.17	6.32	6.45
20	4.03	4.65	5.02	5.30	5.52	5.69	5.84	5.98	6.09
30	3.90	4.46	4.80	5.05	5.25	5.41	5.54	5.66	5.76
40	3.83	4.37	4.70	4.94	5.12	5.27	5.40	5.51	5.60

(Table generated using AS 190 Algorithm, Applied Statistics (1983) vol 32, No. 2)

Functions strng(q,r,v)   q = student range   Example:
         strnginv(p,r,v)   p = confidence level $1-\alpha$

k	v	p
4	16	0.95

v = degrees of freedom df
r = number of samples k
q = 4.0474
p = 0.95

Note: $Q\alpha$, r,v = strnginv($1-\alpha$, r,v)
$Q = R/S$; k = numerator df; v = denominator df.

Answers to Selected Problems

Chapter 1

Search the Internet and refer to other books to prepare your answers.

Chapter 2

2.3. a) $\varphi 20H8/f7$, b) Hole $20^{+0.033}_{+0}$ Shaft $20^{-0.020}_{-0.041}$

2.4. a) Hole basis, b) at MMC Hole 40.000 Shaft, 40.042, at LMC Hole 40.025, Shaft 40.026, c) Locational interference fit, d) Interference Max 0.042, Min 0.001, Mean dia Hole 40.0125 Shaft 40.034, Mean interference 0.0215, Tolerance Hole 0.042, Shaft 0.016, Statistical tolerance 0.045, Interference Max 0.044 Min −0.001

2.5. ± 0.011 for a and b

2.6. 2 ± 0.768

2.7. 12 ± 0.212, 35 ± 0.306, 45 ± 0.334

2.8. 45 ± 0.268

2.9. 30.5 ± 0.006 Unit tolerance approach, 20.1 ± 0.005, 15.2 ± 0.005, 65.7 ± 0.008

2.10. 45H7/g6, Sliding fit, 45H7 (45, 45.025) Mean 45.0125, 45g6 (44.991, 44.9975) Mean 44.9943, Mean clearance 0.0183 (Min 0.0034, Max 0.0331)

Chapter 3

3.1. a) 20.03, b) 0.07

3.2. 25H8 (25.000, 25.033), Bonus Table (Size, Tol) (25.000, 0.010), (25.003, 0.013), (25.006, 0.016), (..., ...), (25.033, 0.043)

3.3. Sketch similar to Fig. P3.2, 25H7 (25.000, 25.021) Bonus Table (Size, Tol), (25.000, 0.016), (25.003, 0.019), (25.006, 0.022), (..., ...), (25.021, 0.037)

3.4. Circularity $= R\left(\frac{1}{\cos 30°} - 1\right) = 6.9615$

3.5. Circularity $= 1.8012$

3.6. Parallelism $= 17 - (-5) = 22$

3.7. Straightness $= 2.530$ using *Straightness.xls*, Angle wrt x axis $= 18.435°$

3.8. Flatness 0.041 using *Flatness.xls*, Normal Direction $(-0.393, -0.405, 0.825)$

3.9. Circularity 0.100 using *CircleSphere.xls*, Center Location (12.000, 5.000)

3.10. Straightness 0.088 using *Straightness.xls*, Angle 17.189° (∼0.3 radians)

3.11. Circularity 0.414 using *CircleSphere.xls*, Center Location (1.000, 1.000)

3.12. Cylindricity 0.150 using *Cylindricity.xls*, Fixed Point (0.000, 0.000, 0.000), Direction of Axis (0.000, 0.000, 1.000)

Chapter 4

4.1. 0.64

4.2. 0.75

4.3. 0.12

4.4. Hint: Draw Venn diagram

4.5. a) 0.429, b) 0.457, c) 0.114

4.6. Mean 6.1, Median 6, Mode 6

4.7. 2.767

4.8. Mean 9.51, Median 9.6, Mode 9.9, skewness -0.189, kurtosis -0.713, IQL 0.95

4.9. 2/3

4.10. 0.3095

4.11. 0.78

4.12. a) binomial distribution, b) 0.1901, c) 0.3231, d) Expected $\mu = 2, \sigma = 1.3416$

4.13. a) 0.7916, b) 0.4933

4.14. 0.2678

4.15. 0.0738

4.16. a) Poisson, b) 0.1339, c) 0.8488

4.17. 0.1152

4.18. a) 0.0067, b) 0.5595

4.19. 3.9199

4.20. 0.9924

4.21. a) $546.71, b) $126.42

4.22. chi-squared distribution

Chapter 5

5.1. $264.0514 \le \mu \le 275.9486$

5.2. $2.5421 \le \mu \le 2.5779$

5.3. $266.4254 \le \mu \le 271.5746$, 273 MPa is not acceptable

5.4. $29.9385 \le \mu \le 30.9615$

5.5. $335.5996 \le \mu \le 342.4004$

5.6. $-1.0543 \le \mu_1 - \mu_2 \le 0.0543$

5.7. Confidence Interval (Normal Approx) $0.0393 \le p \le 0.1321$
Confidence Interval (Clopper–Pearson) $0.0451 \le p \le 0.1449$

5.8. $3.7506 \le \sigma^2 \le 12.9115$

5.9. $0.5903 \le (\sigma_1/\sigma_2)^2 \le 4.47$

5.10. p-value $= 0.0304$, Reject H_0

5.11. p-value $= 0.0053$, Reject H_0

5.12. p-value $= 0.0804$, Insufficient evidence to reject H_0

5.13. p-value $= 0.8032$, Insufficient evidence to reject H_0

5.14. p-value $= 0.00070$, Reject H_0

5.15. 15

Chapter 6

Use chapter computer programs to draw charts and draw conclusions. If data ranges are changed, make changes in the series ranges in charts.

Chapter 7

7.1. a) UCL$_X$ 49.539, LCL$_X$ 49.451, UCL$_R$ 0.160, LCL$_R$ 0.000; b) $\mu = 49.495$, $\sigma = \overline{R}/d_2 = 0.033$

7.2. $LCL_X = 2.890$, $CL_X = 3.001$, $UCL_X = 3.112$; $LCL_R = 0.000$, $CL_R = 0.152$, $UCL_R = 0.347$

7.3. $LCL_X = 2.887$, $CL_X = 3.001$, $UCL_X = 3.114$; $LCL_S = 0.000$, $CL_S = 0.070$, $UCL_S = 0.158$

7.4. a) $UCL_X = 10.225$, $LCL_X = 9.803$, $UCL_S = 0.309$, $LCL_S = 0.000$; b) $\mu = 10.014$, $\sigma = \overline{S}/c_4 = 0.157$

7.5. Use program *EWMA.xls*

7.6. Use program *CUSUM.xls*, Sheet1

7.7. Use program *CUSUM.xls*, Sheet2

7.8. Use program *PNPCharts.xls*

7.9. Use program *PNPCharts.xls*

7.10. Use program *C&UCharts.xls*

7.11. Use program *C&UCharts.xls*

7.12. a) $UCL_X = 9.015$, $LCL_X = 8.985$, $UCL_R = 0.047$, $LCL_R = 0.000$; b) 0.847; c) 0.153

7.13. $UCL_X = 12.107$, $LCL_X = 11.893$, $\beta = 0.970$

Chapter 8

8.1. $C_p = 0.833$, $C_{pu} = 0.556$, $C_{pl} = 1.111$, $C_{pk} = 0.556$

8.2. $C_p = 0.915$, $C_{pk} = 0.686$, $C_{pm} = 0.755$

8.3. $C_{pk} = 4.167$

8.4. Brand A $C_p = 0.800$, $C_{pk} = 0.400$, $C_{pm} = 0.256$, Brand B $C_p = 0.571$, $C_{pk} = 0.476$, $C_{pm} = 0.458$; Brand B is preferable based on C_{pk}, C_{pm}

8.5. $C_{pk} = 1.111$, Confidence Interval (0.844, 1.369)

8.6. %R&R = 25.93, % $P/T = 18.583$

8.7. Mean = 21.333 MPa, Standard deviation = 1.298

8.8. Interval (23.896, 40.504)

8.9. Interval (43.527, 47.473)

8.10. Interval $\geq$ 222.68

8.11. Interval (21.788, 27.212)

8.12. 0.853

8.13. 46

8.14. 0.985

Chapter 9

9.1. 0.294 Binomial

9.2. 0.224 Hypergeometric

9.3. $n = 88, c = 2$

9.4. $P_a = 0.376$, Probability of rejecting on the first sample $= 0.104$

9.5. $n_1 = 62, c_1 = 1, n_2 = 32, c_2 = 3, ASN = 77$. Find other possibilities using $SP2\_ATR.xls$

9.6. $n = 200, c = 2$ (Result from spconline.com)

9.7. $n = 27, k = 1.7372$. Reject if $(U - \bar{x})/\sigma \geq k$ (from $SP\_VAR.xls$)

9.8. $n = 20, M = 6.17\%$ Accept. (Result from spconline.com)

Chapter 10

10.1. Reject null hypothesis

10.2. Reject null hypothesis

10.3. Reject null hypothesis

10.4. Soil has significant effect

10.5. Main and interaction effects are significant

10.6. p-values for A, B, C, AB, AC, BC, ABC; 0.93, 0, 0, 0, 0.43, 0.002, 0.036

10.7. $A -0.875, AB\ 0.375$

Chapter 11

11.1. 265 h

11.2. 0.980

11.3. 0.349, MTTF $= 827.52$ h

11.4. 0.414, MTTF $= 1711.3$ h

11.5. 0.9802

11.6. 0.887, MTTF $= 1838.4$ h

11.7. $R_s = 0.723$, MTTF $= 1538.5$ h

11.8. 0.6706

11.9. $R_s = 0.5806, \lambda_s \approx 0.0072/\text{h}$, MTTF $= 139.85$ h

11.10. 0.7710

11.11. $R_s = 0.9998$, MTTF = 13269 h

11.12. $R_s = 0.9785$, MTTF = 4500 h

11.13. $R_s = 0.9158$, MTTF = 1667 h; $R_s = 0.6188$, MTTF = 833 h for 2 units without standby

11.14. MTTF = 66.67 h, MTTF with 3 standby units = 266.67 h

11.15. $R_s = 0.5488$, MTTF = 1667 h, Number of standby units for R_s of 0.9999 = 5

Chapter 12

12.1. $\alpha = 69.24$, $\beta = 0.9781$, $\gamma = 1.3513$, $MTTF = 71.26$ h, Reliability at 15 hours = 0.8152. Can be modeled using exponential distribution since β is close to 1.

12.2. $\alpha = 580.05$, $\beta = 1.0211$, $\gamma = -0.7376$, MTTF = 574.35 h, Reliability at 50 hours = 0.9203. Can be modeled using exponential distribution since β is close to 1.

12.3. $\alpha = 347.76$, $\beta = 1.5601$, $\gamma = 0.5521$, MTTF = 313.11 h

12.4. $\mu = 7.3036$, $\sigma = 2.8833$

12.5. MTTF = 1669 h, 90% Confidence interval (911.7, 4235.7)

12.6. MTTF = 100000 cycles, 95% Confidence interval (39056, 246383)

12.7. $\hat{\theta} = 4630$ h, $C = 4203$ h, $\hat{\theta} > C$ Acceptable. Probability of accepting a mean life of 2000 h = 0.0138

12.8. $r = 4$, $n = 7$, $\theta/\theta_0 = 0.2594$

12.9. $r = 3$, $n = 5$, $\theta/\theta_0 = 0.2878$

12.10. $h_0 = 1175.15$, $h_1 = -1077.48$, $s = 792.87$

Bibliography

There are many excellent books and articles available on quality and reliability. A selected list of books, on each of the major topics covered in the book, is given below. Readers are encouraged to find other books that are not listed here. Several of these books have been used in the preparation of this book without explicit citation. On completing various chapters from this book, the users should benefit from referring to other books, articles, and Internet resources.

TOLERANCES AND QUALITY SPECIFICATION

COGORNO, G.R., *Geometric Dimensioning and Tolerancing for Mechanical Design*, McGraw-Hill, New York, 2006.

CREVELING, C.M., *Tolerance Design: A Handbook for Developing Optimal Specifications*, Addison-Wesley, Reading, MA, 1997.

FISCHER, B.R., *Mechanical Tolerance Stackup and Analysis*, Marcell Dekker, New York, 2004.

GRIFFITH, G.K., *Measuring and Gauging Geometric Tolerances*, Prentice Hall, Upper Saddle River, NJ, 1993.

GRIFFITH, G.K., *Geometric Dimensioning and Tolerancing: Applications and Inspection*, Prentice Hall, Upper Saddle River, NJ, 2001.

HENZOLD, G., *Geometrical Dimensioning and Tolerancing for Design, Manufacturing and Inspection*, 2Ed, Butterworth-Heinemann, Burlington, MA, 2006.

MADSEN, D.A., *Geometric Dimensioning and Tolerancing*, Goodheart-Willcox Publisher, Tinley Park, IL, 2003.

MEADOWS, J.D., *Geometric Dimensioning and Tolerancing: Applications and Techniques for Use in Design: Manufacturing, and Inspection*, Marcel Dekker, New York, 1995.

MEADOWS, J.D., *Measurement of Geometric Tolerances in Manufacturing*, Marcell Dekker, New York, 1998.

PUNCOCHAR, D., *Interpretation of Geometric Dimensioning & Tolerancing*, 2Ed, Industrial Press, New York, NY, 1997.

WILLHELM, R.G., LU, S.C.Y., *Computer Methods for Tolerance Design*, World Scientific Publishing, Singapore, 1992.

PROBABILITY AND STATISTICS

DeGROOT, M.H., SCHERVISH, M.J., *Probability and Statistics*, 3Ed, Addison Wesley, Reading, MA, 2001.

DEVORE, J.L., *Probability and Statistics for Engineering and the Sciences*, 7Ed, Duxbury Press, Pacific Grove, CA, 2007.

EVANS, M., HASTINGS, N., PEACOCK, B., *Statistical Distributions*, 3Ed, John Wiley, New York, 2000.

HAYTER, A.J., *Probability and Statistics for Engineers and Scientists*, 3Ed, Duxbury Press, Pacific Grove, CA, 2006.

HOEL, P.G., *Introduction to Mathematical Statistics*, 5Ed, John Wiley, New York, 1984.

JOHNSON, R., MILLER, I., FREUND, J., *Miller & Freund's Probability and Statistics for Engineers*, 7Ed, Prentice Hall, Upper Saddle River, NJ, 2004.

LARSEN, R.J., MARX, M.L., *An Introduction to Mathematical Statistics and Its Applications*, 4Ed, Prentice Hall, Upper Saddle River, NJ, 2005.

LEVINE, D.M., RAMSEY, P.P., SMIDT, R.K., *Applied Statistics for Engineers and Scientists*, Prentice Hall, Upper Saddle River, NJ, 2001.

MENDENHALL, W., BEAVER, R.J., BEAVER, B.M., *Introduction to Probability and Statistics*, 13Ed, Duxbury Press, Pacific Grove, CA, 2008.

MILTON, J.S., ARNOLD, J.C., *Introduction to Probability and Statistics*, McGraw-Hill, New York, 2002.

WALPOLE, R.E., MYERS, R.H., MYERS, S.L.,YE, K., *Probability & Statistics for Engineers & Scientists*, 8Ed, Prentice Hall, Upper Saddle River, NJ, 2006.

QUALITY CONTROL AND EXPERIMENTAL DESIGN

ALLEN, T.T., *Introduction to Engineering Statistics and Six Sigma: Statistical Quality Control and Design of Experiments and Systems*, Springer, London, 2006.

ANTONY, J., *Design of Experiments for Engineers and Scientists*, Butterworth-Heinemann, Burlington, MA, 2003.

BARNES, J.W., *Statistical Analysis for Engineers and Scientists*, McGraw-Hill, New York, 1994.

BESTERFIELD, D.H., *Quality Control*, 8Ed, Prentice Hall, Upper Saddle River, NJ, 2008.

BURR, J.T. (Editor), *Elementary Statistical Quality Control*, CRC Press, Boca Raton, FL, 2004.

CHANDRA, M.J., *Statistical Quality Control*, CRC Press, Boca Raton, FL, 2001.

DERMAN, C., ROSS, S.M, *Statistical Aspects of Quality Control*, Academic Press, San Diego, CA, 1997.

DEVOR, R.E., CHANG, T.-h., SUTHERLAND, J.W., *Statistical Quality Design and Control*, 2Ed, Prentice Hall, Upper Saddle River, NJ, 2007.

DIAMOND, W.J., *Practical Experiment Designs for Engineers and Scientists*, 3Ed, John Wiley, New York, 2001.

EVANS, J.R.R., LINDSAY, W.M., *Managing for Quality and Performance Excellence*, Thomson South-Western, Florence, KY, 2007.

GOUPY, J., CREIGHTON, L., *Introduction to Design of Experiments: With JMP Examples*, SAS Institute, Cary, NC, 2007.

GRANT, E., LEAVENWORTH, R., *Statistical Quality Control*, 7Ed, McGraw-Hill, New York, 1996.

GUPTA, B.C., WALKER, H.F., *Statistical Quality Control for the Six Sigma Green Belt*, ASQ Quality Press, Milwaukee, WI, 2007.

HICKS, C.R., TURNER, K.V., *Fundamental Concepts in the Design of Experiments*, 5Ed, Oxford University Press, New York, 1999.

ISHIKAWA, K., *Guide to Quality Control*, 2Ed, Asian Productivity Organization, Tokyo, 1986.

KRISHNAMOORTHI, K.S., *A First Course in Quality Engineering*, Pearson Education, Upper Saddle River, NJ, 2006.

KUME, H., *Statistical Methods for Quality Improvement*, The Association for Oversees Technical Scholarship, Tokyo, 1985.

MITRA, A., *Fundamentals of Quality Control and Improvement*, 2Ed, Prentice Hall, Upper Saddle River, NJ, 1998.

MONTGOMERY, D.C., *Introduction to Statistical Quality Control*, 6Ed, John Wiley, New York, 2008.

OAKLAND, J.S., *Statistical Process Control*, 6Ed, Butterworth-Heinemann, Burlington, MA, 2007.

RYAN, T.P., *Statistical Methods for Quality Improvement*, 2Ed, John Wiley, New York, 2000.

SCHILLING, E.G., *Acceptance Sampling in Quality Control*, Marcel Dekker, New York, 1982.

SMITH, G.M., *Statistical Process Control and Quality Improvement*, Prentice Hall, Upper Saddle River, NJ, 1991.

SUMMER, D.C.S., *Quality*, 3Ed, Prentice Hall, Upper Saddle River, NJ, 2003.

VARDEMAN, S.B., JOBE, J.C., *Statistical Quality Assurance Methods for Engineers*, John Wiley & Sons, New York, 1999.

WHEELER, D.J.J., CHAMBERS, D.S., *Understanding Statistical Process Control*, 2Ed, SPC Institute, Cary, NC, 1992.

ZIMMERMAN, S.M., ICENOGLE, M.L., *Statistical Quality Control Using Excel*, ASQ Quality Press, Milwaukee, WI, 2002.

RELIABILITY

BILLINTON, R., ALLAN, R.N., *Reliability Evaluation of Engineering Systems: Concepts and Techniques*, Kluwer Academic Publishers, Norwell, MA, 1992.

BOX, G.E.P., HUNTER, W.G., HUNTER, S.J., *Statistics for Experimenters*, John Wiley, New York, 1978.

DHILLON, B.S., *Engineering Maintainability: How to Design for Reliability and Easy Maintenance*, CRC Press, Boca Raton, FL, 1999.

EBELING, C.E.E., *An Introduction to Reliability and Maintainability Engineering*, Waveland Press, Long Grove, IL, 2005.

KALES, P., *Reliability for Technology, Engineering, and Management*, Prentice Hall, Upper Saddle River, NJ, 1998.

LEEMIS, L.M., *Reliability: Probabilistic Models and Statistical Methods*, Prentice Hall, Upper Saddle River, NJ, 1994.

LEWIS, E.E., *Introduction to Reliability Engineering*, 3Ed, John Wiley, New York, 2001.

MONTGOMERY, D.C., *Design and Analysis of Experiments*, 6Ed, John Wiley, New York, 2004.

MUKERIEE, R., WU, C.F., *Modern Theory of Factorial Design*, Springer-Verlag, New York, 2006.

MOUBRAY, J., *Reliability-Centered Maintenance*, Industrial Press, New York, 1997.
NELSON, W., *Accelerated Testing*, John Wiley, New York, 1990.
NEUBECK, K., *Practical Reliability Analysis*, Pearson Education, Upper Saddle River, NJ, 2004.
O'CONNOR, P.D.T., *Practical Reliability Engineering*, 4Ed, McGraw-Hill, New York, 2002.
RAMAKUMAR, R., *Engineering Reliability*, Prentice Hall, Upper Saddle River, NJ, 1993.
RIGDON, S.E., BASU, A.P., *Statistical Methods for the Reliability of Repairable Systems*, John Wiley, New York, 2000.
TOBIAS, P.A., TRINDADE, D.C., *Applied Reliability*, 2Ed, Chapman and Hall, London, NY, 1995.

HANDBOOKS AND STANDARDS

ANSI/ASQC Z1.4-1993, *Sampling Procedures and Tables for Inspection by Attributes*.
ANSI/ASQC Z1.9-1993, *Sampling Procedures and Tables for Inspection by Variables for Percent Nonconforming*.
DRAKE, P. (Editor), *Dimensioning and Tolerancing Handbook*, McGraw-Hill, New York, 1999.
IRESON, W.G., COMBS, C.F., MOSS, R.Y., *Handbook of Reliability Engineering and Management*, 2Ed, McGraw-Hill, New York, 1996.
ISO 2859-Parts 0,1,2,3,4,5, *Sampling Procedures for Inspection by Attributes*.
ISO 3951:1989, *Sampling Procedures and Charts for Inspection by Variables for Percent Nonconforming*.
JURAN, J.M., GRYNA, F.M, *Juran's Quality Control Handbook*, 5Ed, McGraw-Hill, New York, 1998.
MIL-HDBK-108 (1960), *Quality Control and Reliability – Sampling Procedures and Tables for Life and Reliability Testing (Based on Exponential Distribution)*.
MIL-STD-105E (1989), *Sampling Procedures and Tables for Inspection by Attributes*.
MIL-STD-414 (1968), *Sampling Procedures and Tables for Inspection by Variables for Percent Nonconforming*.
NIST/SEMATECH *e-Handbook of Statistical Methods*, http://www.itl.nist.gov/div898/handbook/, 2003.

Index

2^2 factorial experiment, 224
2^3 factorial experiment, 224
2^k factorial experiments, 223
2KFactorial.xls, 225, 228, 230, 232

Acceptable quality level (AQL), 184
Accuracy, 166
Alias, 232
Alternative hypothesis, 98
American Society for Quality (ASQ), 3
Analysis of variance (ANOVA), 207, 209
Angularity, 42
ANOVA1B2.xls, 213, 219, 220, 221, 232
Assignable causes, 125
Attribute, 146, 182
Availability, 239
Average outgoing quality (AOQ), 185, 191
Average outgoing quality limit (AOQL), 187
Average run length (ARL), 128
Average sample number (ASN), 191
Average total inspection (ATI), 187, 192

Balanced experiment, 209
Basic dimension, 26
Basic size, 10
Baye's theorem, 57
Bernard's approximation for MR, 256
Bernoulli trials, 70
Beta function, 82
Bias, 167
Binomial distribution, 70, 146, 183, 248
Blocking, 231
Bonus tolerance, 27
Box Muller transformation, 78
BoxMuller.xls, 79, 84
Box plot, 115

BoxPlot.xls, 115, 120
Break-in period, 237

c-chart, 149
C&Ucharts.xls, 150, 154
Cause and effect diagram, 112
Causes of variation, 124
Central limit theorem, 88, 89
Chance causes, 124
Chi-squared (χ^2) distribution, 79
CircleSphere.xls, 35
Circular runout, 47
Circularity, 35
Clearance fit, 13
Concentricity, 46
Conditional probability, 57, 239
Confidence interval, 90
Confidence interval for proportion, 94
Confidence interval for ratio of variances, 97
Confidence interval for the difference between two means, 93
Confidence interval for the mean – variance known, 90
Confidence interval for the mean – variance unknown, 92
Confidence interval for variance, 96
Confidence level, 90
Confounding, 231, 232
Consumer's risk, 102, 184, 264
Continuous distributions, 73
Continuous variable, 59
Contrast, 226
Control chart, 122
Control charts for attributes, 146
ControlChartFactors.xls, 123
Controllable factor, 208

305

C_p, 159
C_{pk}, 162
C_{pl}, 161
C_{pu}, 161
C_{pm}, 163
C_r, 161
Crossby, Philip B., 3, 6
Cumulative distribution function (cdf), 65
Cumulative sum (CUSUM) control chart, 142
Curtailment, 191
CUSUM.xls, 143, 154
Cylindricity, 37
Cylindricity.xls, 37

Datum, 28
Debugging phase, 237
Degrees of freedom (df), 61
Deming, Fourteen points, 4
Deming, W. Edwards, 3, 88
Descriptive statistics, 58
Deviation, 10
Discrete probability distributions, 68
Discrete variable, 59
Distributions, 64
Distributions.xls, 84
Dodge-Romig plans for single sampling, 187
DodgeRomig.xls, 188
Double sampling plan, 189
Double specification limit known σ, 198

Effect, 208
Equal tolerance approach, 20
Errors, propagation of, 172
Event, 54
EWMA.xls, 139, 154
Experiment, 54, 207
Experimental design, 207
Exponential distribution, 238
Exponentially weighted moving average (EWMA) control chart, 138

F-distribution, 82
F-statistic, 212
F-Test, 211
Factor, 207
Factorial Design, 208
Failure, 236
Failure probability density, 237
Failure rate, 236
Failure terminated test, 261, 262
Feigenbaum, Armand V., 3, 6
Fishbone diagram, 112
Fisher, Sir R. A., 82, 207

Fit, 13
Fixed effects model, 209
Flatness, 33
Flatness.xls, 33
Form 2 method, 197
Form tolerances, 29
Fourteen points, 4
Fractional factorial experiments, 232, 235
Full indicator movement (FIM), 47
Fundamental deviation, 10

Gage repeatability, 166, 167
Gage reproducibility, 166, 167
GageR&R.xls, 168, 179, 223, 232
Gamma function, 79
Gaussian distribution, 74
Gaussian random numbers, 78
Generator, 232
Geometric moving average (GMA), 138
Geometric tolerances, 24

Hazard function, 236
Histogram, 111
Histogram.xls, 113, 120
Hole basis system, 14
Honestly significant difference (HSD), 213
Hypergeometric distribution, 68, 183
Hypothesis testing, 98
HypothesisTesting.xls, 101

Independent events, 57
Independent identically distributed (iid), 88
Infant mortality, 237
Inferential statistics, 59
Interaction, 208
Interference fit, 13
Intersection, 55
Ishikawa diagram, 112

Juran, Joseph M., 3, 5

K-of-N system 248

Least material condition (LMC), 11
Least squares, 32
Level of a factor, 208
Level of significance, 90
Life cycle curve, 237
Life testing, 255
LifeTestOC.xls, 264, 266, 267, 270
Limiting quality level (LQL), 184
Line profile, 39
Location parameter, 240

Location, principle of, 29
Location tolerances, 43
Lognormal distribution, 242
Lognormal distribution parameter estimation, 260
Lognormal.xls, 260, 270
Lot, 183
Lot acceptance procedure – preassigned number of failures, 265
Lot acceptance procedure – preassigned time, 267
Lot tolerance percent defective (LTPD), 184
Lower CUSUM, 144
Lower deviation, 10
Lower natural tolerance limit (LNTL), 159
Lower specification limit (LSL), 158

M-method, 197
Malcolm Baldridge National Quality Award, 3, 7
Manufacturing processes, 15
Maximum material condition (MMC), 11
Mean, 59
Mean time between failures (MTBF), 239
Mean time to failure (MTTF), 237, 238, 241, 244, 246
Mean time to repair (MTTR), 239
Median, 60
Median rank (MR), 256
Memoryless property, 239
Minimum zone (MZ), 32
Mode, 60
Multinomial distribution, 72
Multiple sampling plan, 192
Mutually exclusive, 55

Nelder-Mead simplex method, 33
Neyman-Pearson theory, 263
Noncentral chi-squared distribution, 80
Noncentral F-distribution, 82
Noncentral t-distribution, 82
NoncentralDistributions.xls, 84, 105
Normal distribution, 74
Normal inspection, 194
NormalData.xls, 79, 84
Normal probability plot, 116
NormalProbPlot.xls, 118, 120
NormalProbPaper.pdf, 120
np-chart, 149
Null hypothesis, 98

OC Curve for attribute control charts, 153
OC Curve for variable control charts, 151
OC_PChart.xls, 153, 154
OCCurve.xls, 151, 154
OCCurveCh5.xls, 104

Operating characteristic (OC) curve, 104
Order statistics, 122
Orientation Tolerances, 39
Orthogonal column, 226
Orthogonal contrast, 226
Out of control process rules, 127

p-chart, 146
p-method, 99
PNPChart.xls, 148, 154
Parallelism, 40
Pareto chart, 114
ParetoChart.xls, 115, 120
Pareto principle, 6
Perpendicularity, 40
Poisson distribution, 72, 149, 251
Population standard deviation, 61
Pooled estimator for variance, 93
Population, 58
Population mean, 60
Population variance, 60
Position tolerance, 43
Power of the hypothesis, 102
Precision, 167
Prediction interval – known σ, 174
Prediction interval – unknown σ, 174
Preferred fits, 14
Preferred numbers, 8
Primary datum, 28
Principle of location, 29
Probability, 54
Probability distribution function (pdf), 65
Process capability index C_p, 159
Process capability ratio C_r, 161
Process in control, 125
Process out of control, 125
Process spread, 159
Producer's risk, 102, 184
Profile tolerance, 39

Quality definition, 1
Quality historical development, 2
Quality philosophies, 3
Quality trilogy, 5

R chart, 129
Random effects model, 209
Random sample, 58
Random variable, 59
Randomization, 208
RandomOrderIntegers.xls, 208
Range, 60
Range and average method for GR&R, 168

Rational subgroup, 125
Rectifying inspection, 186
Reduced inspection, 194
Redundant system, 245
Regardless of feature size (RFS), 27
Rejectable quality level (RQL), 184
Reliability, 2, 236
Reliability function, 236
ReliabilityOfSystems.xls, 246, 249, 250, 251
Renard, 8
Repeatability, 167
Replication, 209
Reproducibility, 167
Response variable, 207
Roundness, 35
Rule 1, 26
Rule 2, 27
Run chart, 118
Runout, 47

Sample, 58
Sample mean, 59
Sample space, 54
Sample standard deviation, 61
Sample variance, 61
Sampling plans for variables, 195
Sampling procedures for life testing, 263
Scale parameter, 240
Secondary datum, 28
Sequential life test, 268
Sequential sampling plan, 193
Shaft basis system, 14
Shape parameter, 240
Shewhart control chart, 122
Shewhart, Walter A., 3, 122
Single sampling plan, 183
Single specification limit – known σ, 195
Six sigma, 6, 161
SP_VAR.xls, 198, 200
SP1_ATR.xls, 185
SP2_ATR.xls, 190
Specification limit, 158
Standard deviation, 61
Standard sampling plans for attributes, 193
Standard sampling plans for variables, 202
Standby system, 250
Statistical coverage interval, 175
Statistical process control (SPC), 122
Statistics, 54
Stem-and-leaf plot, 110
Straightness, 25, 29
Straightness.xls, 32

Straightness3D.xls, 38
Studentized range distribution, 83, 213
StudentizedRange.xls, 84
Student's t-distribution, 80
Sturgis' rule, 111
Surface profile, 39
Symmetry, 46
System with elements in parallel, 245
System with elements in series, 244
System with series and parallel subsystems, 247

t-test, 101
t-distribution, 80
Tabular approach (CUSUM), 144
Taguchi capability index C_{pm}, 163
Taylor, Frederick W., 3
Taylor's principle, 27
Tertiary datum, 28
Testing with replacement, 261
Testing without replacement, 261
Tightened inspection, 194
Time terminated test, 261, 263
Tolerance, 10
Tolerance grades, 11
Tolerance interval, known σ, 175
Tolerance interval, unknown σ, 176
Tolerance selection, 15–21
Tolerance unit, 11
ToleranceLimits.xls, 177, 179
Tolerances and fits, 9
Tolerances of form, 29
Tolerances of location, 43
Total quality control, 6
Total quality management, 3
Total runout, 47
Transition fit, 14
Treatment, 208
Trimmed mean, 60
Trueness, 166
Tukey intervals, 212
Type A OC curve, 184
Type B OC curve, 184
Type I error, 102
Type II error, 102

u-chart, 150
Union, 55
Unit tolerance approach, 20
Upper CUSUM, 144
Upper deviation, 10
Upper natural tolerance limit (UNTL), 159
Upper specification limit (USL), 158

Variable sampling for single specification – unknown σ, 199
Variance, 60
Venn diagram, 55
Virtual condition, 25, 43
V-mask approach (CUSUM), 142

Waiting time, 266
WECO rules, 128
Weibull distribution, 240
Weibull distribution parameter estimation, 255

Weibull3.xls, 256, 257, 270
Whitney, Eli, 2
WorksheetFunc.xls, 59

x-MR chart, 135
s chart, 133
$\bar{x}$ chart, 129, 133
XbarRChart.xls, 131, 154
XbarSChart.xls, 135, 154

z-test, 101
Zero defects, 6, 7

Printed in the United States
By Bookmasters